INCOMMENSURABI
AND CROSS-LANGUAGE COM.

A dominant epistemological assumption behind Western philosophy is that it is possible to locate some form of commonality between languages, traditions, or cultures – such as a common language or lexicon, or a common notion of rationality – which makes full linguistic communication between them always attainable. Xinli Wang argues that the thesis of incommensurability challenges this assumption by exploring why and how linguistic communication between two conceptually disparate languages, traditions, or cultures is often problematic and even unattainable. According to Wang's presuppositional interpretation of incommensurability, the real secret of incommensurability lies in the ontological set-ups of two competing presuppositional languages.

This book provides many original contributions to the discussion of incommensurability and related issues in philosophy and offers valuable insights to scholars in other fields, such as anthropology, communication, linguistics, scientific education, and cultural studies.

ASHGATE NEW CRITICAL THINKING IN PHILOSOPHY

The *Ashgate New Critical Thinking in Philosophy* series brings high quality research monograph publishing into focus for authors, the international library market, and student, academic and research readers. Headed by an international editorial advisory board of acclaimed scholars from across the philosophical spectrum, this monograph series presents cutting-edge research from established as well as exciting new authors in the field. Spanning the breadth of philosophy and related disciplinary and interdisciplinary perspectives Ashgate New Critical Thinking in Philosophy takes contemporary philosophical research into new directions and debate.

Incommensurability and Cross-Language Communication

XINLI WANG 王新力
Juniata College, USA

Routledge
Taylor & Francis Group

LONDON AND NEW YORK

First published 2007 by Ashgate Publishing

2 Park Square, Milton Park, Abingdon, Oxon OX14 4RN
711 Third Avenue, New York, NY 10017, USA

Routledge is an imprint of the Taylor & Francis Group, an informa business

First issued in paperback 2016

British Library Cataloguing in Publication Data
Wang, Xinli
  Incommensurability and cross-language communication. –
  (Ashgate new critical thinking in philosophy)
  1.Communication in science – Philosophy 2.Science – History
  I.Title
  501.4

Library of Congress Cataloging-in-Publication Data
Wang, Xinli, 1956–
  Incommensurability and cross-language communication / Xinli Wang.
    p. cm. — (Ashgate new critical thinking in philosophy)
  Includes bibliographical references and index.

  1. Comparison (Philosophy) 2. Contrastive linguistics. 3. Comparative linguistics. 4. Translating and interpreting. 5. Communication. I. Title. II. Series.

  BD236.W36 2006
  121—dc22

                                                                    2006011548

ISBN 13: 978-0-7546-3034-0 (hbk)
ISBN 13: 978-1-138-26414-4 (pbk)

To the Memory of
My father, Shi-Rong Wang 王世荣 (1914–1982)

For My Inspirations

My wife, Ling Xu 徐玲
My daughter, Jenny Wang 王絮

Contents

Acknowledgements

Professors Anne Hiskes, Austen Clark, Scott Lehmann, and Samuel Wheeler III, my mentors at the University of Connecticut, Storrs, read early versions of several chapters of the manuscript. They are exceptionally perceptive philosophers and generous friends. Their illuminating questions, detailed written comments, extensive criticisms, and insightful suggestions enabled me to make considerable improvements on the earlier drafts. Without their help and encouragement, I might still be struggling in the 'morass' of the issue of incommensurability. I realize that all those cited above are unlikely to agree with all of my conclusions, but they can pride themselves on having diverted me from erroneous claims that I might have said and helped me to bring my ideas and arguments into clearer focus. I owe an enormous intellectual debt and immense gratitude to them.

I am very fortunate to have the writer Joe Schall as my book editor, who used his literary talent to help me make my expressions more accurate, idiomatic, and readable.

I am grateful to the research leave granted to me by Juniata College in 2005, during which the final version of the book was written. Special thanks go to the college provost, Dr James Lakso, who, besides giving me continuous administrative support and moral encouragement during the past few years, graciously provided a research fund from his office's budget to cover the cost of the manuscript preparation.

I am deeply in debt to an anonymous manuscript reviewer, one of the Ashgate New Critical Thinking in Philosophy advisors, who gave a positive evaluation to the project and provided me with many insightful suggestions.

Many thanks also to Publisher Sarah Lloyd, Humanities Editorial Administrator Anne Keirby, Commissioning Editor Paul Coulam, and editor Jane Fielding at Ashgate Publishing Limited, for their extreme patience and help throughout the process of producing this book.

All errors, either philosophical or linguistic, that may exist in the book are, of course, my sole responsibility.

An earlier version of several chapters of this book appeared in a few philosophy journals. Chapter 2 is an extended version of my article 'A Critique of the Translational Approach to Incommensurability', *Prima Philosophia*, Band 11 / Heft 3 (1998). Chapter 7 is an expansion of my article 'Taxonomy, Truth-Value Gaps and Incommensurability: a reconstruction of the Kuhn's taxonomic interpretation of incommensurability', *Studies in History and Philosophy of*

Science 33 (2002). An earlier version of chapter 8 was published as 'Is the Notion of Semantic Presupposition Empty?' in *Diálogos* 73 (1999). Chapter 12 is a revision based on my article 'Presuppositional Languages and the Failure of Cross-Language Understanding', *Dialogue: Canadian Philosophical Review* XLII (2003). I appreciate the editors and publishers for giving me permission to use them in this book.

Chapter 1

The Many Facets of Incommensurability

1. Incommensurability as Communication Breakdown

In most situations, the speakers of two different scientific languages, i.e., the languages of comprehensive scientific theories, can effectively understand one another's language and successfully communicate with each other. However, in some cases, effective mutual understanding and communication are problematic, difficult, and even in some measure unattainable between two substantially distinct scientific languages.

When Kuhn Encountered Aristotle

A radical conceptual shift in a scientific theory can render its language largely unintelligible to a later age within the same cultural or intellectual tradition. This is what Thomas Kuhn experienced repeatedly when he attempted to understand some out-of-date scientific texts. As Kuhn observed:

> A historian reading an out-of-date scientific text characteristically encounters passages that make no sense. That is an experience I have had repeatedly whether my subject was an Aristotle, a Newton, a Volta, a Bohr, or a Planck. It has been standard to ignore such passages or to dismiss them as the products of error, ignorance, or superstition, and that response is occasionally appropriate. More often, however, sympathetic contemplation of the troublesome passages suggests a different diagnosis. The apparent textual anomalies are artifacts, products of misreading. (1988, pp. 9-10)

Kuhn (1987)[1] records vividly 'a decisive episode' in the summer of 1947, when he, in his struggle to make sense of Aristotle's physics, first encountered what he characterized as the phenomenon of incommensurability fifteen years later—a communication breakdown between two successive competing scientific language communities. Kuhn was deeply perplexed. How could Aristotle have made so many obviously senseless and absurd, not just false, assertions about motion? He discovered that 'Aristotle had known almost no mechanics at all' (Kuhn, 1987, p. 9). But the discovery troubled Kuhn. He wondered 'how could his characteristic talents have deserted him so systematically when he turned to the study of motion and mechanics? Equally, if his talents have so deserted him, why had his writings in physics been taken so seriously for so many centuries after his death?' (Kuhn, 1987, p. 9) Eventually Kuhn suspected that the fault might be his reading of Aristotle (he approached Aristotle from the perspective of Newtonian mechanics).

After changing his way of reading Aristotle and continuing to puzzle over the text, his perplexities suddenly vanished during one memorable hot summer day, as Kuhn recalled vividly 40 years later.

> I was sitting at my desk with the text of Aristotle's *Physics* open in front of me and with a four colored pencil in my hand. Looking up, I gazed abstractly out the window of my room—the visual image is one I still retain. Suddenly the fragments in my head sorted themselves out in a new way, and fell into place together. My jaw dropped, for all at once Aristotle seemed a very good physicist indeed, but of a sort I'd never dreamed possible. Now I could understand why he had said what he'd said, and what his authority had been. (Kuhn, 1987, p. 9)

> My attempt to discover what Aristotle had thought provided my first exposure to incommensurability, and I have encountered it repeatedly since, not only with ancient texts but with texts by figures as recent as Planck and Bohr. Experiences of this sort seemed to me, from the very start, to require explanation. They signal a break between older modes of thought and those now current, and that break must have significance both for the nature of knowledge and for the sense in which it can be said to progress. I changed my career plans in order to engage those issues. (Kuhn, 1999, p. 33)

Guided by these personal experiences, Kuhn found out further that when two successive competing scientific languages, such as the language of Newtonian mechanics and that of Aristotelian mechanics, are separated by a 'scientific revolution', their respective proponents are liable to experience a failure of mutual understanding. They often inevitably talk past one another when attempting to resolve their disagreements. In short, Kuhn was struck by the observation that when rival scientific theories or paradigms clash, we can from time to time identify a communication breakdown between their advocates.[2]

Kuhn's personal experiences with reading old scientific texts are shared by many others. In canonical texts such as Descartes' *Le monde*, Galileo's writings, Bacon's *Advancement of Learning* and *Novum Organum*, and Locke's *Essay Concerning Human Understanding*, we find the writers claiming to be unable to understand some of the fundamental concepts of the Aristotelian. For example, in his *Le monde*, Descartes quoted Aristotle in Latin, claiming to do so because he was unable to understand the sense of his definitions otherwise. Galileo and his supporters were often skeptical about the very possibility of establishing a constructive dialogue with the Tuscan Aristotelians during the debate on buoyancy during 1611-1613 (Biagioli, 1990). I. Hacking (1983) concurs and points out that the medical theory of the well-known sixteenth-century Swiss alchemist and physician, Paracelsus—which exemplifies a host of hermetic interests within Northern European Renaissance tradition—is hardly intelligible to modern Westerners. For example, a Paracelsan assertion that mercury salve is good for syphilis because of the association of the metal mercury with the planet Mercury, the market place, and syphilis makes little sense to the ears of today's physician.

By the same token, substantial semantic and/or conceptual disparities between two comprehensive theories and their languages embedded in two coexistent, distinct, intellectual/cultural traditions can create serious impediments to mutual understanding and communication. For example, Chinese medical theory is hardly intelligible to most Western physicians. They are very skeptical of Chinese medicine and even regard Chinese physicians as something like medicasters. Many Western physicians claim that Chinese medicine sounds strange and alien to them. As one complains, the sentence, 'The loss of balance between the yin and the yang in the human body invites evils which lead to diseases', sounds as nonsensical to him as the utterance, 'ooh ee ooh ah ah'. When a Chinese physician diagnoses a disease as the excess of the yin over the yang within the spleen of a patient, a Western physician would be left in a fog.

A historian or philosopher of science close to the position of logical empiricism would say that these claims about the difficulty of communicating with competitors represent a mere rhetorical strategy. But the question of whether those claims of communication breakdowns are real or rhetorical is beside the point. As Kuhn and others have noticed, such a failure of cross-language understanding and communication breakdown between two scientific languages cannot be simply taken as evidence of the interpreter's limitation of knowledge or lack of interpretative skills. In many cases, difficulty in understanding an alien language and the communication breakdown between the interpreter's and an alien language are experienced by most members of a scientific community, not just by some individuals of the community. It is this failure of mutual understanding and a communication breakdown between two language communities as a whole, rather than between some individual speakers with different dialects, intentions, or conflicting interests, that calls for our attention. The problem clearly involves some deep *semantic and/or conceptual obstructions* between two substantially different languages that make effective mutual understanding and communication difficult and problematic.[3]

Communication Breakdown

Deeply perplexed by apparent communication breakdown between two competing scientific language communities, Kuhn and Feyerabend set out to give a philosophical explication of this phenomenon that they dubbed 'incommensurability'. Both Kuhn and Feyerabend started to use the term 'incommensurability' independently to describe the communication breakdown between two scientific language communities. Kuhn made this point bluntly clear in his first writing about incommensurability.

> We have already seen several reasons why the proponents of competing paradigms must *fail to make complete contact with each other's viewpoints*. Collectively these reasons have been described as the incommensurability of the pre- and post-revolutionary normal-scientific traditions ... . Communication across the revolutionary divide is inevitably partial. (Kuhn, 1970a, pp. 148-9; my italics)

The term 'incommensurability' is borrowed from mathematics. In its original mathematical use, it means 'no common measure' between two irrational numbers. For example, the hypotenuse of an isosceles right-angled triangle is incommensurable with its side or the circumference of a circle with its radius in the sense that there is no unit of length contained without residue an integral number of times in each member of the pairs. The term 'incommensurability' is used metaphorically (Kuhn, 1983b, p. 670) in Kuhn's and Feyerabend's hands in trying to capture their strong intuition that the communication breakdown between two scientific communities is due to lack of some common measure between the two languages used. According to an innocent and widely accepted assumption, any successful communication between two language communities requires some *appropriate* common measure between the languages used;[4] otherwise, communication between the language communities would break down.

However, the controversy arises as to what counts as an appropriate common measure necessary for successful communication between two scientific language communities. Kuhn spent more than thirty years (1960s-1996) trying to pin this measure down. By doing so, Kuhn wished to conceptualize, clarify, and refine his notion of incommensurability, and thus to argue for his celebrated thesis that the phenomenon of incommensurability, i.e., communication breakdown, does exist and that cases of it abound not only in the history of rational thought in general but in the history of sciences in particular. Although Kuhn's position underwent dramatic changes during these years, there was a common thread through all the changes: He sought to identify a significant necessary common measure of cross-language communication, and perhaps primarily to deny that such an identified common measure exists, thereby identifying a certain kind of semantic obstruction between would-be communicants using two competing scientific languages.

Consider briefly the development of Kuhn's thesis of incommensurability since the publication of his *Structure of the Scientific Revolutions* in 1962. Initially, Kuhn specified the common measure as a shared paradigm: the entire constellation of shared metaphysical commitments, shared problems to solve, shared methodological standards of adequacy, and shared perceptions. Faced with extensive criticism, Kuhn realized later that his concept of paradigm was too vague. He thereby concentrated on an essential part of the paradigm, namely, exemplars and similarity relationships among items determined by exemplars.[5] He accordingly specified the common measure as shared similarity relationships (family resemblance) among objects or situations for categorization.[6] However, the major problem with the identification of similarity relationships as a common measure between two competing languages is that a mere shift of similarity relationships does not necessarily cause a communication breakdown.

To further specify the structure of categorical frameworks of scientific languages and to explore their role in successful communication between two competing language communities constituted the central task of Kuhn's explication of incommensurability until the last day of his life. After 1983,

> [t]he phrase 'no common measure' became 'no common language'. The claim that two theories are incommensurable is then the claim that there is no language, neutral or otherwise, into which both theories, conceived as sets of sentences, can be translated without residue or loss. (Kuhn, 1983b, p. 670)

Since 1987, Kuhn gave up the common language requirement as being too broad, and started to focus exclusively on one essential part of a language—its taxonomic structure.[7] At this latest stage, the phrase 'common measure' became 'shared lexical structure or taxonomy' between two competing scientific languages.

It is clear that Kuhn's different formulations of incommensurability evolved in the process of his efforts to specify a significant common measure of successful cross-language communication. The cross-language communication breakdown is the essential sense of Kuhn's notion of incommensurability. To say that two scientific theories are incommensurable is, for Kuhn, to say that a necessary common measure of some sort is lacking between the languages employed by them and that thereby the successful cross-language communication between their advocates breaks down.

2. The Current State of Affairs

No one issue has dominated the landscape of contemporary philosophy of science as has the problem of incommensurability. In fact, any philosopher who takes more than a fleeting interest in the development of science, rational thought, and knowledge must, at some stage, confront the issue of incommensurability in one of its many manifestations. The problem of incommensurability has caught the attention of Anglo-American philosophy in the last quarter of the twentieth century because of its significant implications for many central issues in the philosophy of science, the philosophy of language, ethics, epistemology, and metaphysics. Among them, the problem of theory comparison, the problem of scientific rationality, the problem of scientific progress, and the issue of scientific realism/anti-realism are some important issues that are closely linked to the problem of incommensurability.

It has been widely held that whatever the origins and intentions of Kuhn's and Feyerabend's thesis of incommensurability may be, it is plain that the thesis has made problematic the debate on processes of theory comparison and choice, and has accordingly threatened to undermine our image of science as a rational, realistic, and progressive enterprise. Kuhn and Feyerabend gave fresh respectability to irrationalistic, subjective, and relativistic views about science and knowledge. Consequently, the thesis of incommensurability has met its double fates: On the one hand, the thesis has become 'a modern myth' among many. For many philosophers and theorists in related fields with a relativist bent, the notion often serves as a conceptual foundation for their pet theories, such as social constructionism, multiculturalism, feminism, and postmodernism. On the other hand, the notion and the thesis of incommensurability have become 'a public enemy' in some philosophy circles, especially among philosophers and theorists with an anti-relativist bent and the self-claimed champions of rationality, objectivity of science, and human thoughts. Most notably, Kuhn's and Feyerabend's notion of incommensurability was attacked fiercely by D. Davidson (1984) in conjunction with the very notion of conceptual schemes.

Further, the influence of the thesis of incommensurability has reached far beyond the professional circle of philosophy. Practitioners with a relativistic bent in numerous interpretative fields, such as sociology, anthropology and ethnography, psychology, law, education, political science, economics, cognitive science, decision theory, and linguistics, have been busy discovering similar phenomena in their fields. Through its popularization, the notion of incommensurability has been put on the cultural map, and even becomes part of the weekly glosses in many professional circles.[8]

Because of the above two reasons, the notion of incommensurability has been one of the most revolutionary and influential notions in recent philosophical investigations. Along with its significant impact on the philosophy of science and other related areas, it is one of the most intriguing ideas in recent philosophy. On the one hand, the topic has been so popularized to the extent that one can even claim that in the circle of the philosophy of science 'the doctrine of incommensurability needs no introduction' (Pearce, 1987, p. 1). On the other hand, however, the notion of incommensurability is the most controversial, most often abused notion within contemporary analytic philosophy. Although in the past four decades, philosophers approached it from different directions and presented many historically erudite and conceptually fine-grained analyses of it, we still do not have any clear theoretical conception of what incommensurability is.

The fact that the notion of incommensurability has not been subjected to a satisfactory conceptual clarification explains why hardly any significant progress has been made in the study of the issue of incommensurability in the past.[9] A comment made by Feyerabend 30 years ago can still be used to describe the current research circumstances of incommensurability: 'Apparently, everyone who enters the morass of this problem [referring to the problem of incommensurability] comes up with mud on his head' (1977, p. 363). Kuhn and Feyerabend, two pioneers opening this uncultivated land for us, are no exception. It seems to me that philosophical discussion involving the notion of incommensurability, no matter whether for or against it, tends to come to a deadlock, and is hard to evaluate and often fallacious. There is a danger that, as D. Pearce (1987) pointed out 20 years ago, through popularization and abuse, the term 'incommensurable' would become long in the tooth and the thesis would lose its original bite. This danger has occurred, and it mainly comes from two directions based on two pervasive misconceptions of incommensurability. If incommensurability were interpreted as untranslatability, the thesis would degenerate into a trivial platitude; almost any kind of conceptual difference or conflict could amount to a case of incommensurability. Alternatively, the problem of incommensurability could turn out to be a pseudo-problem for many if incommensurability were reduced to incomparability.

There are many reasons responsible for this slow progress made in the investigation of incommensurability. I would like to mention the following two major reasons: incommensurability as a complex historical-anthropological phenomenon that manifests itself in many facets and ramifications, and the failure of the received interpretation of incommensurability. The first reason will be discussed in the following two sections. The second reason is the topic of chapter 2.

3. The Many Facets of Incommensurability

The Complexity of Incommensurability

It is commonly held that part of the blame for the vagueness of the notion of incommensurability lies with Kuhn himself. Kuhn's notion of incommensurability has often been misinterpreted and abused: partly due to Kuhn's terminological confusion and his constant change of the expression of the notion, partly because many commentators have simply misunderstood Kuhn's point. Kuhn had to clarify himself repeatedly after the publication of his *Structure*. In my opinion, there is a deeper reason responsible for this. When Kuhn and Feyerabend coined the term 'incommensurable' to describe the communication breakdowns that they encountered in the study of the history of sciences, they had nothing so precise in mind. For them, the notion of incommensurability was just a suitable language metaphor to reveal their deep insight gained in their experiences.[10]

The vagueness of the explanation of incommensurability is partially due to the fact that we are dealing with a complex historical-anthropological phenomenon extending beyond the area of philosophy, whose roots are deep in the basic mechanisms of cultures, forms of life, languages, and social institutions. Generally speaking, incommensurability has its natural home primarily in seven disciplinary settings: in intellectual history in general to contrast widely divergent perspectives of understanding different *Weitanschauungen*; in the history of sciences in particular to contrast and understand the conceptually distant explanatory frameworks; in descriptive sociology to contrast kinship systems or other mechanisms for categorizations and explanation of human affairs;[11] in anthropology and cultural studies to contrast and understand totally different modes of justification;[12] in linguistic study to contrast different categorization systems which create 'pattern resistance' to widely divergent points of views;[13] in axiology to contrast and evaluate distinct value (ethical, aesthetical, and others) systems; and in philosophical epistemology to contrast fundamentally diverse perspectives—which start with conceptually disparate presuppositions—of treating explanatory issues. The phenomenon of incommensurability has been and will continue to be rediscovered and enhanced in different disciplinary settings (recently, in cognitive science, psychology, rhetoric, and economics). It is not an exaggeration to say that any philosopher, sociologist, anthropologist, or linguist who takes a comprehensive-historical stand toward the development of rational knowledge and human society would encounter the phenomenon of incommensurability in one way or another at some stage. In Kuhn's words, 'incommensurability has to be an essential component of any historical, developmental, or evolutionary view of scientific knowledge' (1991, p. 3).

Feyerabend explicitly makes an analogy between the clarification of the notion of incommensurability and an anthropological discovery. The term 'incommensurability' is nothing but a 'terminology for describing certain historical-anthropological phenomena which are only imperfectly understood rather than defining properties of logical systems that are specified in detail' (1978,

p. 269). Just like an anthropologist trying to infiltrate an unknown tribe, who must keep in check any eagerness for instant clarity and logical perfection. He or she should not try to make a concept clearer than what is suggested by the available material, keeping key notions vague and incomplete until more information is collected. Feyerabend assumes that such an anthropological method is appropriate for studying the phenomenon of incommensurability. Here, lack of clarity of the notion of incommensurability indicates the scarcity of right information rather than the vagueness of the logical intuitions of it. Therefore, 'the vagueness of the explanation reflects the incompleteness and complexity of the material and invites articulation by further research' (Feyerabend, 1978, p. 270). Feyerabend had even gone so far as to register doubt that, in its present stage of development, the incommensurability thesis was incapable in principle of being given the kind of precise formulation that would serve to satisfy 'analytic' philosophers. In a similar way, Kuhn clearly realized that his attempts to describe the central conception of incommensurability were extremely crude. In his words, 'efforts to understand and refine it have been my primary and increasingly obsessive concern for thirty years' (1993a, p. 315).

The Nature, Sources, and Consequences of Incommensurability

As a complex historical-anthropological phenomenon, the problem of incommensurability manifests itself in many facets and ramifications. The issue of incommensurability in fact is a set of problems, which comprises three interrelated problems: the nature of incommensurability, the sources of incommensurability, and the epistemological and metaphysical implications or consequences of incommensurability. My experience in reading and discussion has been that those either sympathetic or apathetic to the issue of incommensurability often confuse these three problems.

The general question, 'What is incommensurability?' is ambiguous. It can be understood as a question either about the nature or about the sources of incommensurability. For clarity, the question should be divided into two separate questions. First, 'What is *the essential nature* of incommensurability?' The answers will take the format, 'incommensurability as ...' (for example, incommensurability as untranslatability, incommensurability as incomparability, or incommensurability as communication breakdown). Second, 'What are the real sources of incommen-surability?' The answers will take the format, 'incommensurability due to ...' The common alleged sources of incommensurability are, to mention only a few: incommensurability due to radical meaning/reference variance, incommensurability due to value, standard, or problem change, or incommensurability due to lexical structure change.

If we consider the alleged consequences brought about by the thesis of incommensurability, as we have mentioned above, the problem of incommensurability consists of a group of interrelated issues: logical compatibility and semantic comparability between scientific theories; language translation and interpretation; sense and reference of the terms of scientific theories; categorization and taxonomization of scientific language; justification and validity of scientific theories; scientific rationality and progress; value judgment and evaluation criteria; absolutism and relativism; and scientific realism and anti-realism, etc.

The Normative versus the Semantic Dimensions

Much more importantly, the notion of incommensurability is a multiple-dimension concept that involves at least two different dimensions: the normative dimension and the semantic dimension. Accordingly, the concept can be approached from at least two perspectives, which I will call the normative perspective and the semantic perspective. According to the semantic perspective, the problem of incommensurability has something to do with the nature of *a certain kind* of semantic relation between the languages employed by competing scientific theories. Incommensurability can be characterized as a lack of a certain kind of semantic contact between the languages of two competing theories due to changes in either the semantic values (meaning or reference) of the non-logical constituents of sentences or the semantic values (factual meaning, truth-values, or truth-value status) of sentences themselves in these languages. Because of the lack of a certain desirable semantic contact, proponents of incommensurable theories inevitably talk past one another when attempting to resolve their disagreements. Following the convention, I will call the incommensurability identified in the semantic dimension *semantic incommensurability*.

Besides the semantic dimension that addresses linguistic, conceptual aspects of science, the problem of incommensurability arises within a non-linguistic/conceptual dimension also. As far as the sources of incommensurability are concerned, there are, traditionally, at least three alleged non-linguistic/conceptual grounds of incommensurability: (a) the change of problems to solve, (b) incompatible methodological standards of adequacy (such as scientific value judgments on what are sufficient criteria of a good scientific theory, what is an adequate solution to a scientific problem, etc. Some might prefer to call this *methodological incommensurability*), and (c) the *Gestalt* switch between the modes of scientific perceptions. We can observe all these alleged forms of incommensurability in the earlier Kuhn's (before 1980s) exploration of incommensurability.[14] These various interpretations of incommensurability reveal genuine ambiguities and tensions within the earlier Kuhn's understanding of incommensurability. Incommensurability due to the lack of a certain kind of normative contact between two competing scientific theories because of changes in their normative expectations, I will refer to as *normative incommensurability*.

Although incommensurability does have some normative import, I will not be addressing this normative aspect here. My decision is based on the following considerations. First, the linguistic/conceptual aspect of incommensurability still lacks a satisfactory explanation and an adequate solution. Until progress can be made at the conceptual level, there is only a slim chance of achieving success at the normative level. This is because the linguistic/conceptual aspect is more fundamental than the non-linguistic/conceptual aspect. As Pearce points out:

Though an inquiry into values and standards may pay dividends in helping to forge commensurability in the wider sense, it cannot be the sole basis for understanding science as rational enterprise. Even if one would show that science makes a judicious choice of its instruments, if there is a rational gap at the level of concepts then there is a rational gap in science as a whole because the essential process of theory appraisal and choice is, in effect, undermined. (1987, pp. 3-4)

Second, the changes at the non-linguistic/conceptual level can be explained by means of the changes at the linguistic/conceptual level. For example, the perceptual switch observed in perceptual incommensurable cases can be interpreted as the shift between different taxonomies of natural kinds. And the variance in methodological standards of adequacy is caused by the variance in modes of reasoning. The latter is one of the metaphysical presuppositions of scientific languages, as I will argue later. Actually, Kuhn explicitly regards the differences in methods, problem-fields, and standards of solution as necessary consequences of the language-learning process (Kuhn 1983b, p. 648; 1988, p. 10).

Third, the normative interpretation found in the earlier Kuhn's formulation is his premature explication of incommensurability. Kuhn himself gave up his early normative perspective in the 1980s. In Kuhn's own words, 'My original discussion described non-linguistic as well as linguistic forms of incommensurability. That I now take to have been an overextension resulting from failure to recognize how large a part of the apparently non-linguistic component was acquired with language during the learning process' (1988, p. 10).

In conclusion, the linguistic/conceptual aspects of incommensurability are more essential than the normative ones. Although our account of incommensurability may miss some admittedly important features if we leave the normative aspects out, I prefer clarifying one phenomenon in some essential aspects to making only some very broad and vague explications in order to have an inclusive explanation.

Semantic Incommensurability

According to the semantic perspective, the communication breakdown in the case of incommensurability can, and should, be attributed to the lack of *a certain kind* of desirable semantic relationship between the languages employed by two competing scientific theories due to changes of *a certain desirable semantic value(s)* of certain kinds of components (sentences or their constituents) of the languages in question. This seems incontrovertible within the framework of the semantic perspective. But controversy arises with *what kind of semantic relation* is supposed to be the determinant semantic relation between the languages of two incommensurable theories. To see this, we need to identify different kinds of carriers of semantic values and, accordingly, different semantic values associated with these carriers.

First of all, the carriers of semantic values could be some non-logical constituents of a sentence, such as terms—either singular terms including proper names and definite descriptions or general terms including natural kind-terms (water, gold) and concept terms (mass, force)—and predicates. For example, in a Ptolemaic sentence,

(1)   The sun, the largest planet, revolves about the earth, which is a star,

'the sun' and 'the earth' are proper names; 'planet' and 'star' are general terms; 'the largest planet' is a definite description; 'revolves about' and 'is a star' are two-place or one-place predicates. On the other hand, the carrier of semantic values could be a sentence as a whole, for instance, sentence (1).

Secondly, different semantic values are accordingly associated with different kinds of carriers. For a term of a sentence (say, 'the earth' or 'planet'), we can talk about its meaning (sense) or reference; for a predicate (say, 'is a star'), we can talk about its extension. Alternatively, if we take a sentence as a whole as the carrier of semantic values, we can speak of meaning, *factual meaning*, truth-value (Frege's reference of a sentence), or *truth-value status* of the sentence (whether a sentence has a truth-value). For example, the factual meaning of sentence (1) consists in its truth conditions. (1) is either true or false from the point of view of Ptolemaic astronomy, but neither true nor false from the point of view of Copernican astronomy.

Corresponding to the two different kinds of carriers of semantic values and the semantic values associated with them, there are at least two kinds of semantic relations, which could be identified as the determinant semantic relation between the languages of two competing scientific theories in the case of incommensurability. One can focus on parts of sentences and their associated semantic values. In this way, *the meaning/reference relation* between the languages of two competing scientific theories would be the determinant semantic relation in the case of incommensurability. By comparison, one can focus on sentences as a whole and their associated semantic values. Then *the truth-value functional relation* [15] would be the determinant semantic relation between them.

Corresponding to which kind of semantic relation is identified as the determinant relation, there are two possible ways, within the semantic perspective, to characterize the problem of incommensurability. According to my *presuppositional interpretation of incommensurability* presented in this book, the carriers of semantic values in the case of incommensurability are the sentences of the languages of two rival scientific theories. The semantic values that concern us are the factual meanings and the truth-value status of the sentences in question. So it is the truth-value functional relationship between two competing languages that counts as the determinant semantic relationship in the case of incommensurability. To say that two scientific theories are incommensurable is to say that there is a truth-value gap (a substantial number of core sentences of one language is lack of truth-values when considered within the context of a competing language) between the languages employed by the theories, which results in a communication breakdown between their proponents. And such an occurrence of a truth-value gap is in turn due to incompatible metaphysical presuppositions underlying the two languages. In contrast, according to the received *translation-failure interpretation of incommensurability*, terms are the semantic carriers in the case of incommensurability. Accordingly, the semantic values, which play the central role in the incommensurable cases, are the meanings/references of the terms in question. Thus, it is the meaning/reference relationship between the languages of two competing scientific theories that should be identified as the determinant semantic relationship in the case of incommensurability. To say that two scientific theories are incommensurable is to say that the languages of the two theories are mutually untranslatable. The failure of mutual translation in turn is due to the absence of meaning/reference continuity because of the radical variance of meaning/reference.

4. The Presuppositional Interpretation of Incommensurability: A Chapter Review

The received translation-failure interpretation has dominated the discussion of incommensurability for the past four decades. With focus on the meaning/reference relation between two scientific languages, the notions of meaning, reference, and translation have been subjected to many historically erudite and conceptually fine-grained analyses. However, until now no significant progress has been made to clarify the notion of incommensurability. The main reason responsible for such slow progress, in my judgment, is that the translation-failure interpretation is misleading. It cannot establish a tenable and integrated notion of incommensurability. Therefore, we have to go beyond the translation-failure interpretation. This is the conclusion that I draw from chapter 2 after a comprehensive critique of the translation-failure interpretation.

   The doctrine of conceptual relativism and the thesis of incommensurability are twin positions that rise and fall together. Presumably, the notion of conceptual schemes and its underlying metaphysical dualism between scheme and content serve as the conceptual foundation of the thesis of incommensurability. This is why D. Davidson attacks both the notion of conceptual schemes and the thesis of incommensurability together in his influential essay, 'On the Very Idea of a Conceptual Scheme' (1984). Due to such a close affinity of the notion of conceptual schemes with that of incommensurability, we have to respond to Davidson's criticism in defense of the thesis of incommensurability. In chapters 3 and 4, I argue that Davidson's two lines of criticism of the notion of conceptual schemes fail to dismantle the notion of conceptual schemes and therefore do not undermine the thesis of incommensurability. In fact, what Davidson attacks fiercely, even if it were successful, is not 'the very idea' of conceptual schemes, but rather the Quinean notion of conceptual schemes and its underlying Kantian scheme–content dualism. The very notion of conceptual schemes, conceptual relativism, and the thesis of incommensurability escape unharmed from Davidson's attack.

   Incommensurability is typically regarded as a radical conceptual disparity between two competing scientific languages. It is taken to represent a lack of conceptual continuity between them due to some kind of semantic obstruction. A lack of conceptual continuity between two languages is supposed to explain why their proponents inevitably talk past one another. Various attempts to locate the semantic obstruction in the meaning/reference relation between two competing languages fail. The opponents of the thesis of incommensurability applaud the failure of the translation-failure interpretation, and take it tacitly as the failure of the thesis of incommensurability itself. The advocates of the thesis, in contrast, have tried to save it by proposing various remedial measures within the framework of the translation-failure interpretation. Both opponents and proponents of incommensurability, in my opinion, make the same mistake. They both take the translation-failure interpretation as the only valid path to explore incommensurability. To me, it is not the notion of incommensurability that has let us down, but rather a particular conception of how that notion is to be philosophically established, explained, and improved—namely, the received notion of incommensurability—fails us. The failure of the received interpretation of incommensurability does not indicate that there is no such semantic obstruction between two disparate scientific languages, but rather indicates that it could manifest itself in some other more profound way.

To identify such a semantic obstruction, I turn to two case studies in chapter 5, one taken from the history of Western science, i.e., the Newton–Leibniz debate on the absoluteness of space; the other from the comparison between contemporary Western medical theory and traditional Chinese medical theory. I find that those classical conceptual confrontations are not confrontations between two scientific languages with different *distributions of truth-values* over their assertions due to radical variance of the meanings/references of the terms involved so much as they are what is implied by both the translation-failure interpretation and the Quinean notion of conceptual schemes—both notions are rooted in bivalent semantics in which every declarative sentence is either true or false. In contrast, those classical confrontations are the confrontations between two scientific languages with different *distributions of truth-value status* over their sentences due to two incompatible sets of metaphysical presuppositions underlying them. Consequently, the communication breakdown is not signified by the untranslatability between two scientific languages, but is rather indicated by the occurrence of a truth-value gap between them.

A new interpretation of incommensurability emerges from the above case studies. The current discussion of incommensurability and its related topics—such as conceptual schemes, translation, and theory comparison—are based on a tacit assumption that every sentence in a scientific language has a determinate truth-value or is either true or false (the principle of bivalence: Each sentence is assertable). What makes two scientific theories incommensurable, on the received interpretation, is the redistribution of truth-values over their assertions. On the contrary, I argue that in the discussion of incommensurability and related issues, what should concern us is the truth-value status of the sentences used to make the assertions, and not truths or truth-values of assertions. Accordingly, what we should focus on in the study of incommensurability is the truth-value functional relation instead of the meaning/reference relation, since it is the former, not the latter, that is the dominant semantic relation in the case of incommensurability.

In chapter 6, based on the insights we collect from the case studies as well as I. Hacking's and N. Rescher's writings on conceptual schemes, I propose a notion of presuppositional language as an alternative to the Quinean notion of conceptual schemes. A presuppositional language is, roughly, an interpreted language whose core sentences share one or more absolute presuppositions. Those absolute presuppositions, which I call metaphysical presuppositions, are contingent factual presumptions about the world as perceived by the language community whose truth the community takes for granted. Scientific languages are paradigmatic presuppositional languages. According to P. Strawson's notion of semantic presupposition, a sentence would be truth-valueless if one of its semantic presuppositions failed. The notion of presuppositional language is introduced primarily to explain the occurrence of a truth-value gap between two scientific languages, as we will observe in our case studies. When two presuppositional languages conflict with one another, it is likely that many core sentences of one language, when considered within the context of the other language, are truth-valueless due to the failure of some shared metaphysical presuppositions.

In chapter 7, I reconstruct the later Kuhn's taxonomic interpretation of incommensurability. According to my reconstruction, two scientific languages are incommensurable when the core sentences of one language, which have truth-values when considered within their own context, lack truth-values when considered within the context of the other language due to the unmatchable taxonomic structures underlying them. So constructed, Kuhn's mature interpretation of incommensurability does not depend upon the notion of truth-preserving (un)translatability, but rather depends on the notion of truth-value status preserving cross-language communication. Hence, the later Kuhn started to move toward the direction of the presuppositional interpretation.

However, it is a controversial issue as to whether truth-value gaps should be permissible semantically. For many philosophers, the notion of truth-value gap is highly suspect. The theoretical ground of the notion of truth-value gap, namely, the notion of semantic presupposition, has been under constant attack. But it is the very notion of semantic presupposition that is at the heart of my interpretation. To clarify and defend the notions of semantic presupposition and truth-valuelessness, in chapter 8, based on my formally coherent definition of the notion of semantic presupposition, I defend the two notions against some most damaging objections.

Based on the formal treatment of semantic presuppositions, I am ready, in chapter 9, to formalize two crucial notions that I introduce informally in the previous chapters, namely, truth-value gap and presuppositional language. I start with an important but often-ignored distinction between the notions of truth-value (whether an assertion is true or false) and truth-value status (whether a well-formed, meaningful sentence is a candidate for truth-or-falsity), and correspondingly, the distinction between truth conditions and truth-value conditions. Based on my definition of truth-value conditions, truth-value status is relative to a specific language. It is a language that creates the possibility of truth-or-falsehood. Therefore, it is possible for one sentence to be a candidate for truth-or-falsity in one language, but not in the other. In the second part of chapter 9, the formal structure of a presuppositional language is analyzed in detail.

The hallmark of a presuppositional language is its metaphysical presuppositions. In chapters 10 and 11, three kinds of metaphysical presuppositions are identified and illustrated in detail. They are *existential presumptions* about the existing entities in the world, *universal principles* about the existent state of the world, and *categorical frameworks* about the structure of the world around a language community. Since the essence of a presuppositional language consists in its metaphysical presuppositions, two presuppositional languages differ in just this regard, namely, by being laden with different types of metaphysical presuppositions. The metaphysical presuppositions of a presuppositional language determine the truth-value status of its sentences. If two presuppositional languages with incompatible metaphysical presuppositions confront one another, the embodied incompatible metaphysical presuppositions would conflict to the extent that they mutually exclude or suspend each other. Such a mutual violation of each other's metaphysical presuppositions would lead to a rejection of the other language by casting doubt upon whether its ontology is fit to describe reality and suspending all the possible facts associated with it, and consequently causes an ontological gap between the two languages. At the same time, violation or suspension of the metaphysical presuppositions of a presuppositional language would lead to the occurrence of massive truth-valueless sentences when considered within the context of a competing language, and results in a truth-value gap between the two languages.

Incommensurability is a semantic phenomenon closely related to the issue of how two presuppositional language communities can effectively understand and successfully communicate with one another. We have found that the advocates of the two distinct presuppositional languages often experience a communication breakdown between them. One primary purpose of introducing the notion of presuppositional language is to explain those communication breakdowns, which Kuhn and Feyerabend dubbed as the phenomenon of incommensurability. To do so, I need to locate some essential semantic and/or conceptual obstructions between two competing presuppositional languages and thus to identify a significant *necessary condition* of effective cross-language understanding and that of successful communication between the two language communities. This becomes a main focus of the rest of the chapters.

I start with, in chapter 12, one extreme of the communication breakdown, i.e., failure of cross-language propositional understanding. I argue that the comprehension of the metaphysical presuppositions of an alien presuppositional language is necessary for understanding it effectively. This shows that truth-value status plays an essential role in effective understanding: An interpreter can effectively understand an alien language only if the sentences of the language, when considered within the context of the interpreter's language, are (conceptually) true or false. When an interpreter encounters an alien language, the most common approach to understand it is to try to consider it within the context of his or her own language. In this case, the interpreter might easily and facilely project his or her own well-entrenched language (form of life, tradition, worldview, or framework) onto the alien's. When the metaphysical presuppositions of the two languages in question are compatible, this projective way of understanding usually does not cause a problem that would hinder mutual understanding; but when they are incompatible, the would-be communicators will experience a *complete* communication breakdown due to the failure of effective understanding of the other's language.

However, it is a mistake to assert that the would-be-communicators of two incommensurable languages cannot understand one another *per se*. Anything that can be said in one human language can be, with imagination and effort, understood by the speaker of another human language. The as-of-now communication breakdown only shows us that propositional understanding does not work in the case of incommensurability. It calls for a different way of understanding. However, what we need in the incommensurable cases is not the opposite approach to the projective understanding, namely, the adoptive way of understanding—to suppose that one can understand others only when one thinks, feels, and acts like others. The thesis of incommensurability does not necessarily lead to radical relativism. H.-G. Gadamer's hermeneutic understanding provides us a way of going beyond both projective and adoptive ways of understanding, both of which are propositional understanding in nature. In chapter 13, I discuss the possibility of restoring mutual understanding between two competing presuppositional languages in terms of Gadamer's hermeneutic understanding, and explore some hermeneutic dimensions of the thesis of incommensurability.

Even if hermeneutic understanding can overcome *complete* communication breakdown due to the failure of mutual propositional understanding, there still exist some much more significant cases of communication breakdowns between two competing presuppositional languages with incompatible metaphysical presuppositions. When the speakers of two competing languages can understand but cannot successfully communicate with one another, they experience a *partial* communication breakdown. In chapters 14, 15, and 16, I argue that such a partial communication breakdown is inevitable between two competing presuppositional languages with incompatible metaphysical presuppositions. It is the existence of such partial communication breakdown, i.e., communication breakdown *per se*, that establishes the metaphysical significance of the phenomenon of incommensurability.

Chapter 14 focuses on the received transmission model of linguistic communication (informative communication). Propositional understanding should be distinguished from informative communication. Understanding is necessary, but not sufficient for successful communication between two incommensurable presuppositional languages. Successful cross-language communication requires much more than mutual understanding. Even in the case that both sides are bilinguals who can understand one another's language, they still cannot communicate successfully with each other if the metaphysical presuppositions of the two languages are incompatible. This is because successful informative communication between two presuppositional languages requires shared or compatible metaphysical presuppositions. Otherwise, the communication between them is inevitably *partial*.

Chapters 15 and 16 discuss the two versions of the dialogical model of communication, which uses conversation or discourse, rather than transmission of information, as the central metaphor. One is Gadamer's conversation model based on his philosophical hermeneutics, and the other is J. Habermas's discourse model derived from his theory of communicative action. For both Gadamer and Habermas, cross-language communication is essentially the process of coming to hermeneutic understanding oriented toward agreement or consensus—not simply comprehension as propositional understanding entails—in terms of either genuine conversation (Gadamer) or through argumentation in dialogue (Habermas). In chapter 15, through analysis and criticism of Gadamer's common language requirement through a full fusion of horizons, I conclude that cross-language communication between two incompatible presuppositional languages is inevitably partial. After identifying three essential conditions of communication according to Habermas's discourse model in chapter 16, I argue that those conditions can be met only if the metaphysical presuppositions of two presuppositional languages are compatible. Therefore, chapters 14, 15, and 16 eventually reach the same conclusion: Shared or compatible metaphysical presuppositions between two presuppositional languages are necessary for successful communication between two distinct presuppositional languages. In other words, the communication between two presuppositional languages with two incompatible metaphysical presuppositions is inevitably partial. It is such *partial communication breakdown* between two distinct presuppositional languages that gives the real theoretical thrust of the thesis of incommensurability as communication breakdown.

A communication breakdown between two language communities is semantically indicated by the occurrence of a truth-value gap between them. I contend that this kind of communication breakdown due to the occurrence of a truth-value gap signifies that the two languages are incommensurable. In light of the above considerations, it seems to me that it is much more appropriate to use the occurrence of a truth-value gap between two rival presuppositional languages, signified by a communication breakdown and caused by incompatible metaphysical presuppositions, as a touchstone of incommensurability. For two presuppositional languages to be incommensurable is for the core sentences of one language to lack truth-values when considered within the context of the other language. More precisely, corresponding to two degrees of communication breakdown—that is, partial or complete communication breakdown—we can identify two degrees of incommensurability: moderate versus radical incommensurability. The moderate incommensurability relation between two competing languages associated with partial communication is the incommensurability of real metaphysical significance. The above is what I argue in the final chapter.

The presuppositional interpretation is inspired by a few philosophers' works on some related topics. H. Gaifman's (1975, 1976, and 1984) works on

Wittgenstein's concept of ontology rescues me from the 'morass' of the problem of incommensurability and puts me on the right track. I. Hacking's (1982, 1983) and N. Rescher's (1980) emphases on the distinction between the notion of truth-value and the notion of truth-value status as well as their clarification of the notion of conceptual schemes on the basis of truth-value status set a basic semantic framework for my project. My new interpretation emerges from reading the above works. Taking such a new perspective, I first puzzled over Kuhn's and Feyerabend's mature works on incommensurability. To my surprise, I found out that both pioneers have already hinted at the new interpretation in many profound ways.

However, although those philosophers has expressed ideas of the presuppositional interpretation, in different ways, most of these are expressed implicitly when they deal with other related issues. There are only some scattered insights here and there, which have not been made as a full case. This is part of the reason why these insights on incommensurability have not gained their deserved attention up to now. My purpose is to develop these insights into a full case, to give them a clear and coherent formulation. I hope that this treatment will help the presuppositional interpretation of incommensurability gain the acceptance that it deserves. This would be not only good in itself—to establish the tenability of the notion of incommensurability—but the effect on other related issues would be quite beneficial, such as the notion of conceptual schemes, the issue of cross-language understanding and communication, and the notion of truth-value status and truth-value conditions.

5. Scientific Language, Trivalent Semantics

Before we start our adventure of dipping into 'the morass' of the problem of incommensurability, let us clarify a few crucial concepts and related terminology conventions that I have been using above and will continue to use throughout the book.

Scientific Theory and Scientific Language

If semantic incommensurability has something to do with the semantic relations between two radically disparate scientific texts, then which of the following candidates is/are the appropriate term(s) of such a relation: scientific theories or scientific languages?

The terms 'scientific theory' and 'scientific language' are often used interchangeably by many analytical philosophers, as in the writings of T. Kuhn, P. Feyerabend, W.V. Quine, D. Davidson, H. Putnam, N. Rescher, due to the conceptual affinity between the two notions. Although it does not matter much to many how one construes the notions of scientific theory, scientific language, and their interrelation, many confusions in the discussion of the issue of incommensurability and related matters, such as the notion of conceptual schemes, do arise because of a lack of reasonable distinction between scientific theory and scientific language. Therefore, a certain distinction between them is called on here.

The term 'scientific theory' can be used to refer to a specific individual theory whose core is a very specific set of related doctrines (hypotheses, axioms, or principles) that can be utilized for making specific experimental predictions and for giving detailed explanations of natural phenomena. Examples are Maxwell's theory of electromagnetism, Wegener's theory of continental drift, and Einstein's theory of the photoelectric effect, etc. On the other hand, the term can be used to refer to a whole spectrum of individual theories. The core of such a scientific theory consists of much more general, much less easily testable, sets of assumptions or presuppositions. Following P. Feyerabend, I use the term 'comprehensive scientific theory', which roughly correspond to P. Feyerabend's 'background theories', T. Kuhn's 'paradigms', I. Lakatos' 'research programmes', or L. Laudan's 'research traditions', such as quantum theory (including quantum field theories, group theories, S-martix theory, and renormalized field theories), Newton's mechanics, Aristotelian physics, Einstein's relativity theory, traditional Chinese medical theory, etc. What we are concerned about in the discussion of incommensurability, as Kuhn and Feyerabend emphasized, is not restricted individual scientific theories, since restricted theories rarely lead to the needed conceptual revisions, but comprehensive scientific theories.

Feyerabend once pointed out that comprehensive scientific theories are sufficiently general, sufficiently 'deep', and have developed in sufficiently complex ways. They can be considered, to some extent, along the same lines as well-developed natural languages (Feyerabend, 1978, pp. 224-5). As I see it, we can link scientific theory with scientific language on the basis of a modified 'semantic' approach of Beth and van Fraassen's (Fraassen, 1970, 1989). We can think of a scientific theory as a set of theoretical definitions plus a number of theoretical hypotheses. Divorced from the 'syntactic' approach of the classic view—the idea that underlying any scientific theory is a purely formal logical structure captured in a set of axioms formulated in an appropriate formal language—the semantic approach shifts the focus from the axioms (as linguistic entities) to the models of axioms (non-linguistic entities—any physical or conceptual entities and processes that satisfy the axioms). The important distinction between the two approaches consists in the idea that the semantic account takes models as fundamental while the syntactic account takes statements, particularly laws, as fundamental. Contrary to the syntactic approach, which identifies a theory (i.e., Newtonian physics) with a definite set of statements (i.e., Newton's three laws plus the law of universal gravitation), scientific theories are, according to the semantic approach, not linguistic entities. Rather, theories must be some extra-linguistic structures standing in mapping relations to the world.

If so, a scientific theory has to be formulated in some theoretical language with a specific lexicon (such as Kuhn's lexical structure), plus syntax and logic. The theoretical language of a scientific theory consists of a consistent set of sentences or statements while the theory formulated in the language is either these sentences marked as 'believed', or a distribution of degrees of beliefs, in Bayesian style, over the sentences. Adopting such a semantic approach leaves wide latitude in the choice of languages for formulating particular scientific theories. In principle, any language could be used to formulate a theory, including extensions of everyday natural languages constructed through pragmatic observations of the linguistic usage within a scientific community, not just formal languages. We can call the theoretical language employed by a scientific theory the language of that theory, such as the language of modern chemistry, the language of the Copernican theory of astronomy, or the language of the phlogiston theory of combustion. In general, I will call accompanying languages of scientific theories *scientific languages*.

We need to notice that the concept of language in the discussion of incommensurability has several different meanings, which sometimes leads to confusion, especially in the discussion of the untranslatability thesis. Some distinctions are necessary here. First, 'a language' can refer to either a scientific language or a natural language. Obviously, scientific language should be distinguished from natural language. On the one hand, scientists express their ideas and theories in extensions of natural languages. The very same scientific language (say, the language of Newtonian physics) used to *formulate* a scientific theory (Newtonian physics) could be embedded in, expressed by, or coded in different extended natural languages (say, English or German). On the other hand, two different scientific languages (say, the language of Newtonian physics and that of Aristotelian physics) can be couched in or recorded by the same natural language (English). For our later discussions, 'language' usually refers to 'scientific language' when the context is clear. Whenever it is necessary, I will make it clear which language, natural language or scientific language, is under consideration. Second, sometimes it is necessary to distinguish the object language (such as the Newtonian language versus the Aristotelian language) that we are discussing from the metalanguage that we use to do the discussion—in our case, it will be English.

These two distinctions are especially important when we discuss the issue of possible translation between two texts. We have to be clear about the terms of translation relation: translation between two scientific languages, between two natural languages, or between one scientific language and one natural language. According to the accepted translation-failure interpretation, the languages of two alleged incommensurable theories cannot be mutually translated into one another. However, this does not mean without further qualification that an out-of-date scientific language (say, the language of phlogiston theory) cannot be translated into a contemporary natural language (say, English).[16]

We are now in a position to identify the terms of incommensurability relations. According to common usage, incommensurability is supposed to be about the relation between two conceptually disparate scientific theories. We usually claim that two scientific theories are incommensurable. Whereas, it is the *languages* of two scientific theories that are incommensurable. Theories are incommensurable only in a derivative sense. According to my presuppositional interpretation, two scientific theories are incommensurable if there is a truth-value gap between the languages of the theories. Furthermore, according to the received translation-failure interpretation, to say that two theories are incommensurable is actually to say that they are formulated in languages that are not mutually translatable. Again, it is languages, not theories, that are the terms of incommensurability relations. Therefore, the proper terms of incommensurability relations should be scientific languages. For this reason, I will confine my investigation to the analysis of the notion of incommensurability between scientific languages. Although I will be talking about incommensurability between two scientific theories from time to time, this is only a conventional way of speaking.

The notion of incommensurability seems to deal exclusively with scientific theorizing. But this notion can be extended to apply to a broader linguistic phenomenon, such as possible incommensurability between two conceptually disparate natural languages. Many philosophers (Feyerabend, R. Rorty, and others) extend the domain of incommensurability even further to cover much broader non-linguistic entities, such as traditions, forms of life, worldviews, cultures, or cosmological points of view. Although I am sympathetic with such a broad use of the notion of incommensurability, I will primarily focus on the linguistic dimension of incommensurability in the following discussion.

Bivalent Semantics vs. Trivalent Semantics

Our discussion of the presuppositional interpretation of incommensurability will proceed within the framework of trivalent semantics, which will be presented and formalized in chapter 8. Nevertheless, the use of a very broad brush here on the difference between bivalent and trivalent semantics may be helpful for the reader.

In logic, a proposition that has a definite truth-value, either true or false, is called bivalent. In standard bivalent semantics, a declarative sentence has, by default, a definite truth-value. It could only be either true or false. No truth-valuelessnesss (neither-truth-nor-falsity) is permissible. That means that any declarative sentence is assertable as a statement (by definition, a statement or assertion is either true or false).

In contrast, within trivalent semantics, a declarative sentence (such as 'the present king of France is bald') could be true, false, or *neither-true-nor-false*. The standard bivalent semantics only deals with truth-or-falsity, not neither-truth-nor-falsity. Within a three-valued semantics, besides the classical truth-values (truth or falsity), we have to add one more *kind* of truth-related semantic value of declarative sentences, which I will dub 'the truth-value status'. The notion of truth-value status concerns whether a declarative sentence has a (classical) truth-value or whether a sentence is a candidate for truth-or-falsity (this is never a question within standard bivalent semantics). A sentence has two different truth-value statuses: either-true-or-false versus neither-true-nor-false. If a sentence has a truth-value, we say that it has a positive truth-value status (being true-or-false); otherwise it has a negative truth-value status (being neither-true-nor-false).

Accordingly, the truth-related evaluation of a declarative sentence S involves two connected stages: First, we have to determine the truth-value status of the sentence based on some given truth-value conditions. The question is whether S has a truth-value when considered within a certain language L. Second, if the answer to the first question is positive (*S* has a truth-value), then we can go ahead to determine what *S*'s truth-value is (it is either true or false) based on some given truth conditions. If the answer to the first question were negative, *S* would be truth-valueless (having no classical truth-value) from the viewpoint of language L. In this case, there is a truth-value gap regarding *S*. If a substantial number of core sentences of a language L_1, when considered within the context of a competing language L_2, lack classical truth-values, or vice versa, then there is a truth-value gap between the two languages L_1 and L_2.

Notes

1   T. Kuhn recalled this life-altering event again in his later publication in 1999.
2   Kuhn, 1983b, p. 669; 1987, pp. 8-12; 1988, pp. 9-11; 1991, p. 4; 1999.
3   Kuhn, 1983b, p. 669; 1987, pp. 8-9; 1988, pp. 9-10; 1991, p. 4.
4   This assumption is a basic methodology of communication theory that is forced on us when we try to communicate with others who employ different languages. Our expression of the basic assumption is innocent since it does not specify the content of agreement. But at the same time, it is too vague to be useful in the process of communication.
5   Kuhn's early position is represented in Kuhn, 1970a, 1970b, 1976, 1977b, and 1979.
6   See Kuhn, 1970a, pp. 200-201; 1970b, pp. 275-6; 1976, p. 195; 1979, p. 416; 1987, pp. 20-21.
7   Kuhn, 1983b, p. 683; 1988, pp. 9, 16; 1991, pp. 4-5, 9; 1993a, pp. 323-6.
8   To have a taste of the popularity and pervasiveness of the notion of incommensurability in other areas, simply go to Amazon.com and search for books on incommensurability. You will find a paper collection from *Symposium: Law and Incommensurability* (University of Pennsylvania, 1998), a psychology book entitled *Toward a Unified Psychology: Incommensurability, Hermeneutics and Morality* (Institute of Mind and Behavior, 2000), a book on political economy, *Ordinary Choices: Individuals, Incommensurability and Democracy* (Routledge, 2005), two books on rhetoric and communication, *Rhetoric and Incommensurability* (Parlor Press, 2005) and *Judgment, Rhetoric, and the Problem of Incommensurability* (University of South Carolina Press, 2001), and so on.
9   Attempts at further explications and even precise formulations of the semantic notion of incommensurability different from the received interpretation to be presented in chapter 2 can occasionally be found in the literature, such as W. Stegmülar, 1979, H. Hung, 1987, J. Hintikka, 1988, W. Balzer, 1989, B. Ramberg, 1989, among others. These attempts are not satisfactory for many different reasons.
10   See Kuhn, 1983b, pp. 669-70; 1988, p. 10; 1991, p. 4; 1999.
11   Please see P. Winch, 1958 for the manifestation of incommensurability in social sciences.
12   For a prize case study, please read Feyerabend's analysis of the cosmology shift from archaic to classical Greece in his 1978, chapter 17.
13   See B. Whorf, 1956.
14   Kuhn, 1970a, pp. 103, 148, 111-20, 150, 109-10.
15   By a truth-value functional relation between two languages, I mean the semantic relation regarding the truth-value status of the sentences in these languages. This is intended to be distinguished from the familiar logical relation regarding truth-values of

the sentences—namely, the truth functional relation between two languages. It is my task to clarify such a truth-value functional relation between the languages of two competing scientific languages in the case of incommensurability in the following chapters.

16 H. Putnam's argument against the thesis of untranslatability is based on confusion between scientific language (as object language) and natural language (as metalanguage). See Putnam, 1981, p. 114.

Chapter 2

Incommensurability as Untranslatability

1. The Received Interpretation: Incommensurability as Untranslatability

Even though adequate analyses of the notion and the thesis of incommensurability are conspicuously absent from the literature, a somewhat more specific sense is nevertheless often associated with the notion—namely, incommensurability as untranslatability due to radical variance of meaning[1] (sense and/or reference) of terms occurring in competing scientific languages.

It is widely accepted that Kuhn's and Feyerabend's thesis of incommensurability hangs on their contextual theories of meaning, which were advanced as a result of their rejection of the standard empirical account of the meaning of scientific theory. In brief, the traditional empirical theory of meaning has two main components: the reductionist account of meaning (reduction of meanings to sensory experience) and the 'double-language' model of scientific theory (the distinction between theoretical language and observation language). Feyerabend finds that such an 'orthodox empiricist' theory of meaning is untenable, for it depends on an *a priori* assumption of a special relationship between the human observer and an observer-independent physical reality. This *a priori* assumption is at the heart of the theoretical-observational distinction, which holds that sensations or sensory stimulations of the human observer have significance for meaning. Feyerabend contends that a truly empirical philosophical position must see the interaction between the observer and the 'world' as one of the processes that needs to be examined and cannot be taken as *a priori*. In particular, we need to examine it in the light of other hypotheses concerning the relation between observers and the 'world'. Feyerabend proposes just such a hypothesis: All observation sentences are theory-laden in the sense that experience and empirical meaning, if any, must occur within a comprehensive 'world-view' or a theory. Based on this hypothesis, Feyerabend advances his version of the contextual theory of meaning, according to which, it is a theory that gives meaning to observation sentences and not the other way around.[2] The following quotation from Feyerabend summarizes his position best:

> After all, the meaning of every term we use depends upon the theoretical context in which it occurs. Words do not mean something in isolation; they obtain their meanings by being part of a theoretical system. Hence if we consider two contexts with basic principles that either contradict each other or lead to inconsistent consequences in certain domains, it is to be expected that some terms of the first context will not occur in the second with exactly the same meaning. Moreover, if our methodology demands

the use of mutually inconsistent, partly overlapping, and empirically adequate theories, then it thereby also demands the use of conceptual systems that are mutually irreducible (their primitives cannot be connected by bridge laws that are meaningful and factually correct), and it demands that the meanings of all terms be left elastic and that no binding commitment be made to a certain set of concepts. (1965c, p. 180)

The early Kuhn concurs: 'Successive theories are incommensurable (which is not the same as incompatible) in the sense that the referents of some of the terms which occur in both are a function of the theory within which those terms appear' (1979, p. 416).

According to the above readings of Kuhn's and Feyerabend's contextual theories of meaning, the meanings of some apparently similar terms employed by two competing scientific languages would be different. Moreover, it is hardly possible to identify and is impossible to formulate the meanings of the terms occurring in a competing language within the home language. This is because the meaning of a theoretical term is supposed to be determined by the theory in which it occurs. If a theoretical term t (say 'phlogiston') is taken out of its appropriate theoretical background L_P (say, the phlogiston theory), it would lose its original semantic value in that theory. So, in this view of meaning, when t is examined from the point of view of a competing theory L_C (say, modern theory of chemistry), even if the original semantic value of t in L_P might be *identifiable* in terms of some expressions within L_C, the value cannot be expressed or *reformulated* in L_C without loss. In this case, we would say that the full semantic value of t in L_P is *unrecoverable* within L_C. In particular, the meanings of the shared terms in two competing languages (such as 'mass' in both Newtonian and relativity theory) would be different. If this is the case, we say that there is a lack of a certain desirable semantic relation—that is, *meaning continuity*—between L_P and L_C. Consequently, the proponents of two competing scientific languages would inevitably talk past each other when they attempt to resolve their disagreement. It is not even clear whether there is any genuine logical disagreement at all between them. In this case, the two languages are incommensurable.

Presumably, a lack of meaning continuity between two competing scientific languages has one closely related chief consequence for inter-language translation between them. It seems self-evident that mutual translation between two languages is possible only if at least two conditions can be met: (a) the meanings of shared terms (such as 'mass' in both Newtonian physics and relativity theory) can be preserved; (b) the meanings of different terms in the competing language (such as 'phlogiston' in phlogiston theory) can be identified and reformulated in the home language (for example, the modern theory of chemistry). The lack of meaning continuity between the terms occurring in two languages would inevitably render the two languages untranslatable. As such, it is believed that untranslatability turns out to be the very nature of incommensurability (*incommensurability as untranslatability*). As a matter of fact, as early as at the end of 1960s and during 1970s, Kuhn[3] realized that 'untranslatable' (Quine's term) is a better word than 'incommensurable' to describe a communication breakdown between two alien contexts: 'In applying the term "incommensurability" to theories, I had intended only to insist that there was no common language within which both could be fully expressed' (Kuhn, 1976, p. 191). Thus, instead of saying that Aristotle's physics and Newton's physics are incommensurable, one should say that many substantial Aristotelian sentences could not be translated into Newtonian sentences.

As far as the sources of incommensurability are concerned, we can identify two alternatives within the framework of the translation-failure approach with respect to which semantic value, meaning (sense) or reference, plays the essential role in the explication of incommensurability. *The meaning alternative* regards incommensurability as the result of a radical change in the meanings (senses) of the non-logical terms occurring in two competing languages in question (incommensurability due to radical meaning variance). *The reference alternative* attributes incommensurability to radical change in the references of the terms employed in two competing languages (incommensurability due to radical reference variance).

To sum up, according to the received translation-failure interpretation, to say that two scientific theories T_1 and T_2 are incommensurable is to say that the languages employed by T_1 and T_2 do not exhibit meaning continuity due to the different semantic values attached to their key terms. Consequently, the two languages are mutually untranslatable.

2. An Influential Line of Criticism

The above standard reading of Kuhn's and Feyerabend's thesis of incommensurability has been the target of an influential line of criticism. It is not surprising that almost all the criticisms have focused on Kuhn's and Feyerabend's alleged contextual theories of meaning. Critics argue that Kuhn's and Feyerabend's thesis of incommensurability based on their contextual theories of meaning faces many difficulties.

First, critics argue that Kuhn's and Feyerabend's treatment of meaning is too general. Kuhn does not specify precisely how he thinks the referents of scientific terms are fixed or established by the theoretical context, and how referents of these terms can change with shift in the theoretical context in which these terms occur. For example, the content of 'context' in a contextual theory has to be specified. Otherwise, it will become absurd: If any change in a theory radically alters the meanings of its terms, then any two theories with any slight difference would be incommensurable. However, there is no appropriate way to specify 'context' no matter whether we identify it as the whole theory or as a part of it. Kuhn's and Feyerabend's positions are too ambiguous so no unequivocal analysis can be given.[4] This is the reason why many complain that Kuhn and Feyerabend owe us a clearly formulated theory of meaning that is specific enough to permit extensive scrutiny.

Second, critics claim that there are no good reasons, nor does Kuhn or Feyerabend provide any, for maintaining that all concepts necessarily have their meanings wholly determined by their contexts of distinctive theoretical concepts. A contextual theory, like any other proper theory of meaning, should concern two aspects of 'the problem of meaning', namely, the determination of meaning on the one hand, and the change or likeness of meaning on the other. Let us assume, for the sake of argument, that the meanings of the terms of a theory T were fully determined by T. The fact that the meanings of the terms in T were fully determined by T does not compel us to conclude that the meanings of these terms would change radically if T were modified radically. In other words, from the fact that the meanings of the terms of a theory T were established by T it does not follow that there would be a one-to-one relationship between the meanings of the terms in T and theory T. It is very likely that some shared terms in two different theories T_1 and T_2 have the same meanings. Advocates of the contextual theory do not provide us with any satisfactory criteria of meaning change. Therefore, the thesis of incommensurability does not necessarily follow from the contextual theory of meaning.[5]

Third, critics question in what sense can two theories (such as Newtonian physics vs. relativity theory) be rivals or compete with each other if they are so semantically disparate that there is no meaning continuity between them. For example, if the meaning of a shared term 'mass' is different in relativity theory and Newtonian theory, how can we say of a statement in relativity theory that

(2)   The mass of a particle increases with the velocity of the particle

is contradictory with a corresponding statement in Newtonian physics that

(3)   The mass of a particle does not change with the change of the speed of the particle.

Last but also most significantly, according to some critics, not only is the contextual theory of meaning refuted by actual scientific practice, but also it theoretically stems from a confused concept of reference, namely, the traditional descriptive theory of reference.[6] According to this theory, each term is logically associated with a more or less certain number of descriptions. The referent of a term is the object that fits most of these descriptions. These associated descriptions are the meaning (Frege's sense) of the term. So, the reference of a term is determined by its meaning (sense). In this way, terms have meanings (senses) essentially and references only contingently. Consequently, whenever meaning (sense) changes reference changes accordingly.

Critics also contend that Kuhn and Feyerabend neglect to consider the possible separation between the change of meaning (sense) and that of reference. To defend this contention, the critics have been busy finding a way to detach the change of reference from meaning (sense) in order to keep either meaning (sense) or reference fully or partially unchanged during theory change. The believers of informal semantics[7] take an externalistic view on meaning and separate meaning (sense) from reference. It is believed that full references of scientific terms remain stable while their meanings (senses) might change with the variance of the theory containing these terms. Kripke–Putnam's causal theory of reference (Kripke, 1972; Putnam, 1973), I. Scheffler's notion of co-reference (1967), M. Prezelecki's relativized full-identity of reference (1979), and P. Kitcher's full-identity of token-reference (1978) are some representatives of informal semantics. In contrast, the followers of formal semantics[8] admit that the references of scientific terms change with the theory in which they occur. However, they try to locate some unchanged common core or commensurable part of the two references of the same term before and after the change of the theory in which the term occurs. For example, M. Martin (1971) argues that although the references of the same term in two competing theories are not identical, they somehow overlap. H. Field (1973) agrees, contending that we can identify the shared partial reference between the same terms occurring in two competing theories.

In comparison with the above reference alternative, those following the radical meaning alternative argue that what changes with the change of a theory is the references of terms while the meanings (senses) of the terms remain stable. Others, such as D. Shapere (1984) and N. Nersessian (1984, 1989), who follow the modest meaning alternative, grant meaning variance by accepting the basic principle of the contextual theory of meaning, but argue that there is a continuity of meaning (sense) during meaning variance on the basis of the chain-of-reason.

In sum, critics conclude that, contrary to Kuhn and Feyerabend, there is no such radical change in the meanings of scientific terms corresponding to any substantial change in the theory in which the terms occur. No matter how much the meanings of the terms change with the variance of the theory containing them, we can always locate some sort of meaning continuity between the terms in two theories. This continuity manifests itself either as a common core of semantic values shared by the same terms or as the recoverable semantic values between the different terms in two competing theories. It is the very existence of such a meaning continuity between the terms of two competing theories that makes mutual translation possible. The thesis of incommensurability as untranslatability fails.

* * * * *

My objections to the above influential line of criticism of the thesis of incommensurability fall into two parts. In the first part, I will assume, for the sake of argument, that the basic premise of the received translation-failure interpretation, namely, incommensurability should be identified as untranslatability. Even so, the influential line of criticism fails for two major reasons. First, Kuhn's and Feyerabend's expositions of the contextual theory of meaning are not as vague as the critics think. On the contrary, Kuhn and Feyerabend do offer us two quite comprehensive theories of meaning, which deal with all three aspects that any adequate theory of meaning is supposed to address: determination, change, and formulation of meaning. Second, the critics' attempts to locate meaning continuity is either unsuccessful or fails to render mutual translation possible between two alleged incommensurable languages.

In the second part of my response, the basic premise of the influential line of criticism, namely, the identification of incommensurability with untranslatability, is questioned and rejected. In fact, Feyerabend's and Kuhn's explication of the thesis of incommensurability is different from the standard interpretation. As will become clear later in chapters 7 and 10, Feyerabend's and the later Kuhn's expositions of incommensurability have moved away from the translation-failure interpretation.

3. Feyerabend's and Kuhn's Contextual Theories of Meaning

In some general philosophical positions concerning the theory of meaning, the problem of meaning is usually taken to have two thrusts. One is the significance or the having of meaning; the other is the likeness of meaning or synonymy. I would like to add another aspect to any adequate theory of meaning: the formulation of meaning—i.e., how to formulate the meaning of a term employed by one scientific language in a competing language. This is because whether the meaning of a term in one scientific language can be identified by the users of another scientific language is a different issue from whether the meaning of that term can be *formulated* or expressed fully within the second language. For example, the term 'dephlogisticated air' in phlogiston theory, if treated as a token expression, sometime refers to oxygen itself, sometimes refers to an oxygen-enriched atmosphere. However, although the language of modern chemistry can be used to identify the reference of 'dephlogisticated air' (as a token) employed by phlogiston theory, it is a totally different issue whether the meaning of the same term 'dephlogisticated air' can be expressed without loss within the language of modern chemistry.[9] Therefore, a full-fledged theory of meaning needs to deal with three separate aspects of meaning: (a) the determination of meaning; (b) the change of meaning; (c) the formulation of meaning.

Feyerabend's Contextual Theory of Meaning (Sense)

Examining Feyerabend's contextual theory of meaning (sense) from these three aspects, we find that Feyerabend's view, while it is certainly contextualistic, is not simply that the meanings of scientific terms are determined by the theoretical context in which they occur and vary with change of the context as usually construed.

The first issue is the determination of meaning. According to the standard reading, Feyerabend's account of meaning may initially suggest that the meanings (senses) of terms are determined by the entire theoretical context in which they occur. A theory as a whole is somehow constitutive of the meanings of the terms occurring in the theory. However, according to Feyerabend, it is not a theory as a whole, but only some specific part of it, that determines the meanings of the terms within it.

> Now it seems reasonable to assume that the customary concept of meaning is closely connected, not with definitions which after all work only when a large part of a conceptual system is already available, but with the idea of *fundamental rule*, or a *fundamental law*. Changes of fundamental laws are regarded as affecting meanings. ... There exists therefore a rather close connection between meanings and certain parts of theories. (Feyerabend, 1965a, p. 14, fn. 17; *italics* as original)

Very briefly put, by 'fundamental rules or laws'—which Feyerabend later also calls 'universal principles'—Feyerabend means some fundamental factual presuppositions underlying a scientific language. A fundamental rule of a scientific theory determines the meanings of the terms occurring in the theory in a way that suspending the rule will suspend the meanings of the terms. For instance, the notion of impetus depends upon the Aristotelian principle that all motion is the result of the continuous action of some force, which constitutes the fundamental rule of the impetus theory (Feyerabend, 1962, pp. 52-5).

Second, the issue of meaning determination is not the issue of meaning variance. From the mere fact that meaning depends on theoretical context it does not at all follow that meaning therefore bears a one-to-one relationship to theoretical context. It is possible for two terms in different theories to have the same meaning even if the meaning of each of them is fully determined by the theory in which they occur. As a matter of fact, Feyerabend himself explicitly denies a charge of extreme meaning variance based on a reading of his account. According to this reading, Feyerabend commits himself to an extreme view that, with any change of theory, no matter how slight, the meanings of the terms in this theory must change.[10] However, for Feyerabend, not every change of theory involves change of meaning, but only some more restricted forms of change do. In the following passages, Feyerabend distinguishes changes not affecting meaning from those doing so:

> The change of rules accompanying the transition T → T' is a fundamental change, and that the meanings of all descriptive terms of the two theories, primitive as well as defined terms, will be different: T and T' are incommensurable theories. (1965a, p. 115)

> A diagnosis of a stability of meaning involves two elements. First, reference is made to rules according to which objects or events are collected into classes. We may say that such rules determine concepts or kinds of objects. Secondly, it is found that the changes brought about by a new point of view occur within the extension of these classes, therefore, leave the concepts unchanged. Conversely, we shall diagnose a change of meaning either if a new theory entails that all concepts of the preceding theory have extension zero or if it introduces rules which cannot be interpreted as attributing specific properties to objects within already existing classes, but which change the system of classes itself. (1965b, p. 268)

Actually, Feyerabend here specifies two criteria of meaning change with respect to theory change: The meanings of terms in a theory T change if T undergoes substantial change to the extent either that the scheme of categorization of T is affected, or that the fundamental rules are changed.

Third and much more importantly, there is another sense in which Feyerabend's account of meaning is not merely contextualistic. Not all meaning variances lead to incommensurability in Feyerabend's sense. Feyerabend repeatedly stresses that incommensurability is not simply a matter of difference of meaning between theories:

> Mere difference of concepts does not suffice to make theories incommensurable in my sense. The situation must be rigged in such a way that *the conditions of concept formulation* in one theory forbid the formulation of the basic concepts of the other. (1978, p. 68, fn. 118; my italics)

Ten years later, Feyerabend, while denying the semantic interpretation given by H. Putnam (1981) of his thesis of incommensurability, emphasizes two differences between Putnam's interpretation and his own:

> First, incommensurability as understood by me is a rare event. It occurs only when the *conditions of meaningfulness* for the descriptive terms of one language (theory, point of view) do not permit the use of the descriptive terms of another language (theory, point of view); mere difference of meanings does not yet lead to incommensurability in my sense. Secondly, incommensurable languages (theories, points of view) are not completely disconnected—there exists a subtle and interesting relation between their *conditions of meaningfulness*. (1987, p. 81; my italics)

According to Feyerabend, neither the failure of meaning identification nor even meaning variance constitutes a sufficient condition for incommensurability. Instead, two theories are incommensurable if some central concepts in one theory cannot be formulated in the language of the other because the conditions of meaningfulness (the conditions of concept formulation) for these concepts are incompatible with each other. Furthermore, as we have mentioned above, Feyerabend identifies two basic conditions of meaningfulness: one relates to the universal principles and the other to the schemes of categorization underlying the language in which the concepts are to be formulated. According to this reading, Feyerabend, in saying that the conditions of concept formulation of one theory forbid the formulation of the concepts of another theory, is actually offering us two criteria of incommensurability:

(F1)   Two theories T_1 and T_2 are incommensurable if the respective *schemes of categorization* are incompatible in the sense that the concepts used in one scheme cannot be formulated in the other.

(F2)   Two theories are incommensurable if the respective universal principles are incompatible in the sense that the principles of one theory forbid the expression of the concepts that are expressible in terms of the principles of the other theory.

Kuhn's Lexical Theory of Reference

In the later 1980s, Kuhn introduced a new bearer of incommensurability, i.e., the lexicons/lexical structures of scientific languages. Notice that Kuhn uses 'lexicon', 'lexical structure', 'taxonomy', or 'taxonomic structure' interchangeably in his writings. By 'lexicon', Kuhn does not mean a lexicon in a linguistic sense, but rather refers to the categorical structure of a scientific language. By focusing on the lexical structure of language, the later Kuhn modifies his early contextual theory of reference by associating the determination and the change of reference of kind-terms with the nature of taxonomization. In my reading, Kuhn, by doing this, presents a modified, finer-grained, and coherent theory of reference, which I will call the lexical theory of reference of kind-terms (hereafter LTRK).[11]

The main concern of Kuhn's LTRK is about the determination and change of the reference of a kind-term. The question is: How do kind-terms actually attach to nature? What are the references of kind-terms? Here Kuhn is concerned with actual referents themselves, namely, with the set of objects or situations to which kind-terms actually attach. According to Kuhn's logic, the basic process of how the references of kind-terms are established in a language community can be summarized as follows.

Kind-terms, as category terms, are attached by a language community to the existing categories of objects or situations in a world to which they apply. It is natural to use the extensions of these corresponding categories of objects or situations as the actual referents of the terms.

Furthermore, kinds are taxonomic categories. Kinds, especially scientific kinds, can be arranged taxonomically. Each taxonomy has its unique structure that can be specified by a process of taxonomization. Usually, taxonomization is a categorizing process that can be divided conceptually into two stages: (i) individuation: both individuating entities into distinguishable items and categorizing the domain into taxonomic categories; (ii) distribution: distributing items into pre-existing categories. Corresponding to these two stages, the structure of taxonomy consists of two parts: taxonomic categories or kind-terms and relationships (similarity/dissimilarity) among these categories. More formally, a taxonomic or lexical structure is the vocabulary structure shared by all members of a language community that provides the community with both shared taxonomic categories/kind-terms and shared relationships between them.[12]

Existing in a taxonomy, interrelated taxonomic categories or kind-terms in a local area cannot be isolated from one another because 'those categories are interdefined', must be learned or relearned together, and laid down on nature as a whole (Kuhn, 1983b, p. 677). More precisely, kind-terms with normic expectations (the expectations with exceptions), which we call 'high-level theoretical kind-terms',

> must be learned as members of one or another contrast set. To learn the term 'liquid', for example, as it is used in contemporary non-technical English, one must also master the terms 'solid' and 'gas'. The ability to pick out referents for any of these terms

depends critically upon the characteristics that differentiate its referents from those of the other terms in the set. (Kuhn, 1993a, p. 317)

On the other hand, the kind-terms with nomic expectations (the expectations without exceptions), which we can call 'low-level empirical kind-terms', must be learned with other related terms and from situations exemplifying laws of nature. For example, 'force' has to be learned both with 'mass' and 'weight' and with recourse to Hooker's law and Newton's three laws of motion.[13] Therefore, taxonomy is necessarily locally holistic. Change in one category would affect other categories of a taxonomy in a local area. This means that references of kind-terms, as the extensions of the corresponding categories in a taxonomy, are determined by the taxonomic structure in some local part of the language. In this sense, 'the criteria relevant to categorization are *ipso facto* the criteria that attach the names of these categories to the world' (Kuhn, 1987, p. 20). Categorization is a coinage with two faces, one looking outward to the world to divide it in different ways, the other inward to the world's reflection in the referential structure of the language to determine the references of kind-terms.

In conclusion, according to Kuhn's LTRK, the extensions of kind-terms are determined by taxonomic structures of languages, and will change with change of such structures. These reference changes are confined to some selected high-level theoretical kind-terms and leave other terms referentially stable. Therefore, a necessary condition of sameness of reference of shared kind-terms between two languages is that they share the same taxonomic structure.[14]

> Unlike two members of the same language community, speakers of mutually translatable languages need not share terms: 'Rad' is not 'wheel'. But the referring expressions of one language must be matchable to coreferential expressions in the other, and the lexical structures employed by speakers of the languages must be the same, not only within each language but also from one language to the other. Taxonomy must, in short, be preserved to provide both shared categories and shared relationships between them. (Kuhn, 1983b, p. 683)

If so, when two languages with incompatible taxonomic structures confront each other, then the kind-terms from one language cannot be translated into the corresponding terms in the other without violating the non-overlap principle (No two kind-terms in the same level of a taxonomy may overlap in their extensions. For example, there are no dogs that are also cats in the mammal taxonomy). That means that the mutual translation between the two languages is impossible. This combination of a logical theory of taxonomy and a linguistic theory of translation constitutes the later Kuhn's taxonomic interpretation of incommensurability. 'Incommensurability thus becomes a sort of untranslatability, localized to one or another area in which two lexical taxonomies differ' (Kuhn, 1991, p. 5).

It is clear that, as we have shown above, there does not exist, nor does Kuhn or Feyerabend claim, an alleged one-to-one correspondence relationship between a theory and the meanings of its terms. It is not that all changes of a theory would affect the meanings of its terms. The fact that some shared terms in two different theories might have the same meaning is perfectly consistent with Kuhn's and Feyerabend's versions of the contextual theory of meaning. Kuhn's and Feyerabend's contextual theories of meaning, if correct, really lead to the conclusion that there exists some sort of meaning discontinuity between two competing languages. This sort of discontinuity would lead to untranslatability.

4. Meaning Continuity and Translatability

The critics contend (a) that it is possible to locate some kind of meaning continuity between two competing languages, and (b) that the kind of meaning continuity identified would render mutual translation between them possible. However, the critics' efforts to identify meaning continuity are either unsuccessful, or, if successful, the kinds of meaning continuity identified cannot be used to disprove the thesis of untranslatability.

The Meaning (Sense) Alternative

Do the critics succeed in identifying the desired meaning continuity between two competing languages? For the answer, let us examine the meaning (sense) alternative first. D. Shapere (1984) contends that although meaning (sense) will change with the variance of beliefs or theories, we can still locate meaning (sense) continuity between the languages of two theories by identifying a chain-of-reasoning connecting the two theories. According to Shapere, the rationality of science can be preserved as long as we can locate such a chain-of-reasoning during the development of a scientific theory. However, the attempts to retain the rationality of science in terms of explaining the rationality of conceptual change or meaning change are doomed to failure for the following reason. It is true that a historian of science can make the shift of scientific belief systems intelligible by tracing back the chain of change for each concept. However, historical explanation itself is not a philosophical justification. There is nothing the philosopher can add to what the historian has already done to show that this intelligible and plausible course of development is a rational one. The result would be that everything that has happened in the history of science would be rational, as long as we can reconstruct its actual process of change. This inference is valid only if we accept a *presupposition that science itself is a rational enterprise.* It begs the question; since whether or not science is rational is the issue at stake that is what Shapere wants to prove through the introduction of the concept of a chain-of-reasoning in the first place.

In addition, it is reference variance rather than meaning (sense) variance that has the relativistic and anti-realistic implications allegedly promoted by the thesis of incommensurability. Historical relativism and scientific anti-realism cannot follow from meaning variance alone. What motivates many philosophers to oppose the thesis of incommensurability is the defense of scientific realism. Scientific realism is a doctrine about reference (the notion of reality and truth), not about meaning (sense). Even we assume that meaning (sense) can be separated from reference, meaning continuity still cannot avoid the consequences brought about by radical change of reference (Leplin, 1979). Therefore, even if the opponents following the meaning alternative succeed in proving the existence of stable meaning or continuity of meaning, they still cannot defeat the thesis of incommensurability and avoid its undesirable implications.

The Reference Alternative

The critics who take the route of informal semantics within the framework of the reference alternative argue that the meanings (sense) of terms may change with the variance of a theory in which the terms occur, but that their references remain stable. This alternative seems to be more promising than the meaning alternative. Even if the meaning (concept) of a term in a theory cannot survive the change of a whole belief system, since they are intertwined inextricatively, it is still possible that the reference of the term survives the change. One possible way to explain the stability of reference is to follow G. Frege's initiative. I. Scheffler (1967) argues that by separating sense from reference following Frege's approach, we can keep reference unchanged while meaning varies. This is a strange proposal considering that Frege explicitly adopts and promotes the description theory of meaning. From the fact that some terms different in meaning (sense) have the same reference, it does not follow that a term can always keep its reference unchanged while changing its meaning. Frege's theory of meaning provides no basis for expecting such stability of reference at all. On the contrary, as long as we accept Frege's theory, the task of identifying stable reference is hopeless. According to Frege, the only way we have of identifying the reference of a term is by the criteria formulated in its sense; so even if it were true that reference remained the same when sense changed, we would have no (*a priori*) means of knowing it.

   Another possible way to explain the stability of reference is more revolutionary, namely, to apply S. Kripke's (1972) causal theory of names to general terms, as H. Putnam (1973) does. According to this theory, there is a causal chain of reference-preserving links between the naming ceremony of a term and its use on some occasion. The reference of a term, either an individual or a general term, cannot change after it is fixed in the initial naming ceremony by ostension. However, it is well known that the causal theory of reference faces many difficulties. As G. Evans (1973) points out, the existence of a causal chain is neither sufficient nor necessary for an individual to identify the reference of a name. Especially, the references of many scientific terms (such as 'positron', 'photon', 'heat', or 'water') were not fixed by any so-called naming ceremony by ostension. Actually, their initial references were fixed by their roles in a certain theoretical context, rather than on the basis of some experience attributed to the causal agency of the alleged referent (Leplin, 1979). So the causal theory at most tells us what a theory of reference *could* be, not what a theory of reference *actually* is. More importantly, the references of many scientific terms do change during the development of scientific theories, as Kuhn argues convincingly in many places, such as the difference in the references of the kind-term 'compound' in pre-Dalton and post-Dalton chemistry. Besides, it is perfectly conceivable that the reference of a name will change in some circumstances (imagine that two babies are switched without notice right after their mothers give them names). Therefore, any theory of reference that cannot accommodate the possibility of reference variance is essentially inadequate.

The failure of informal semantics seems to encourage a move toward formal semantics. According to formal semantics, although the reference of a term does change with the variance of the theory in which it is embedded, there exist some sort of commensurable parts between two different references of the same term embedded in two alleged incommensurable theories. This reminds us of Shapere's chain-of-reasoning within the meaning alternative. However, formal semantics is not in better shape than informal semantics. To see the chief drawback of formal semantics, we need to consider the second issue concerning referential continuity: Does the existence of referential continuity identified by formal semantics between two alleged incommensurable languages disprove the thesis of incommensurability as untranslatability?

As I have mentioned, the motivation behind the attempts to identify commensurable reference is to restore mutual translation between two alleged incommensurable languages. However, formal semantics, if successful, fails to give us translatability. The reason is simple. The thesis of incommensurability as untranslatability is a negative thesis. To argue for untranslatability between two competing languages, one only needs to show that some necessary conditions of intertranslation cannot be met between two alleged incommensurable languages. Although controversy exists about what should count as necessary conditions of translation, it is safe to say that a successful translation requires one to *formulate* any expression e_s in the source language (the language to be translated) into an appropriate expression e_t, a semantic equivalent of e_s, in the target language (the language into which the translation is made). Sameness of reference and meaning (sense) are two obvious desirable semantic equivalents that should be preserved during translation. As long as Kuhn shows that it is impossible to formulate some crucial expressions in the source language (say, the term 'mass' in Newtonian theory) into an appropriate expression in the target language (say, the term 'mass' in relativity theory) while preserving the same reference and meaning of the original term, he succeeds in arguing for the thesis of untranslatability. But in order to show the failure of the thesis, the critics have to demonstrate either that mutual translation between two alleged incommensurable languages is possible or, at least, that fully identical reference or meaning of expressions e_s in the source language (say, 'dephlogisticated air' in phlogiston theory) can be preserved by formulating e_s into the corresponding expression e_t in the target language (say, 'oxygen' in modern chemistry). Unfortunately, the advocates of formal semantics fail to do either. On the one hand, formal semantics only shows us the possibility of preserving partial or overlapping references, instead of identical references, between the corresponding terms in two alleged incommensurable languages. On the other hand, the existence of overlapping references or partial identical references (even the existence of full identical references) between the shared terms in two alleged incommensurable languages is not sufficient for translation. Therefore, even if the critics succeeded in showing that there exists some kind of overlap between the references of the shared terms between two alleged incommensurable languages, they would still not disprove the thesis of untranslatability.

A similar argument can be applied to informal semantics as well. According to informal semantics, it is possible for references of scientific terms to remain stable while their senses change with the variance of the theory containing these terms. Therefore, it is possible to identify the full reference of an expression in the source language in terms of an appropriate expression in the target language. For instance, P. Kitcher (1978) argues that the language of modern chemistry can be used to identify the referents of the terms or expressions of phlogiston theory, at least to the extent that those terms or expressions actually do refer in each specific context (context-sensitive reference or token-reference). For example, 'dephlogisticated air' refers in some contexts to oxygen itself, or refers in some contexts to an oxygen-enriched atmosphere. However, as Kuhn (1983b) points out, the crucial issue concerning translation is not whether the referent of a term in the source language can be *identified* by the appropriate expression in the target language, which is often possible; but whether the term can be *formulated* by the corresponding expression(s) in the target language to the extent that those expressions can capture the beliefs, intentions, and implications conveyed by the original terms (such as the whole belief system associated with the terms 'phlogiston'). Reference-identification alone is not a sufficient condition of translation. Hence, reference-identification is not translation.

To sum up, the critics are either unsuccessful in providing us with the meaning continuity that they promise us or fall short of what they intend to achieve; namely, mutual translation between alleged incommensurable languages.

5. Untranslatability and Incommensurability

The previous two sections might give readers the impression that I am trying to defend the standard interpretation of incommensurability as untranslatability. This is not the case at all. It is not my intention to make a judgment about which side, Kuhn's and Feyerabend's thesis of incommensurability or the above influential criticism of it, holds more water. Instead, I believe that both sides are on a wrong track: It is unproductive and futile to identify incommensurability as untranslatability. The notion of incommensurability (commensurability) does not depend on the notion of untranslatability (translatability).

There is a common assumption underlying both the arguments for and against the thesis of incommensurability as untranslatability: Incommensurability has something to do with a specific kind of semantic relation, i.e., the meaning relation, between the terms of two radically disparate languages:

(MR) It is the meaning relationship between the terms (rather than sentences) of two competing languages that constitutes the determinant semantic relationship in the case of incommensurability.

(MR) is supported by the following fallacious reasoning:

(M1) Mutual translation is necessary (and sufficient) for effective mutual understanding and successful communication between two scientific language communities.

(M2) Meaning continuity between the terms of two competing languages is necessary for mutual translation between them.

Therefore,

(M3) Meaning continuity is necessary for effective mutual understanding and successful communication between any two distinct language communities.

Attributing the premise M2 to both the advocates and opponents of incommensurability as untranslatability is obvious; for no matter what notion of translation is adopted, mutual translation between two languages is possible only if at least two basic semantic values of expressions (sense and reference) in the translated language can be preserved in the translating language. To justify the premise M1, we need to realize that although there is no general agreement concerning the nature of incommensurability, there is one basic intuitive sense of incommensurability which is likely to be acceptable to all; namely, incommensurability as the communication breakdown between two competing languages. To say that two languages are incommensurable is another way of saying that successful cross-language communication breaks down between them.

Based on this shared line of reasoning, the views of the advocates and opponents of incommensurability as untranslatability diverge due to conflicting views about the possibility of the existence of meaning continuity between the terms occurring in two allegedly incommensurable languages. The advocates of incommensurability argue, in terms of the contextual theory of meaning, that there is a lack of the meaning continuity between the terms of two competing languages due to radical meaning variance. The two languages are hence not mutually translatable. Consequently, a communication breakdown between the two language communities is inevitable. The two languages are therefore incommensurable. On the contrary, the critics argue that some degree of meaning continuity must exist between two allegedly incommensurable languages. Furthermore, if the meaning continuity is not only a necessary, but also a sufficient condition for translation, then the thesis of untranslatability simply fails.

However, although the critics argue against the untranslatability thesis, they agree with the advocates that the issue of incommensurability is conceptually linked with the issue of translation: Incommensurability is equivalent to untranslatability and commensurability to translatability.[15] In particular, the critics accept also that it is the link between untranslatability and incommensurability that makes the meaning relation the determinate semantic relation in the case of incommensurability. In this sense, the critics agree with the advocates on the translational-failure approach to incommensurability.

My criticism of the translational-failure approach to incommensurability focuses on two questions. First: Can the notion of untranslatability carry 'the weight' of clarifying the nature of incommensurability? The answer to this question depends on the answers to the following two related questions: Is there a tenable and integrated notion of translation that we can use to clarify the notion of incommensurability? Is mutual translation necessary (and sufficient) for effective cross-language understanding and communication? Second: Is the meaning relation between the terms of two competing languages the determinant semantic relation between two allegedly incommensurable languages? The answer to this question depends in turn upon the answer to the following question: Is meaning continuity necessary for cross-language translation?

6. The Very Notion of Translation

Presumably, whether or not translation could be used to clarify the notion of incommensurability depends on whether we can have an adequate notion of translation to work with. It is noteworthy that what counts as an acceptable translation is vague and full of confusion. Here are just a few examples. (a) A distinction could be made between *literal* (strict, or exact) translation into a home language as it is already constituted—i.e., a translation without change at all to the home language on the one hand and *liberal* (loose) translation that achieves a degree of success by modifying the existing usages, and hence the existing meanings of words of the home language, on the other. (b) The previous distinction roughly corresponds to the distinction between translation into the home language in the normal sense and interpretation, or a far looser sort of reinterpretative reconstruction of the alien language into the home language.[16] (c) Intensional translation focusing on meaning (sense) could be distinguished from extensional translation focusing on references. W. V. Quine (1981, p. 42) fully realizes that translatability is a flimsy notion, unable to fulfill its overloaded mission of distinguishing alternative languages and thereby 'unfit to bear the weight of the theories of cultural incommensurability that Davidson effectively and justly criticizes'. In fact, Quine (1970) tries to construct a very strict notion of translation to measure the conceptual remoteness of two alternative languages: i.e., a translation in a purely extensional sense, or word-by-word isomorphic mapping between the extensions of all respective general terms and the designata of all corresponding singular terms. Nevertheless, at the same time, he clearly allows awkward and baffling translation between languages embodying alternative conceptual schemes (Quine, 1981, p. 41).

Kuhn on Liberal versus Literal Translation

To illustrate Quine's point, let us take a close look at the notion of translation used by Kuhn in his explication of incommensurability. Kuhn has employed two different notions of translation during the development of his thoughts on the thesis of incommensurability as untranslatability. The earlier Kuhn employs a concept of translation that is used by professional translators in practice. According to this concept, the final goal of a translator, like Quine's radical translator or interpreter, is to achieve a better understanding of an alien text. For this purpose, the translator 'must find the best available *compromises* between incompatible objectives'. This is because 'translation, in short, always involves compromises which alter communication. The translator must decide what alterations are acceptable' (Kuhn, 1970b, p. 268).

Translation in this practical sense, so-called *liberal translation*, has the following features: (a) The target language is permitted to be changed by introducing new concepts and more or less by subtly changing the old concepts. (b) It does not require a systematic replacement with or mapping of words or word groups in the source language to the corresponding words or word groups in the target language. (c) The translation is a matter of degree; no exact translation is available. (d) Most importantly, the translation is closely connected with the process of language learning, and thereby involves a strong interpretative component. Actually, liberal translation is the enterprise practiced by historians and anthropologists when they try to understand an old text or break into an alien culture. In such cases, in order to make the source language intelligible, the translator (strictly speaking, the interpreter) has to learn the source language in the process of translation, and afterward to look for the closest counterparts of expressions of the source language in the target language (Kuhn, 1983b, pp. 672-3).

An immediate problem with the explication of incommensurability in terms of liberal translation is that if we accept this concept of translation, then it is always possible to translate any different text, no matter how alien it is to the translator's home language; for it is always possible theoretically to make an alien context intelligible by language-learning. Thus, there would be no incommensurable texts at all eventually. The notion of untranslatability leads to the dissolution of the issue of incommensurability. This line of reasoning makes it too easy for the opponents of incommensurability to claim a victory.

A further problem with the notion of liberal translation is related to Kuhn's treatment of the relation between translation and the existence of a common language. The earlier Kuhn explicitly regards the existence of a common language as a prerequisite for commensurability or translatability.

> In applying the term 'incommensurability' to theories, I had intended to insist that there was no common language within which both could be fully expressed and which could therefore be used in a point-by-point comparison between them. (Kuhn, 1976, p. 191)

As Kuhn and Feyerabend argue repeatedly, there is not any common language available between two incommensurable languages. In this way, Kuhn makes his argument for the nonexistence of any neutral language the foundation of his thesis of untranslatability. However, untranslatability of two languages L_1 and L_2 into a common language L is not equivalent to mutual untranslatability between L_1 and L_2. On the one hand, the requirement for a neutral language as a necessary condition for mutual translation or commensurability is too strict. It cannot be met. On the other hand, the requirement for a common language as a sufficient condition for mutual translation is too loose. We can represent various mathematical theories T_1 and T_2 in set theory, but we may not be able to represent T_1 in T_2. The problem is that a translation *into* set theory does not necessarily give us a translation *from* set theory, which is what we are going to need to get from T_1 to T_2 via set theory.[17] Especially, this notion of translation based on the existence of common languages is directly contradictory with the concept of loose translation Kuhn employs at this stage; for the former is so strict as to render the translatables untranslatable while the latter is so loose as to render the untranslatables translatable.

After 1983, by virtue of introducing the new tool of lexicon and its structure, Kuhn continues to rely heavily on the notion of untranslatability in his explication of incommensurability, and further links untranslatability closely with change of the taxonomic structures of scientific languages. At this stage, the concept of incommensurability is directly connected with the notion of untranslatability without resource to the notion of a neutral language.[18] 'Incommensurability thus equals untranslatability' (Kuhn, 1988, p. 11). 'If two theories are incommensurable, they must be stated in mutually untranslatable languages' (Kuhn, 1983b, pp. 669-70).

More significantly, at this stage, Kuhn consciously operates with a different notion of translation, *literal translation*: Translation must be taken in a strict sense as the *formulation* of expressions within the translating language (the target language) that are semantic equivalents of expressions of the translated language (the source language). Semantic equivalence is not word-to-word synonymy, but it does require that the translator systematically substitute the appropriate expressions in the target language for the corresponding expressions in the source language to produce an equivalent text in the target language. In comparison with the notion of liberal translation, literal translation has the following features: (a) During the process of translation, the target and the source languages should remain unchanged. No addition of new kind-terms and no change of semantic values (sense, reference, and truth-value) of the expressions are permissible. (b) The translation requires a systematic mapping of concepts (kind-terms especially) in the source language to the corresponding concepts in the target language such that each concept is mapped to a concept with exactly the same semantic values. (c) The translation is an all or nothing issue: The translation between two languages is either possible or impossible. (d) The translation is a totally different linguistic activity from language learning or interpretation. The purpose of interpretation is to make an alien text intelligible. We can achieve this by language learning. For this purpose, it suffices to identify the related semantic values of the expressions in the source language. However, the purpose of literal translation is not only to identify the semantic values of the expressions in the source language, but also to *formulate* semantic equivalents of these expressions within the target language. Translatability in this sense is a function of what can be said or expressed in the target language. It depends upon the potential ability of the target language to produce semantic equivalents of the expressions in the source language without changing its taxonomic structure and without extending its semantic resources. However, the target language's ability of translating the source language is limited and constrained by the capacity of its taxonomic structure. A translation fails if formulating the semantic equivalents of the expressions of the source language in the target language requires either change of the target language's taxonomic structure or an extension of its semantic resources by semantic enrichment.[19]

What constitutes semantic equivalency as sought in a literal translation? Sameness of sense or intension, sameness of references or the extensions of shared kind-terms (including the sameness of the way in which the reference of a term is determined as well), and sameness of truth-values of shared sentences are obvious desiderata. As far as the sufficient conditions of literal translation are concerned, some pragmatic aspects of language, such as conversational implicatures, the speaker's intentions, the speaker's meaning, the speaker's reference, and context specifiers have to be included as well. Unfortunately, it is almost impossible to list all the sufficient conditions for literal translation (in fact, it is doubtful whether there exists such a complete set of sufficient conditions).

According to Kuhn's notion of literal translation, translation is an all-or-nothing issue. Such a notion of the perfect translation presupposes that there exists, if any, one and only one right translation for an expression in question. However, this presupposition is questionable. Quine has argued convincingly that there can be many mutually incompatible systems of translation consistent with all possible behavioral data. We have no way to determine which of many available translations is the only right translation. Hence, translation is indeterminate. Of course, we can follow Davidson's suggestion to put enough constraints—both the constraints from the bottom, such as formal constraints and empirical constraints (tested as true), and the constraints from the top, namely, the methodological perception or the principle of charity—on the language to be translated to reduce multiple acceptable translations for the language to a small amount of translations or even to the one best translation. However, a translation so constructed is no longer a literal translation. As Davidson contends, such a translation does not (should not) preserve the detailed propositional attitudes of the speakers (such as the speaker's beliefs or intentions) and does not, and should not, make use of unexplained linguistic concepts (such as the speaker's meaning, synonymy). Besides, due to inextricability between meaning and belief, we have to put the speaker in general agreement on beliefs with the interpreter and try to maximize such an agreement. This often inevitably distorts the speaker's belief system. Even so, there will still be no perfect mapping from sentences held true by the speaker in the source language onto the sentences held true by the translator in the target language.

Recall an example given by Davidson: It is sometimes even impossible to give a literal translation in English of an English speaker's expression, such as, 'There's a hippopotamus in the refrigerator'.

> Hesitation over whether to translate a saying of another by one or another of various nonsynonymous sentences of mine does not necessarily reflect a lack of information: it is just that beyond a point there is no deciding, even in principle, between the view that the other has used words as we do but has more or less weird beliefs, and the view that we have translated him wrong. (Davidson, 1984, pp. 100-101)

The inextricable intertwining of the speaker's belief and the speaker's meaning establishes that there is no such literal translation available in any case. Therefore, Kuhn's notion of literal translation is too strict and could never be exemplified. If we accept such a notion of translation, we cannot even translate the speech of others who speak our own language, not to mention translation between two distinct languages.

The conclusion from the above examination of Kuhn's two notions of translation is that neither the notion of literal translation nor that of liberal translation can help us clarify the notion of incommensurability. In fact, the thesis of incommensurability as untranslatability becomes trapped in a fatal dilemma: If 'translation' is construed narrowly and strictly to mean 'literal translation', then it indeed follows that two distinct languages (such as Newtonian vs. Einsteinian languages) are untranslatable. But incommensurability in this sense would degenerate into a trivial platitude since there is no literal translation available between any two languages anyway. In this sense, some degree of partial untranslatability marks the relationship of every language to every other. Consequently, it would make the commensurables incommensurable. On the other hand, if 'translation' is construed broadly and loosely to mean 'liberal translation', then untranslatability will not qualify as a criterion of incommensurability since any two languages would be translatable. Consequently, it would make the incommensurables commensurable. Either way, the thesis of incommensurability would turn out to be an illusion.

The Truth-Preserving Translation

Some might reply, 'Yes, maybe Kuhn's two notions of translation are trapped in the above dilemma. But this does not exclude the possibility of a more tenable notion of translation that could be used as a criterion of commensurability'. It is true that there might be some more appropriate notion of translation than Kuhn's two radical notions of translation. But I doubt that we could have a 'desirable' notion of translation as the advocates expect, which could be used to clarify the notion of incommensurability. To show this, I am not going to consider every possible notion of translation presented or to be presented by the advocates, such as D. Pearce's (1987) notion of context-sensitive translation as reduction.[20] For my limited purpose, it suffices to point out that the issue of incommensurability, as I will argue later, is essentially associated with the possibility of truth-or-falsehood. It is the issue about the distribution of *truth-value status* among the sentences of the competing languages, not about the distribution of different *truth-values* among the sentences of the competing languages. Two languages are incommensurable when both sides disagree about the truth-value status of some substantial number of core sentences in the other language. Such a truth-value gap indicates a linguistic communication breakdown between the two language communities. However, the traditional notion of translation (whether Kuhn's notion of literal translation or any other alleged more 'appropriate' notion of translation), Quine's indeterminacy of translation, and Davidson's notion of radical interpretation all start from an idea that an adequate translation should preserve the truth-values of the sentences translated and gain the truth-value matching of sentences between the target and the source languages.

Traditional extensional semantics uses truth-value preservation as a condition for an adequate translation. Kuhn, who adopts intensional semantics on translation, certainly does not regard truth-value preservation as a sufficient condition for literal translation. But the truth-values of the translated sentences are, for Kuhn, an important semantic value that should be preserved in literal translation. Similarly, there is an assumption underlying Davidson's notion of radical interpretation: Although the speaker and the interpreter in discourse may assign opposite truth-values to some core sentences of a language, they all agree that all the sentences in the competing language are either true or false. The task of the radical interpreter is to design a translation (more precisely, a theory of truth for the language in question), based on Davidson's principle of charity, that can preserve as much truth as possible. It is my conviction that this kind of truth-preserving translation is irrelevant to the incommensurable texts because what really matters is not *redistribution of truth-values* between the sentences of two compatible texts, but whether or not both sides still accept the other's sentences as *candidates for truth-or-falsehood*. Therefore, using the notion of a truth-preserving translation to clarify the notion of incommensurability is on the wrong track. For this reason, I conclude that there is no tenable and integrated notion of translation that can be used to clarify the notion of incommensurability.

7. Translation and Cross-Language Communication

There is still another way to connect the notion of translation with the notion of incommensurability as communication breakdown. Many contend that translation is either sufficient or necessary for effective cross-language understanding and communication. An interpreter can understand an alien language through or only through the relation of the translated language to the interpreter's own home language. If so, the failure of mutual translation between two languages would indicate the breakdown of cross-language communication between the speakers of two languages, and hence the two languages would be incommensurable. This alleged connection between cross-language communication and translation is a hidden motivation for many philosophers, advocates and opponents of the thesis of incommensurability alike, who identify commensurability with translatability.

Davidson has already observed that a translation manual is to be contrasted with a theory of meaning or interpretation. A theory of interpretation for an unknown language directly describes the way in which the language functions by providing an account for the essential role of each sentence in that language. Since the meaning of each sentence in a language is determined by the essential role of the sentence in the language as a whole, a theory of interpretation for a language does help us understand the language. For Davidson, a theory of interpretation is a theory of understanding. A translation manual tells us only that certain expressions of the translated language mean the same as certain expressions of the translating language, without telling us what, specifically, the expressions of either language mean. It is theoretically possible, Davidson contends, to know of each sentence of a given language that it means the same as some corresponding sentence of another language, without knowing at all what meaning any of these sentences has. Therefore, a translation manual itself does not constitute a theory of meaning or understanding. A translation manual leads to an understanding of the translated language only via the translating language, an understanding that it does not itself supply. In other words, a translation itself is not sufficient for understanding.[21]

Furthermore, according to the truth-value conditional theory of understanding to be presented in chapter 9, the notion of truth-value status, instead of the notion of truth, plays an essential role in understanding; an interpreter can understand a sentence in a given language only if the sentence has a truth-value from the point of view of the interpreter. Shared notion of truth-value status between the interpreter and the speaker is necessary for the interpreter to understand fully the sentences in the speaker's language. In this way, understanding an alien text is a matter of recognizing new possibilities for truth-or-falsehood, and of how to appreciate and follow other ways of thinking embedded in other languages that bear on these new possibilities. What we need to capture in communication is not what each word in a sentence or the whole sentence means in an alien language, which can be done by constructing an adequate translation manual, but the factual commitments underlying each language that create new possibilities for truth-or-falsehood. For example, even if a sentence in traditional Chinese medical theory, 'The combination of yin and rain makes one drowsy', could be translated into some appropriate expressions in Western medical theory, Western physicians are still at a loss if they do not know the mode of reasoning underlying the sentence. What is true-or-false in the way of reasoning used by Chinese medical theory may not make much sense in another until one has learned how to reason in that way.[22] Hence, a truth-preserving or meaning-matching translation, if any, is not sufficient for effective understanding of an alien text.

On the other hand, translation is certainly not necessary for understanding. There is a good reason to think that the whole focus on actual translation in understanding is misguided. The key category with which we should be concerned in understanding is surely not translation but interpretation or language learning. Interpretation is a very different linguistic activity from translation. By interpretation, following many others, I mean the enterprise practiced by historians and anthropologists when they try to understand an old text or break into an alien culture. The interpreter, at the very beginning, only masters his or her own home language. The source language is totally unknown to him or her. The purpose of interpretation is to make an unknown alien text intelligible. The most effective means to make an old or alien text intelligible is to learn the language, instead of to translate it. For this purpose, it suffices to identify the related semantic values of the expressions in the source language. For instance, to identify the referent of an out-of-date kind-term (say, 'phlogiston' in phlogiston theory) and the role of the term in the theory is sufficient for understanding the term. However, to understand the theory, the interpreter has to learn the source language by identifying the related semantic values of all the expressions in the source language. By comparison, a translator is supposed to master both the source and the target languages at the very beginning. The purpose of translation is to *formulate* semantic equivalents of expressions of the source language in the target language. Translatability in this sense is a function of what can be said or expressed in the target language. It depends upon the potential ability of the target language to produce semantic equivalents of the expressions in the source language without change of the taxonomic structure of the target language and without extending its semantic resources. A target language's ability to translate a source language into a specific language is limited and restrained by the capacity of its underlying taxonomic structure, descriptive universal principles, and mode of reasoning. There can be no genuine translation where the underlying descriptive principles, taxonomic structures, or the modes of reasoning of two languages involved are substantially different.

After understanding an alien context by learning it, interpreters may be able to make an alien text intelligible to fellow speakers through paraphrasing, 'explaining' or the like, the alien text into the nearest 'corresponding' (not exact mapping) expressions in their own home language. For example, an English speaker, after learning Chinese, may attempt to interpret the Chinese sentence, 'Jenni hen fung maan', as, 'Jenny is voluptuous, has a full figure, or is full and round, well-developed, full-grown, plump, and smooth-skinned'. But as every Chinese speaker knows, the above English expressions not only do not convey all evaluative elements contained in the Chinese adjective 'fung maan', but also cannot even describe exactly the physical appearance of a 'fung maan' woman. This is because Chinese speakers categorize the woman's body differently from the way English speakers do, using different discriminations in doing so. Under this circumstance, 'fung maan' remains an irreducibly Chinese term, not translatable into English. But this does not block understanding. An English speaker can understand the term very well by learning Chinese. In general, the process of learning an alien language or text may well involve a complex process of theory building or a loose sort of reinterpretative reconstruction of the alien language. But this is not a process of translation. Translation might be a desideratum, but not a *sine qua non* necessity for understanding. Interpretation can serve perfectly well in understanding.

Therefore, translation is neither necessary nor sufficient for understanding. To understand is to learn, not to translate.[23] In addition, translation does not help understanding, but often sets obstacles. Linguistic studies show that the best way to learn a foreign language is not to learn it by means of word-by-word translation, but by living in the community of native speakers and learning the language from scratch as a child does. The more you forget your native language, the more effectively you learn, and the deeper you understand the foreign language.

8. Meaning Continuity and Cross-Language Translation

Whether or not there is meaning continuity between the terms of two competing languages in the case of incommensurability is supposed to determine whether mutual translation between the two languages is possible. For the opponents of incommensurability as untranslatability, mutual translation is possible because of the existence of a sort of meaning continuity between two languages. For the advocates, mutual translation is impossible due to a lack of meaning continuity. Now, after we have detached the notion of untranslatability from the notion of incommensurability, it is not clear at all how much weight the meaning relation can carry regarding the explication of incommensurability, not to mention how the relation can remain a determinant semantic relation between incommensurable languages.

I would like to point out further that, in general, there is no necessary connection between meaning continuity and translation. Translation does not presuppose sameness of meaning. Whether two languages are mutually translatable depends upon the target language's capacity for encapsulating substantial expressions of the source language. Such a formulation does not require sameness of meaning. For illustration, let us consider the following artificial case about color classification. Different languages may divide the spectrum in different ways and thereby have different color predicates. Let us imagine two sets, S_1 and S_2, of color predicates in two different languages L_1 and L_2. S_1 and S_2 divide the spectrum in such a way that none of the color predicates of them match up with the color predicates of the other set. Therefore, we have two different color category systems. Let us further suppose that S_1 has 'red', 'orange', 'yellow', 'green', 'blue', and 'purple' as its finest-grained color predicates, while S_2 has 'Orange', 'Green', 'Purple' at the corresponding level of discrimination. The extensions of S_2 can be imagined to be distributed along the spectrum so that 'Orange' matches up with the shades of both 'red' and 'orange', 'Green' with both 'yellow' and 'green', 'purple' with 'blue and purple' in S_1. The conceptual mismatch between S_1 and S_2 is adjustable since each color predicate of S_2 can be defined in S_1. The concept 'Orange' in S_2 can be expressed as 'both red and orange' in S_1, and so on. More generally, all the color predicates in L_2 are translatable into corresponding expressions in L_1. Hence, the two languages are translatable. This case shows us that translation does not presuppose the sameness of reference.

Interestingly enough, based on the rule of contraposition, we can easily infer from the above conclusion that meaning continuity is not necessary for translation that the discontinuity or variance of meaning between two competing languages is not sufficient for untranslatability between two languages. If so, Kuhn's and Feyerabend's arguments for meaning variance, based on their versions of the contextual theory of meaning, do not entail without further ado the thesis of untranslatability. If this argument against untranslatability holds water (which I think it does), then the same argument can be used against translatability also. Ironically, appeal to the meaning relation (continuity or discontinuity) is a double-edged sword that can be used to hurt both proponents and opponents of the thesis of incommensurability as untranslatability. This establishes that there is no necessary conceptual connection between meaning continuity and the possibility of mutual translation between two languages.

Finally, I would like to point out that untranslatability does not establish anything interesting about many theoretical issues that are supposed to be implicated by the thesis of incommensurability, such as the issues of the rationality of science and scientific progress as well as the debate between relativism and realism. For example, the debate between relativism and rationalism simply does not depend on the possibility of translation. On the one hand, the ineliminable possibility of mutual translation (liberal translation) between two languages does not guarantee either that the speakers of the two languages share criteria of truth, rules of inference, or have a common core of empirical knowledge, or that the two languages enable their speakers to refer to some fundamental common set of referents. On the other hand, the impossibility of mutual translation between two incommensurable languages does not exclude the possibility that the speakers of the two languages may have shared criteria of truth, shared domain of referents, or common criteria of theory evaluation. The alleged semantic obstruction implied by the thesis of incommensurability does not arise from untranslatability.

* * * * *

For the past four decades, the translational-failure interpretation has dominated · much of the discussion of incommensurability. Questions of meaning (sense), reference, and translation have been examined both from the standpoint of general philosophical theories as well as from case studies in the history of science. However, in spite of tremendous efforts, little progress has been made toward the goal of understanding and clarifying the thesis of incommensurability. The notion of incommensurability still remains mysterious to us.

Too much attention paid to the meaning relation between incommensurable theories is, I believe, to a large extent responsible for the slow progress that has been made toward establishing the integrity and tenability of the notion of incommensurability. This is because the translational approach is not an *effective*[24] way of exploring the essence of incommensurability. It cannot provide us with a tenable and integrated notion of incommensurability. This is mainly because the notions of translatability and untranslatability are sterile without any independent interpretative power in the case of incommensurability. Reference to the notion of translation neither identifies nor resolves the problem of incommensurability.

Therefore, the received interpretation of incommensurability as untranslatability due to radical meaning change has to be rejected. Of course, the failure of the received notion of incommensurability does not indicate the failure of the very notion of incommensurability. It is not the notion of incommensurability that lets us down, but the *received* notion of it that does so. In fact, the rejection of the translational-failure approach paves the road to a more promising interpretation of incommensurability, namely, the presuppositional interpretation of incommensurability.

Notes

1   The term 'meaning' could be used in a narrow sense to refer to 'sense' or in a broader sense to refer to 'sense and reference'. In our following discussions, the term 'meaning' will be used in the broader sense in most cases. When it is necessary, I will specify whether it refers to sense or reference.

2   See Feyerabend, 1962, pp. 29, 57, 59, 68; 1965a, pp. 114-15; 1965b; 1965c, p. 180.

3   Kuhn, 1970a, 1970b, 1976, 1979.

4   H. Sankey identifies three distinct interpretations of Kuhn's early position on reference. See Sankey, 1994, pp. 153-61.

5   See C. Kordig, 1971, pp. 35-48.

6   Here the description theory refers to the so-called cluster theory of reference defended by John Searle and others. See J. Searle, 1958.

7   Informal semantics gives a realistic interpretation of the references of scientific terms. According to it, (a) a scientific theory is not a logically reconstructed formal system; (b) the reference of a term is identified with the set of 'real world' objects to which it applies. Therefore, the intended references of central concepts of a scientific theory are fixed absolutely, not relative to some specific theoretical framework. The references of the terms of a scientific theory should remain unchanged during theory change.

8   Formal semantics gives a relativistic interpretation of reference. A scientific theory is believed to be a logically reconstructed formal system. The issue of meaning and reference can be dealt with within a formal (model-theoretic) framework. Meaning and reference are determined by and relative to a formal framework, instead of being fixed absolutely.

9   For the distinction between determination of the meanings (sense or reference) of terms and formulation of the meanings of those terms and its significance in the explication

of incommensurability, please read Kuhn's (1983b) response to P. Kitcher's (1978) strategy of translation.

10 J. English (1978, pp. 57-8) attributes this extreme view to Feyerabend.

11 Kuhn's LTRK is a restricted thesis which mainly concerns the referents of selected kind-terms, especially the high-level theoretical categorical terms, such as 'planet', 'mass', 'force', 'compound', 'element', 'phlogiston', etc., and less concerns the referents of low-level empirical categorical terms such as 'alloy', 'metal', 'physical body', 'salts', 'gold', 'water', etc. LTRK does not concern the referents of names and other singular object designators, such as 'Earth', 'Moon', etc. For Kuhn, the references of names and the references or extensions of kind-terms (especially, natural kind-terms) are two separate issues. Kuhn partially accepts causal theory of reference for determination of the referents of names (Kuhn, 1979; 1988, p. 25; also see Sankey, 1994, pp. 163-71). However, Kuhn uses his version of a contextual theory of reference to determine the referents of kind-terms.

12 Kuhn, 1983b, pp. 682-3; 1991, pp. 11-12; 1993a, pp. 325, 329.

13 Kuhn, 1983b, pp. 676-9; 1988, pp. 14-21; 1993a, p. 317.

14 Kuhn, 1983b, p. 683; 1993a, pp. 325-6, 329.

15 Many critics of the thesis of incommensurability explicitly accept the identity of incommensurability with untranslatability. D. Pearce (1987) contends that translatability is necessary, and sufficient under suitable conditions, for commensurability. W. Balzer (1989) claims that incommensurability is intimately linked with a particular kind of potential translation. It is well known that D. Davidson (1984) equates incommensurability with untranslatability and argues that it does not make sense to talk about two untranslatable conceptual schemes. In a similar way, H. Putnam (1981, p. 114), M. Devitt (1984), and many others regard incommensurability as being or implying the untranslatability of languages. H. Sankey devotes a book (1994) to defend the thesis of incommensurability as untranslatability.

16 For some analysis of details, please see N. Rescher, 1980, pp. 326-8, M. Forste, 1998, pp. 137-9, and D. Henderson, 1994, section II.

17 The second argument is suggested by S. Lehmann.

18 Kuhn, 1983b, pp. 669-70; 1988, p. 11; 1991, p. 5.

19 Kuhn, 1983b, pp. 672-6, 679-82; Sankey, 1994, pp. 76-7.

20 D. Pearce contends that translatability is necessary and sufficient under suitable conditions for commensurability. A translation can be considered adequate in the semantic sense if it respects an acceptable model-theoretic reduction of the structuralistic sort. Science as a whole is so rich in concepts and so diversified in its patterns of conceptual change and development that one cannot expect to find one general method of translation that can be used to produce adequate specific translation between any two languages. A successful response to the challenge of the doctrine of incommensurability as untranslatability consists in examining concrete examples of theory change in science and providing translation of the right kind that applies there.

21 It seems to me that Davidson's above argument does not presuppose any specific notion of translation. It provides a general objection to regarding translation as a sufficient condition of understanding.

22 This argument is essentially I. Hacking's (1982).

23 For a good defense of the distinction between understanding and translation, see H. Sankey, 1997, chapter 6.

24 I intentionally choose the expression 'not effective' because I do not wish to imply that the translational-failure approach is completely wrong, although it does lead to misinterpretation.

Chapter 3

Incommensurability and Conceptual Schemes

1. Conceptual Relativism and Incommensurability

In essence, the thesis of incommensurability is a philosophical doctrine of what D. Davidson calls radical conceptual relativism. Briefly put, conceptual relativists believe that there are distinct conceptual schemes (either between two distinct intellectual or cultural traditions or over the course of history within the same intellectual or cultural tradition) to schematize our experience such that meaning, truth, cross-language understanding and communication, and human perceptions of reality are relative to conceptual schemes. Radical conceptual relativists contend that conceptual schemes or the languages associated with them can be and actually are, in many cases, *radically* distinct or *massively* different without any significant overlap, even to the extent of being incommensurable and leading to *massive*, even complete, communication breakdown between two language communities. For modest conceptual relativists, on the other hand, even if radically distinct schemes without any significant overlap are hard to come by, partially distinct conceptual schemes with some shared common parts are pervasive, which often leads to partial communication breakdown between the language communities associated with them.

According to Davidson's interpretation of the Quinean notion of conceptual schemes, a conceptual scheme is identical with a sentential language held to be true by its believers. To say that two conceptual schemes or languages are distinct amounts to claiming that they are untranslatable into each other. Similarly (it is not by coincidence), based on the received translation-failure interpretation of the thesis of incommensurability adopted by Davidson, "'incommensurable' is, of course, Kuhn's and Feyerabend's word for 'not intertranslatable'" (Davidson, 1984, p. 190). Thus, for Davidson and many others, the thesis of incommensurability as untranslatability and Quinean conceptual relativism are two sides of the same coin. They rise and fall together.

Even if one does not accept the Quinean notion of conceptual schemes and the thesis of incommensurability as untranslatability, one still cannot deny conceptual affinity between the two doctrines. In fact, the thesis of incommensurability presupposes conceptual relativism. Without the possible existence of two conceptually distinct conceptual schemes or languages associated with them, there is simply no issue of incommensurability. On the other hand, the belief that two distinct schemes or languages could be incommensurable is at the very heart of conceptual relativism. To say that two scientific languages are incommensurable is to say that they embody two radically distinct conceptual schemes—whether they are B. Whorf's grammar of natural languages, R. Carnap's linguistic frameworks, W.V. Quine's sentential languages, or T. Kuhn's paradigms and lexicons. This is why most conceptual relativists (B. Whorf, W.V. Quine, T. Kuhn, C.I. Lewis, and P. Feyerabend among them) accept the thesis of incommensurability.

However, both the thesis of incommensurability and conceptual relativism can be intelligible only if we can make sense of 'the very notion of a conceptual scheme'. The notion of conceptual schemes and its underlying metaphysical dualism between scheme and content serve as the conceptual foundation of the thesis of incommensurability. Dismantlement of the notion would lead to the fall of both conceptual relativism and the thesis of incommensurability. This dismantlement is exactly what Davidson sets out to achieve in his celebrated essay, 'On the Very Idea of a Conceptual Scheme' (1984).

With a strong faith in the universality of cross-language understanding and communication, Davidson firmly believes that mutual understanding and communication across different languages, cultures, and traditions is theoretically possible and practically achievable. It is not simply because we can somehow manage, through tremendous efforts in some cases, to fully understand and communicate with others whose thoughts, ideologies, moral systems, ways of thinking, or language schemes associated with them, are so remote from us, but rather because others are after all not that different from us. 'There are limits to how much individual or social systems of thoughts can differ' (Davidson, 2001a, p. 39). All human beings are alike, or at least we have no good reasons to think otherwise in order to engage in productive dialogue with each other.

Conceptual relativism seems, Davidson believes, to threaten to break our cherished hope of cross-language/scheme understanding and communication by 'imagining' the existence of radically different mental schemes or mind-sets associated with different cultures, traditions, and languages. Those distinct mental schemes, mediated between the world and the thoughts, create serious impediments to cross-scheme understanding and communication.

Of course, Davidson is not so naïve as to deny the existence of different systems of thoughts that often make mutual understanding and communication difficult. He admits that 'there are contrasts from epoch to epoch, from culture to culture, and person to person, of kinds we all recognize and struggle with; but these are contrasts which with sympathy and effort we can explain and understand' (2001a, p. 40). The debate between Davidson and his opponents is about whether those massive differences in thoughts and ideologies are caused by fundamental semantic and/or conceptual obstructions, namely, radically distinct conceptual schemes, or simply because of differences in beliefs.

Instead of taking on different forms of conceptual relativism one by one, Davidson sets to dig out its root, i.e., the very notion of conceptual schemes. To discredit the notion, Davidson presents a host of different kinds of arguments, some simple, some complex, some straightforward, some carrying plenty of heavy controversial theoretical baggage of their own, such as his truth-conditional theory of meaning/understanding/translation. As complicated as they appear to be, all of Davidson's arguments actually follow two distinct lines of reasoning: One focuses on the verification conditions of alternative conceptual schemes, which targets the Quinean notion of conceptual schemes; the other attacks directly the Kantian schemes–content dualism. Before examining those two lines of criticism of Davidson's in the next chapter, we need to clarify and define the notion of conceptual schemes as comprehensively as possible first.

2. The Very Notion of Conceptual Schemes Defined

Philosophically, the notion of conceptual schemes begins its intellectual life with Kant's transcendental philosophy. To answer his ingenious question of how experience is possible, Kant divides the mind into active and passive faculties, i.e., sensibility or unsynthesized *a priori* sensible intuitions of space and time on the one hand and understanding with endowed pure concepts or categories on the other. All possible experience, in order to be the object of our experience at all, is constructed *a priori* and subjectively by our minds through the joint work of these two basic mental faculties; i.e., understanding uses *a priori* concepts to 'interpret' what 'the world' (Kant's noumenal world) imposes on the sensibility. Without those two basic kinds of mental schemes, no human experience is possible. Of course, for Kant, such mental schemes are universal and unchanged. There are no distinct mental schemes.

However, once the distinction between the data of experience and the schemes for conceptualizing them—the Kantian scheme–content dichotomy—is granted, it is not hard to imagine the existence of distinct alternative schemes. In rejecting Kant's transcendental idealism, the 'Historical School' in late eighteenth- and nineteenth-century Germany (including Hegel and others who possessed an impressively broad and deep understanding of the history of ancient ideas, languages, and texts) turned Kant's grand *a priori* scheme on its head by making it related to specific historical and cultural contexts, and thereby setting the foundation for the development of conceptual relativism.

Historically, conceptual relativism has been associated with two models. According to the Kantian categorical model, some set of basic or categorical *a posteriori* concepts are necessary for any possible experience within a context, and such schemes of concepts can change with contexts. By comparison, the Quinean linguistic model takes schemes to be identical with sentential languages composed of sentences accepted as true. Since there can surely be radically different languages, there could be radically distinct conceptual schemes as well.

Although it is controversial as to what counts as a conceptual scheme, a conceptual scheme is, figuratively speaking, commonly considered to be a specific way of organizing experience and/or a way of 'carving up' the world under consideration. It determines a unique conceptual way of contacting and describing reality. Briefly put, a conceptual scheme is supposed to be the conceptual framework used to schematize our experience in terms of its metaphysical presuppositions of existents, states of affairs, modes of reasoning, and categorization. Accordingly, conceptual relativism contends that there exist some substantial semantic and/or conceptual disparities between two distinct conceptual schemes. Such a conceptual disparity could create serious impediments to mutual understanding between the respective believers, which, in many cases, could even render one side incommensurable to the other. For instance, C. I. Lewis (1929) asserts that our cognitive experience is composed of two distinguishable elements: the immediate data of the senses presented to the mind, and the form or construction that the mind imposes upon these data. Functioning as schemes of interpretation, different forms of the mind yield different descriptions of reality. If two schemes are not translatable, then their corresponding descriptions of reality are incommensurable.

What is at stake here is not a trivial empirical fact that others could be or actually are different from us in their beliefs, thoughts, worldviews, even normal concepts, but whether those differences in the contents of thoughts are originated from and shaped by the schemes or the forms of thoughts themselves, which may be radically different sets of categorical concepts used to categorize possible experience, different linguistic frameworks of describing reality, different forms of explanation and interpretation for certain domains of discourse, different modes of reasoning/justification, or different ways of thinking. So defined, a conceptual scheme is a much more narrow notion than a worldview, a form of life (L. Wittgenstein), a tradition (H.G. Gadamer), or a culture (R. Rorty), which appear to be the combination of the content and the scheme. A conceptual scheme refers instead to the essential conceptual core or the ontological presuppositions of a worldview.

The thesis of conceptual relativism is not just a philosophical speculation. It has been emphatically supported by many empirical studies from many interpretative disciplines. For example, in the twentieth century, anthropologists gradually realized that the difference between the beliefs of two distinct cultures cannot simply be reduced to the fact that those two cultures are at different developmental stages of a common path, i.e., aiming to achieve a causal, scientific explanation and understanding of the natural world. Other cultures actually think of, interpret, and categorize reality in quite different ways from ours (Evans-Pritchard, 1964.) Therefore, to understand fully the thoughts of alien cultures or distinct historical thoughts within the same cultural tradition, we must understand their categories of thought, forms of explanation, as well as modes of reasoning and justification, which they impose upon reality as they conceive it.

Advocates of conceptual relativism need to be more specific about the essential parameters of a conceptual scheme if they want to avoid the charge of engaging in only 'empty metaphorical talking'. In fact, behind the above apparently different metaphorical ways of description we can identify three basic doctrines of conceptual relativism.

Scheme–Content Dualism

It is believed that there is an underlying *conceptual or linguistic modality* that determines a specific way of conceptualizing/perceiving the world or of constructing possible experience. Such a metaphysical distinction between the world/experience and schemes for conceptualizing them to form a set of beliefs or a theory is necessary if the notion of conceptual schemes can serve as a metaphysical notion at all. C. I. Lewis, for example, believes the scheme–content distinction to be an almost self-evident philosophical truth; similarly, P. Strawson, W. V. Quine, J. Searle, and many others embraced the dichotomy wholeheartedly. Davidson identifies correctly the dichotomy as the cornerstone of conceptual relativism, but regards it as the third dogma of empiricism that we had better discard. A variety of scheme–content dualisms have been proposed so far:

> The primary bearer of a conceptual scheme: (S1) a set of concepts, such as Kant's *a priori* categorical concepts or Neo-Kantian *a posteriori* basic concepts, such as P. Strawson's contextual interconnected basic concepts (the Kantian categorical model); or (S2) a sentential language with all the sentences accepted as true (the Quinean linguistic model).

> The empirical contents which a conceptual scheme frames: (W) the world/reality for the scheme–world distinction); or (E) experience for the scheme–experience distinction.

> The relation between a scheme and its empirical content: (R1) A scheme could be a categorical framework (usually either a set of categorical concepts or the lexicon of a language) to *categorize* its empirical content. Alternatively, (R2) a scheme could be a system of interpreting or a way of perceiving/describing (usually a set of basic assumptions about existence or fundamental principles on basic structure of the world) to *construct*, not represent, its empirical content. Alternatively, (R3) a scheme could be a form of representation (usually a theory or a sentential language) to fit (face, predict and account for) experience.

Alternative Schemes

A sound scheme–content distinction is necessary, but is not sufficient for conceptual relativism. Kantian built-in *a priori* schemes, for instance, do not allow for the existence of alternative schemes. We need a second doctrine of conceptual relativism:

> It is possible to have two radically different conceptual schemes that could even differ massively and be incommensurable to each other.

To justify the above doctrine, a few questions need to be addressed: How could we have two distinct conceptual schemes? In which way can a scheme be relative to language, tradition, culture, and history? The answers to those questions lie in another doctrine of conceptual relativism.

A Non-Fixed, Fuzzy 'Analytic–Synthetic' Distinction

Scheme–content dualism is metaphysical in nature; namely, the distinction between our conceptual apparatus and the world/experience. However, it is very tempting to confuse the distinction with a closely related semantic distinction, the analytic–synthetic distinction (the so-called first dogma of empiricism), that is, the distinction between sentences being true in virtue of their meanings/concepts alone (these being the analytic sentences) and sentences being true in virtue of both their meanings/concepts and their empirical content (these being the synthetic sentences). Despite such differences, many have tried to reduce one distinction to the other. Some argue that the duality of scheme–content commits one to the duality of analytic–synthetic.[1] Others suggest that the scheme–content distinction relies on or is motivated by the analytic–synthetic distinction. Hence, the former should be seen as a variant of the latter. The dualism of scheme–content either leads to or presupposes the dualism of analytic–synthetic. R. Rorty argues that the analytic–synthetic distinction is necessary for conceptual relativism, and accordingly, necessary for scheme–content dualism (Rorty, 1982, p. 5). C. I. Lewis (1929, p. 37) also constructs his version of scheme–content dualism based on the analytic–synthetic distinction. Therefore, if we follow Quine in abandoning the first dogma of empiricism, we have to abandon the concept of meaning or the scheme of concepts going with it, which will lead to the fall of the third dogma.

There are some obvious conceptual connections between the two distinctions. Imagine that a scheme functions as a conceptual filter (a thought processor) between the world/experience (the empirical content of a scheme, the input) and the beliefs/thoughts/propositions/theory (the cognitive content of a scheme, the output) in the way that it organizes its empirical content to form its cognitive content. Following this way of thinking, the analytic–synthetic distinction does presuppose scheme–content dualism since the very analytic–synthetic distinction depends upon the scheme–content distinction. However, the contrary is not true. Scheme–content dualism neither necessarily leads to nor presupposes a sharp analytic–synthetic distinction as Quine construed. In the first place, the scheme–content distinction does not entail that the sentences used to describe a scheme have to be analytical. We may hold, following Quine's holism, that all beliefs form an interconnected 'web of beliefs', thus all sentences used to express them have empirical contents and are subject to revision. Therefore, scheme–content dualism will not necessarily lead to a *fixed*, *sharp* analytic–synthetic distinction. Secondly, after we abandon the absolute analytic–synthetic distinction, we are not compelled to abandon the scheme–content distinction. We may still hold the very ideas of empirical content and revisable conceptual schemes. In fact, within the Quinean framework, we can retain the dualism of scheme–content even after we have abandoned the analytic–synthetic distinction; as Davidson has admitted, 'the scheme–content division can survive even in an environment that shuns the analytic–synthetic distinction' (1984, p. 189). This is the path followed by Quine, Kuhn, Feyerabend, and many other patrons of conceptual relativism.

Although Davidson realizes correctly that scheme–content dualism could well survive after the fall of the analytic–synthetic distinction, he is wrong to allege that 'giving up the analytic–synthetic distinction has not proven a help in making sense of conceptual relativism' (1984, p. 189). I will argue that it is exactly the denial of a fixed and absolute analytic–synthetic distinction that makes alternative conceptual schemes possible.

Abandoning the analytic–synthetic distinction leads to abandoning the rigid distinction between concept, meaning, or language on the one hand and belief, thought, or theory on the other. It is no longer a novel idea today that all concepts themselves are empirical and none *a priori* as argued by American pragmatists against Kant's transcendental philosophy. Concepts we deploy upon experience are themselves the products of empirical inquiries. In other words, concepts are theory-laden and change with theories. According to Quine's holism, our conceptual schemes (concepts or meanings) are no longer considered as being separated from their cognitive contents (beliefs). We should think of them instead as a whole language standing in relation to the totality of experience. The Kantian 'fixed framework of concepts' model was replaced by the model in which meanings/concepts and beliefs are intertwined.

Based on their extensive knowledge of the historical development of scientific thought, Kuhn and Feyerabend argue that the traditional belief that we describe the world through constructing scientific theories about it in terms of a fixed system of concepts distorts the actual procedure of scientific theory construction. There is no sharp distinction between language (concept and meaning) and theory (belief system). Rather, 'meaning is contaminated by theory'. The concepts/meanings of scientific terms, such as the concepts of space and time, are not fixed; instead, they are adjusted and redefined by new scientific principles within the framework of emergent theories such as the theory of relativity. Consequently, a new conceptual scheme or paradigm emerges with a new theory. Following a similar line, J. Searle (1995, p. 60) argues that conceptual schemes, as a subspecies of 'systems of representation', are influenced by cultural, economic, historical and psychological factors.

However, even if it is true that scheme–content dualism neither necessitates nor presupposes the analytic–synthetic distinction, a total abandon of all forms of the analytic–synthetic distinction seems to render the very notion of a conceptual scheme unintelligible. If meaning is contaminated by theory as Kuhn and Feyerabend try to convince us, then 'to give up the analytic–synthetic distinction as basic to the understanding of language is to give up the idea that we can clearly distinguish between theory and language' (Davidson, 1984, p. 187). However, if a scheme is defined as a framework of concepts or language to form its cognitive content (beliefs, propositions, theory) by organizing its empirical content, we must be able to distinguish somehow one's scheme from one's beliefs and to discuss them separately. This in turn naturally implies that there is a distinction between the language used to describe a scheme and the theory used to describe experience if we can meaningfully talk about a conceptual scheme at all. Furthermore, without some kind of analytic–synthetic distinction in place, there is no way to tell whether or not two alleged conceptual schemes contain different concepts or whether they simply embody different beliefs. In other words, without some sort of analytic–synthetic distinction, there is no way to tell whether or not any communication failure between two linguistic communities is due to their words having different meanings or due to their having different beliefs since, in radical interpretation, meanings and beliefs are always intertwined. Therefore, the complete abandoning of the analytic–synthetic distinction would lead to the self-destruction of the very notion of conceptual schemes.

It may be true that the meanings of expressions used in the formulation of a theory are introduced, changed or redefined by the theory itself, but it does not follow that no distinction can be made between the language used to formulate a theory and the theory couched within the language. After we abandon the fixed, sharp analytic–synthetic distinction, the organizing role that was exclusively attributed to analytic sentences and the empirical content that was supposedly peculiar to synthetic sentences are now seen as shared and diffused by all sentences of a language. But it does not mean that all sentences play equal roles in forming our beliefs. Still using Quine's metaphor of 'a web of beliefs', sentences in the center of the web are those we are most reluctant to give up and are primarily used to describe the scheme of concepts. Those sentences play primarily the organizing role in the formation of beliefs. Confronted with the conflict of experience, we would prefer to keep those sentences fixed by comparison to the sentences on the fringes, which we would more easily revise in the light of experience. We can still call them 'analytic sentences' in a modified sense that the truths of the sentences are widely accepted and fully protected by their users, but they are subject to revision also.

To illustrate such a non-fixed, fuzzy analytic–synthetic distinction, Wittgenstein's riverbed metaphor comes in handy. Wittgenstein asks us to imagine our worldview as a riverbed, where the bed of a river represents certain 'hardened propositions', which is the scheme or the essential conceptual core of the worldview, and the river running on the bed represents the mass of our ever-changing belief systems. 'The river-bed of thoughts may shift. But I distinguish between the movement of the waters on the river-bed and the shift of the bed itself; though there is not a sharp division between one or the other" (Wittgenstein, 1969, p. 15e). On the one hand, our conceptual schemes, as riverbeds, are relatively fixed and firm over a certain period. They form and guide our beliefs. On the other hand, our beliefs, as the rushing waters of the river could slowly change the shape of the riverbed and alter the course of the river, could change our schemes over time. Thus the distinction between the scheme (or language) and its cognitive content (beliefs and theory) can be distinguished relatively.

Although such a distinction is fuzzy and changeable, it is a necessary distinction we use all the time. By the same token, even if there is no hard and fast criterion for determining when a difference between languages is a difference of concepts rather than beliefs, it does not follow that no distinction can be drawn. This is why Quine does not abolish the analytic–synthetic distinction totally; rather he shows it to be one of degree rather than kind. This leads to our third doctrine of conceptual relativism:

> There is a non-fixed, fuzzy distinction between the statements about the scheme of concepts, i.e., the language on the one hand and the statements expressing beliefs of reality/experience, i.e., the theory, on the other.

3. The Quinean Linguistic Model of Conceptual Schemes[2]

In spite of its lasting popularity among its advocates, the notion of conceptual schemes remains murky and widespread confusion lingers over its meaning. Even so, a somehow popular notion of conceptual schemes dominates the discussion— namely, the Quinean linguistic model of conceptual schemes. To have a clearer picture of the Quinean notion, let us see where the Quinean linguistic model stands against the three doctrines of conceptual schemes identified earlier.

Quinean Scheme–Content Dualism

> The primary bearer of a conceptual scheme (S2): For Quine, a conceptual scheme is a set of intertranslatable sentential languages. 'It is a fabric of sentences accepted in science as true, however provisionally' (Quine, 1981, p. 41). That is, a sentential language with its referential apparatus, such as predicates, quantifiers, terms, and its descriptive apparatus, i.e., whole sentences held to be true by the speaker.

> The empirical content of a conceptual scheme (E1): uninterpreted sensory experiences, which are often associated with different names, such as sense data, 'uninterpreted sensation', 'surface irritations', or 'sensory promptings'. It refers to what falls under an umbrella term, 'the sensuous given', or briefly, the given, such as the patch of color, the indescribable sound, the fleeting sensation.

> The relation between a scheme and its empirical content: The scheme or language is used to predict future experience in the light of past experience by providing a manageable structure into the flux of experience. More accurately, (R1) through its referential apparatus the language individuates, organizes, and categorizes experience (such as the posits, physical objects, forces, energy, etc.). [The organizing model] (R3) More importantly, the language as a whole *fits* (faces, predicts, and accounts for) experience [the fitting model].[3]

It is this misleading figure of a language-fitting experience (which is different from R2) that leads Davidson to focus on sentences, rather than predicates, as a primary element of the Quinean notion of conceptual schemes.

The Analytic–Synthetic Distinction

A conceptual scheme or language is like an interconnected 'web of beliefs', 'a man-made fabric which impinges on experience only along the edges', so much so that 'the total field is so underdetermined by its boundary conditions, experience'. 'A conflict with experience at the periphery occasions readjustments in the interior of the field'. No sentence, not even the logical rules, is immune to revision (Quine, 1980, pp. 42-3). Consequently, there is no fixed, absolute distinction between analytic sentences used to describe the language and synthetic sentences used to describe experience. However, this does not exclude a relative, contextualized distinction between beliefs at the center of the web that we are most reluctant to give up and the beliefs on the fringes that we are more ready to revise in light of experience. A fuzzy, non-fixed analytic–synthetic distinction is still possible.

Alternative Conceptual Schemes

One natural consequence of the above non-fixed, contextualized analytic–synthetic distinction is that any language (the riverbed) is subject to change through its interaction with our ever-changing belief system (the river). Meanings or concepts could be contaminated by theory. Therefore, alternative languages or conceptual schemes are not just possible, but are reality, either as real cultural differences due to both posited cultural universals and posited radical differences in conceptual schemes or as conceptual shifts within the same cultural tradition such as from Ptolemy to Kepler, Newton to Einstein, and Aristotle to Darwin.[4]

Notes

1   In some places, Davidson (1984, p. 189) seems to suggest that the duality of scheme–content commits one to or supports the analytic–synthetic distinction as construed by Quine.
2   By the label 'the Quinean linguistic model', I do not mean 'Quine's model' as such, namely, what Quine formulates and defends as a well-established theory. Rather, I intend to use the phrase to refer to one pervasive notion of conceptual schemes based on the works of Quine and many other similarly-minded conceptual relativists including T. Kuhn, P. Feyerabend, B. Whorf, and C. Lewis. In general, it is a notion of conceptual schemes identified and criticized by Davidson.
3   Davidson, 1984, pp. 191-4; 2001a, pp. 40-41.
4   Quine, 1960, p. 77; 1970, pp. 9-11; 1980, p. 43.

Chapter 4

In Defense of the Very
Notion of Conceptual Schemes

1. Inverifiability of Alternative Conceptual Schemes

It seems imaginable that there might be intelligent creatures whose minds operate within a different framework of concepts and an unfamiliar mode of thinking from our own such that their experience of the world could be substantially remote from our own. In fact, we are never in short supply of creative thought experiments from philosophers on such totally alien conceptual schemes. For instance, imagine that the Martians are equipped with very different sensory apparatus and their minds operate in terms of a conceptual scheme bearing no similarity to ours. Would they be disappointed to find that we, the alleged intelligent creatures with whom they disparately want to establish contact, are just noisemakers? Does not this, and other similar cases, prove that totally disengaged conceptual schemes are at least theoretically possible?

Nevertheless, Davidson is not impressed at all with those thought experiments for a good reason: They cannot be verified *semantically*. Given, they may be possible; but anything is possible in a logical sense. The questions with which Davidson is concerned are how to verify semantically the existence of a conceptual scheme, and how to distinguish semantically between two alternative conceptual schemes. Presumably we need a semantically verifiable criterion for the existence of a conceptual scheme (a criterion of schemehood/languagehood for the Quinean conceptual schemes: whether some form of sounds, writings, and recorded symbols made by some creatures represent the use of a language at all) and a criterion for distinguishing between two alternative conceptual schemes. As Davidson discovers, after examining a few available criteria of identity of conceptual schemes allegedly associated with three different forms of conceptual relativism, the Quinean notion of a conceptual scheme simply cannot pass the verifiability test.

Extreme Conceptual Relativism: The Criterion of Intelligibility

For an extreme conceptual relativist, an alien conceptual scheme could be extremely remote from ours to the extent of being 'mutually unintelligible' or 'forever beyond our grasp' and 'rational resolve'.[1] Consequently, cross-language understanding between those schemes is unattainable in principle. Davidson quickly points out that such a criterion simply does not make sense. In order to identify an alternative conceptual scheme, it has to be somehow intelligible to us to the extent that we can recognize it as a state of mind. Since such a scheme does not even make any sense to us, what justification could we possibly have for believing it to exist as a state of mind? What would stop us to deny it as a state of mind at all?

To respond, we need to clarify the key assumption of the argument, i.e., the *unintelligibility to us* of an alien conceptual scheme. It could mean that it is unintelligible *contextually* or *hypothetically* for us if we stick to our own framework of concepts and way of thinking; or it could mean that it is unintelligible *categorically* for us even if we try to learn to think in the way of the alien conceptual scheme by grasping its concepts. Davidson is highly ambiguous as to what he really means by 'unintelligibility' here. In one place, he seems to mean the former sense when he complains that conceptual relativism 'seems (absurdly) to ask us to take a stance outside our own ways of thought' (Davidson, 2001a, p. 40). But in other places he clearly means the latter interpretation since unintelligibility is something 'forever beyond our grasp'. Going along with Davidson, I believe that extreme conceptual relativism in the latter sense is untenable, but not for the logical reason given by Davidson. Even if there exists an alien human language that cannot be made intelligible through interpretation by our semantic and conceptual apparatus, we can always make it intelligible, if it qualifies as a *human language* at all, by learning it from scratch as a child does. If it is unlearnable in principle, we have no other way but to conclude that it is not a human language (the language of bats, for example).

Davidson's ambiguity over the notion of intelligibility reveals a fundamental flaw in his reasoning: He often confuses alleged extreme conceptual relativism with a more promising form of conceptual relativism. What Davidson's argument shows us is that total categorical unintelligibility of another language, culture, or tradition is not an option. 'It slays the terrifying mystical beast of total and irremediable incomprehensibility. But what we suffer from in our encounters between peoples are the jackals and vultures of partial and (we hope) surmountable noncommunication' (Taylor, 2002, p. 291). To the best of my knowledge, not any of the conceptual relativists targeted by Davidson, i.e., B. Whorf, W. V. Quine, T. Kuhn, C.I. Lewis, P. Feyerabend, or P. Strawson, really holds such an extreme position. Most of them actually belong to the camp of radical conceptual relativists who contend that conceptual schemes or the languages associated with them can be and actually are, in many cases, *radically* distinct or *massively* different without any significant overlap, even to the extent of being incommensurable and leading to *massive*, even complete, communication breakdown. But such a failure of cross-language understanding is contextual and can be *partially* overcome, although it remains questionable whether full communication could be carried out without residue between two conceptually remote language communities. Davidson unfairly paints extreme and radical conceptual relativism with the same brush and treats them both as a case of complete failure of translatability (1984, p. 185; 2001a, p. 40). However, they are actually two totally distinct breeds; one holds that some cross-scheme understanding is in principle unattainable while the other believes it is attainable. It is necessary to make the distinction in order to defend radical conceptual relativism.

Radical Relativism: The Criterion of Translatability

How do we effectively identify a radically distinct conceptual scheme? How do we distinguish and compare radically distinct conceptual schemes? The intelligibility criterion does not work here since intelligibility is not an issue for radical conceptual relativism. A different criterion is needed. The identification of the Quinean conceptual schemes with languages prompts us to consider cross-language translation. As Quine realized, it is 'a measure of what might be called the remoteness of a conceptual scheme but what might better be called the conceptual distance between languages' (Quine, 1981, p. 41). But what else can be used to measure such conceptual distance between languages better than the possibility and degree of intertranslation between them? (Quine, 1969, p. 5). Thus in the case of radical conceptual relativism, intertranslatability between languages becomes, as Davidson construes and attributes to the Quinean conceptual relativism, a criterion of conceptual schemes, as both a criterion of languagehood and one of different conceptual schemes. Lack of shared common parts between two radically distinct conceptual schemes makes intertranslation between the two associated languages impossible and leads to total translation failure. 'No significant range of sentences in one language could be translated into the other' (Davidson, 1984, p. 185). Thus, the complete failure of intertranslatability between two languages becomes a necessary and sufficient condition for difference of conceptual schemes associated with them, respectively. Thus we have the following criterion of identity and difference of conceptual schemes:

> The criterion of intertranslatability: If a set of languages can be translated into one another, then they share the same conceptual scheme; otherwise they have different conceptual schemes. Accordingly, a form of activity represents the use of a language for an interpreter if and only if it could be translated, at least partially, into the interpreter's own language.

The belief that there are some distinct alien languages untranslatable into a known language simply does not make sense if translatability is a necessary criterion of languagehood.[2] If we cannot translate a language at all, how can we even be able to recognize it is a language, not just noises? Radical conceptual relativism apparently falls into the similar dilemma faced by extreme conceptual relativism. Of course, Davidson fully realizes that this objection in terms of translatability is too much of a 'cheap shot' that does not have any substantial persuasive power to conceptual relativists. They can quickly offer rebuttal that translatability is not, and should not, be a necessary condition for languagehood. To answer these possible objections, Davidson asks, could we make sense of there being a language that we cannot translate at all but that we could still recognize as a language by other criteria? In other words, could we have 'a criterion of a languagehood that did not depend on, or entail, translatability into a familiar idiom'? (Davidson, 1984, p. 192) Where do we look for such a criterion? The opponents can easily supply a bundle of criteria of languagehood independent of translatability. At least interpretation in a loose sense or other criteria focusing on communicative functions will do the trick.[3] But Davidson will not give in to these suggestions. For Davidson, the problem at hand is not whether there are other creditable criteria of languagehood available (I do not think Davidson would deny that there are such criteria) but whether Quinean radical conceptual relativism itself can supply such a criterion. The question becomes, could we derive a criterion of languagehood independent of translatability from Quinean scheme–content dualism?

As Davidson observes, Quinean scheme–content dualism is associated with two popular metaphors: A language either (R1) organizes its empirical content through its referential apparatus or (R3) fits its content in terms of its descriptive apparatus. The organizing model R1 cannot be made intelligible and thus cannot supply the criterion needed. The fitting model R3 simply means that a language fits all potential experiences. But 'the point is that for a theory to fit or face up to the totality of possible sensory evidence is for that theory to be true. … [t]he notion of fitting the totality of experience … adds nothing intelligible to the simple concept of being true'. Therefore, 'our attempt to characterize languages or conceptual schemes in terms of the notion of fitting some entity has come down, then, to the simple thought that something is an acceptable conceptual scheme or theory if it is true' (Davidson, 1984, pp. 193-4). Then the proposed criterion of languagehood becomes that a form of activity represents the use of a language for an interpreter if and *only if* it is largely true from the interpreter's perspective. However, according to Davidson's truth-conditional theory of meaning/translation, to be true and to be translatable always go hand in hand. Hence, the proposed truth-criterion slips right back to the translatability criterion: A form of activity represents the use of a language for an interpreter if and only if it is translatable into the interpreter's language. Davidson concludes that Quinean conceptual relativism cannot supply a criterion of languagehood independent of translatability.

By comparison, the translatability criterion of alternative conceptual schemes appears to be logically coherent, if not tenable, from a third-person perspective, simply imagining that one is observing two competing languages, somehow recognized as such, which cannot be translated into one another. But remember that Quinean content–scheme dualism can be boiled down to the claim that a conceptual scheme different from the interpreter's is largely true. Then 'the criterion of a conceptual scheme different from our own now becomes: largely true but not translatable' (Davidson, 1984, p. 194). But again according to Davidson's truth-conditional theory of meaning/translation, truth cannot be divorced from translation. In conclusion, the translatability criterion of conceptual schemes, both as a criterion of languagehood and as a criterion of alternative conceptual schemes, is incoherent.

A radical conceptual relativist can respond to Davidson on two fronts, either to defend the Quinean translatability criterion—for a Quinean conceptual relativist, many fundamental assumptions of Davidson's arguments are either unsound or questionable—or to separate radical conceptual relativism from the Quinean relativism as Davidson construes by removing the translatability criterion out of the equation. The first route is a well-worn path by many critics[4] that I will not belabor except to remark on briefly, below. The second route, to me, is more effective and will be discussed in the next section.

One can and should ask the question: Why does translation have to be a necessary condition of languagehood? The basic assumption lurking behind such a rather strict requirement is that having justification for believing in the linguistic/conceptual character of someone's behavior requires being able to understand their language/concepts. At first glance, the assumption seems to be quite innocent since if we cannot make another's language intelligible, how could we even know whether a certain form of activity represents a linguistic act. However, whether such an assumption is acceptable depends upon the kind of cross-language understanding involved here. Among many different notions of cross-language understanding, the translatability criterion of languagehood assumes the translatability notion of cross-language understanding. That is, we can understand an alien language only by mapping it onto our own language.

What makes the case in hand more provocative is the fact that such a translatability notion of cross-language understanding is actually Davidson's own stand, not just the Quinean conceptual relativist position as Davidson construes and criticizes. To see why, one needs only to recall Davidson's truth-conditional account of understanding. According to it, to know the (Tarskian) truth conditions of the sentences of a language is both necessary and sufficient to understand it. But since truth and translation always go hand in hand, it amounts to a translatability notion of cross-language understanding: To be able to translate a language (in the sense of truth-functional translation) it is necessary to understand it. Ironically, what Davidson attacks is actually his own position. In other words, Davidson's criticism itself is incoherent. More importantly, as I will argue in chapter 12, it is the knowledge of *truth-value* conditions (under what conditions a sentence has a truth-value), not the knowledge of *truth* conditions (under what conditions a sentence is true), that plays an essential role in cross-language understanding.

In addition, Davidson's arguments against the translatability criterion do not work without help from his interpretation of Quinean content–scheme dualism and his own truth-conditional theory of translation; for only the combination of both can make the notion of a conceptual scheme, which is supposed to be largely true but untranslatable, incoherent. Recall that Davidson concludes that the distinction can be boiled down as follows: 'Something is an acceptable conceptual scheme or theory if it is true' or 'largely true'. Being largely true from whose perspective? Davidson answers, from the interpreter's notion of truth! This is simply another application of his principle of charity. As to be argued below, the principle is untenable. More to the point, as I will argue later, it is a false assumption that an interpreter would hold the sentences of an alien language embodying a radically distinct conceptual scheme as true.

Modest Conceptual Relativism: The Criterion of Interpretability

Modest conceptual relativism recognizes the existence of the common part between two partially distinct conceptual schemes. Because of this, partial translation between the two associated languages is always possible. As such, translatability would not suit for the identity criterion in the case of modest conceptual relativism.

The basic supposition of modest conceptual relativism must be, Davidson contends, that the difference between two conceptual schemes can be identified by reference to the common part shared by both. The common part that is intertranslatable must represent shared concepts and beliefs/cognitive contents, and the other parts are presumed to constitute the difference between two schemes, namely, difference in meanings or concepts, not just difference in beliefs. Since those parts are not translatable, we can only hope to discover the difference in meanings and concepts through interpretation. In addition, since those parts are not shared, the interpretation cannot presuppose shared beliefs, meanings, or concepts. This is actually the case of so-called radical interpretation in which the interpreter has as evidential basis for interpreting a subject only the subject's physical behavior in their environmental context. Thus, the identity criterion of alternative partially distinct conceptual schemes becomes:

> The criterion of interpretability: An alien conceptual scheme is different from an interpreter's when the interpreter can identify the difference in meanings or concepts in terms of radical interpretation.

Obviously, the above interpretability criterion assumes that the interpreter, in the case of radical interpretation, could decisively separate the speaker's concepts from his or her beliefs and could determine, when the speaker thinks and speaks differently from the interpreter, whether the difference lies in his or her concepts rather than his or her beliefs. If such a difference were to occur between the two languages, 'we should be justified in calling them alternations in the basic conceptual apparatus' (Davidson, 1984, p. 188), not simply alternations in beliefs. But the assumption is, Davidson argues, not justified given the underlying methodology of radical interpretation—the principle of charity.

According to Davidson, a central source of trouble in interpreting others in the case of radical interpretation is the way beliefs and meanings conspire to account for utterances. There is no way to disentangle completely what aliens means from what they believe. We do not know what alien believe unless we know what they mean; we do not know what they mean unless we know what they believe. Therefore, we are always captured in such an interlock between meanings and beliefs. This is especially the case when we encounter an alleged alternative conceptual scheme (Davidson 1984, pp. 27, 142, 196). To break into such a vicious circle of radical interpretation, we have to hold one end steady while studying the other end. Following Quine, Davidson suggests that we should fix beliefs constant as far as possible while solving for meanings. In fact, this is actually not much of a choice for the interpreter since meanings are unobservable for the interpreter while some belief-related behaviors are observable within specific contexts. Of course, we cannot hold all beliefs constant for it is neither possible nor feasible. Fortunately, we do not need to do so. Among the behavioral facts ascertainable by observation prior to interpretation are notably the speaker's beliefs of *holding* certain token-sentences *to be true* through consent when those token-sentences are actually true according to the interpreter's truth conditions. For Davidson, it is the only legitimate evidential basis for radical interpretation. Therefore, what we need in radical interpretation is to fix a specific part of beliefs, namely, holding constant sentences of the speaker's language to be true by the speaker from the interpreter's perspective while deriving meanings from it.

The notion of 'holding true' plays the most essential role in Davidson's project. But why does the interpreter have to determine whether the speaker holds a sentence in his or her own language to be true by attributing the interpreter's own holding-true conditions of the sentence to the speaker? The answer lies in Davidson's truth-conditional interpretation of the principle of charity, i.e., in radical interpretation, putting the speaker in general agreement on beliefs with the interpreter according to the interpreter's point of view. But in order to complete a radical interpretation, only holding *some* of the speaker's sentences true is not sufficient. Radical interpretation can only be accomplished through more generous 'charity', that is, through the interpreter ascribing to the speaker's beliefs as largely true by the interpreter's own lights. In other words, most sentences in an alien language that are held to be true by the speaker must be actually true according to the interpreter's truth conditions.

A main reason for this strict interpretation of the principle of charity is this: Radical interpretation can only be accomplished if the interpreter knows what the speaker's subject matter is. If the speaker's saying is *about* anything at all, it is necessary for the speaker to possess a *massive* network of true beliefs about the subject matter in question. So the interpreter has to assume that the speaker's beliefs are largely true of the speaker's own environment as the interpreter believes them to be; otherwise if the speaker's beliefs bear little or no relation to reality, then the interpreter could not even start the first step of radical interpretation, which is purely based on the speaker's reaction to his or her immediate environment. Similarly, in order for the interpreter to identify what the speaker's subject matter is, he or she has to share with the speaker a massive network of true beliefs relating to that subject matter (Davidson, 1984, pp. 155-70). Furthermore, for communication to be possible there must not only be a high degree of agreement in factual judgments, i.e., the shared beliefs of holding true, there must also be agreement in definitions, meanings, or concepts. Davidson thus concludes that radical interpretation requires the discovery of *massive* agreement between the interpreter and the speaker in both beliefs and concepts. This is why Davidson is convinced that we should prefer a theory of interpretation that maximizes the agreement and minimizes disagreement between the speaker and the interpreter. In order to understand others, we have to suppose that others are like us, to attribute our own beliefs to others, and to interpret others' utterances based on our own beliefs.

If the charity principle is a methodological prescript forced on us in radical interpretation and not an option as Davidson strongly believes, nothing can force us to allocate a difference in respect to holding a sentence true to a disagreement in beliefs, rather than to a disagreement in concepts. 'We could not be in a position to judge that others had concepts or beliefs radically different from our own', for 'when others think differently from us, no general principle, or appeal to evidence, can force us to decide that the difference lies in our beliefs rather than in our concepts' (Davidson, 1984, p. 197). It is always possible to conclude, through a variety of semantic maneuvers, that others share our scheme but have different beliefs than ours. Therefore, as with other previous criteria, the interpretability criterion fails to identify the existence of alternative conceptual schemes.

Typical responses to Davidson's above argument are to focus on its shaky foundation, namely, the principle of charity. I agree with those critics that the principle of charity is untenable, theoretically flawed, and practically unproductive.[5] However, my interest here is not in the principle of charity in general, but in Davidson's truth-functional interpretation of the principle, which will be discussed in detail below.

2. Two Assumptions of the Quinean Model

Davidson's arguments from inverifiability against the Quinean notion of conceptual schemes are fully loaded with many unwarranted assumptions, theories, principles, or notions, such as his truth-conditional theory of meaning/translation, the translatability criterion of alternative conceptual schemes, the principle of charity, and the so-called fitting model of conceptual schemes. They are at least as shady and controversial as the target notion he wants to overthrow. Instead of examining those assumptions one by one, my strategy is to examine and challenge the two basic assumptions of the Quinean linguistic model on which Davidson's arguments rely. One assumption is explicit; that is, the identification between conceptual schemes and sentential languages, which prompts Davidson to focus on sentences, truth-values of sentences, and sentential translation in distinguishing alternative conceptual schemes. His alleged fitting model of conceptual schemes, which is indispensable to his inverifiability argument, is also based on such an interpretation. Nevertheless, if the thesis of scheme–language identity is unfounded, the translation criterion goes out of the window and his argument of inverifiability becomes baseless. This is, I will argue, exactly the case. The other assumption is hidden, tacitly subscribed to by Quine, Davidson, and many others, conceptual relativists and anti-relativists alike, according to which we should focus on, in the discussion of alternative conceptual schemes, the notion of truth (whether truth is relative to a conceptual scheme or relative to a language) and redistribution of truth-values crossing two distinct languages. Application of Davidson's truth-functional theory of interpretation/translation and his version of the principle of charity presuppose the second assumption. Its removal would undermine Davidson's project.

Conceptual Schemes as Sentential Languages

Presumably, conceptual schemes are about concepts. However, the notion of concept is notoriously murky and slippery, often referring to a shadowy and problematic entity of some obscure sort, such as Platonic mind-independent abstract entity, disposition, mental ability to perform certain activities, or whatever composes the propositional content of one's thought or belief. 'The linguistic turn' in the last century seems to point a way out of those confusions. After all, it is widely recognized that thinking is essentially a linguistic activity. Language is intimately associated with categorization and conceptualization, the two primary functions of concepts. Thinking in terms of language can certainly help us better grasp concepts, better detect conceptual variation, and better measure conceptual distance between alternative conceptual schemes. Through the association with concepts, conceptual schemes have commonly come to be associated with languages. For these reasons, many friends of conceptual relativism as different as R. Carnap, B. Whorf, and W.V. Quine have been thinking of the notion of conceptual schemes along such a linguistic line.

The exact relationship between conceptual scheme and language depends mainly upon two factors: What the linguistic counterparts of concepts are thought to be on the one hand and kinds of language involved on the other. Since concepts are linked to meanings, the primary linguistic vehicle of meaning would presumably be the primary bearer of a conceptual scheme. In the first half of the twentieth century, we witnessed an important reorientation in semantics, seen in Frege's and Russell's works as well as in the verification theory of meaning of logical positivism, whereby the primary vehicle of meaning came to be seen no longer in the term but in the sentence. Along with such a reorientation in semantics came an ontological shift from concepts to sentences as the primary elements of conceptual schemes. Through the link to sentences, conceptual scheme becomes more closely connected to language. This is partly why R. Carnap started to speak of linguistic framework when others talked about conceptual framework/scheme.

However, Carnap's linguistic framework, which mainly refers to the meaning postulates of a language or theory, is still not a language itself. Quine pushes us further down the path to assimilate conceptual scheme into language. Quine contends that a conceptual scheme is not merely associated with a language, but is, rather, *identical* with it. Based on Quine's holistic semantics, even in taking individual sentences as the unit of meaning, we have drawn our grid too finely since 'our statements about the external world face the tribunal of sense experience not individually but only as a corporate body' (Quine, 1980, p. 41). Instead, the unit of linguistic meaning with empirical significance is neither terms nor sentences, but the whole language. In some sense, once meaning is linked to language as a whole, meanings themselves, as obscure intermediary entities, may well be abandoned. One might wonder whether the notion of conceptual schemes should go by the wayside along with the dissolution of meaning and its counterpart, the concept, in the language. But Quine, as an empiricist who is unwilling to part with the scheme–content dichotomy (Quine, 1981, pp. 38-40), continues to use the term 'conceptual schemes or frameworks for science' and regards it as 'a tool' or 'a device for working a manageable structure into the flux of experience'. Thus, 'science has its double dependence upon language and experience' although this duality is not significantly traceable into the duality of analytic sentences used to describe a scheme on the one hand and synthetic sentences used to describe experience (the dogma of the analytic–synthetic distinction) on the other (Quine, 1980, pp. 42-6). Thirty years later, Quine, in response to Davidson's criticism, admitted 'where I have spoken of a conceptual scheme I could have spoken of a language. Where I have spoken of a very alien conceptual scheme I would have been content, Davidson will be glad to know, to speak of a language awkward or baffling to translate' (1981, p. 41). Considering the possibility that not all languages are conceptually distinct, for Quine, a conceptual scheme eventually becomes a set of languages that share the same conceptual make-up, whatever it may be.

Furthermore, in Quine's mind, the language identical with the conceptual scheme is not a language in a special technical sense, such as the logical positivist theoretical language, but rather 'ordinary sentential language, serving no technical function' (Quine, 1981, p. 41). Obviously not all declarative sentences in an ordinary language are true. However, for a conceptual scheme to organize experience and face reality its sentences have to be true, at least from the speaker's perspective. Therefore, a conceptual scheme eventually becomes identical with a sentential language with all its sentences held to be true. This is exactly the way in which Davidson construes the Quinean notion of conceptual schemes: 'We may identify conceptual schemes with languages, then, or better, allowing for the possibility that more than one language may express the same scheme, sets of intertranslatable languages' (Davidson, 1984, p. 185).

While thinking in terms of language may promise to clarify our understanding of the notion of conceptual schemes, it is not immediately clear how that clarification should be carried out. In other words, while it is relatively unproblematic and widely accepted now that conceptual schemes are and should be somehow associated with languages, it is controversial as to how to associate conceptual schemes with languages; since there are certainly many different ways in which a conceptual scheme could be connected to a language. To identify a conceptual scheme with a sentential language is certainly not the right way.

Which kind of language do Quine and Davidson have in mind when they identify a conceptual scheme with a sentential language? Strictly speaking, a notion of *sentential* language itself does not make much sense. Any linguist would tell us that it is obviously misguided to think of a language as a totality of sentences, not to mention as a totality of all declarative sentences held to be true. A language is a grammar and a vocabulary, not just sentences. But let us suppose, for the sake of argument, that the notion makes sense. What kinds of sentential languages then? Among the advocates of conceptual relativists identified by Davidson, Whorf exclusively focuses on natural languages while Kuhn and Feyerabend primarily concentrate on scientific languages. In his criticism, Davidson mixes and treats them both as sentential languages held to be true by their speakers without further distinction. In Quine's naturalized epistemology, it is 'the total science' or a network of comprehensive scientific theories that is supposed to be a conceptual scheme facing the totality of experience. Thus, we have two possible candidates for sentential languages that Quine and Davidson have in their mind: natural language and scientific language.

A natural language *per se* such as English or Chinese is in no sense a conceptual scheme; otherwise, it would be a bizarre way to construe the notion of conceptual schemes. Does any conceptual relativist seriously think that all Chinese would inherit a different conceptual scheme from that of all English simply because they speak different natural languages? To put it another way, in what sense is the Chinese language throughout its 5,000 years of history a conceptual scheme and over time the same conceptual scheme? A natural language is not a theory either. A natural language like English does not schematize experience or even metonymically predict or fit reality. Although part of a natural language, i.e., its grammar, does in some sense determine the logical space of possibilities (Whorf, 1956), it is the theoretical assertions made in the language that predict and describe reality and in so doing assert which logical spaces are occupied, i.e., which logical possibilities are actualized in the world. In addition, a natural language as such is not even a worldview as many are convinced. 'If every language is a view of the world, it is so not primarily because it is a particular type of language (in the way that linguists view languages) but because of what is said or handed down in this language' (Gadamer, 1989, p. 441). What constitutes the so-called worldview proper is not natural language as language, whether as grammar or as lexicon; the worldview consists in the coming into language of what has been handed down from the cultural tradition associated with the language. Only in this sense do we say that a language-view is a worldview.

Many have treated a scientific language as a conceptual scheme. Although it is more closely related to a conceptual scheme than a natural language is, a scientific language construed as the totality of all its sentences is not a conceptual scheme. A conceptual scheme is, briefly put, supposed to be a conceptual framework which schematizes our experience in terms of its metaphysical presuppositions of existents, states of affairs, modes of reasoning, and categorization to form a theory. Many parts of a conceptual scheme, such as a categorical framework, usually a lexical structure of a scientific theory, are simply not a set of sentences, but rather a categorical system. In addition, a conceptual scheme, which serves as the conceptual framework of a theory, cannot in itself be the theory or the language expressing the theory. Nor will it improve matters to stipulate that a conceptual scheme is the totality of sentences held to be true by its speaker. A conceptual scheme does not describe reality as the Quinean fitting model R3 suggests; it is rather the theory it formulates that describes reality. A conceptual scheme can only 'confront' reality in a very loose sense, namely, by becoming in touch with reality in terms of a theory. Therefore, a conceptual scheme cannot be said to be true or largely true. It is simply not the bearer of truth. Only the assertions made in a language and a theory couched in the language can be said to be true or largely true.

To treat a conceptual scheme as a sentential language is to confuse language with language scheme, i.e., the conceptual core of language. When Whorf compares a natural language to a scientific theory, he actually means that it is the grammar of a natural language (construed as the rules for the use of expressions to determine what states of affairs are expressible and what expressions make sense) that functions just like a scientific taxonomy. Grammar or taxonomy fixes the logical space of possibilities that the world may or may not occupy, in as much as grammar determines what it makes sense to say. Any sentences couched in the language with that grammar or taxonomy will have truth-values, and they are true or false depending upon whether things are as they are asserted to be. It is exactly the role of the later Kuhn's notion of lexical structure played out in a scientific theory. Similarly, for Carnap, a linguistic framework is not a language, but mainly refers to the meaning postulates of a scientific language. We can call those different conceptual apparatus of a language the language scheme. The language scheme of a scientific language is the conceptual scheme of the corresponding theory. Strictly speaking, neither a language nor a theory can be a conceptual scheme.

Truth-Values Cross Alternative Conceptual Schemes

If a conceptual scheme were identical to a sentential language, construed as the totality of sentences held to be true by its speaker, then it would seem to be natural to deal with a conceptual scheme by the notions of truth and meaning in Tarski's style. This is exactly the strategy used by Davidson in his inverifiability argument against the notion of conceptual schemes. The Tarskian notions of truth and meaning are based on the classical bivalent logic, in which a declarative sentence has a definite truth-value; that is, it can only be either true or false and cannot be neither true nor false (having no truth-value at all). If this is so, why do we not distinguish alternative conceptual schemes in terms of the redistribution of truth-values?

According to Davidson's reading of Quinean content–scheme dualism, especially the fitting model R3, this dualism can be boiled down to the claim that a form of activity represents the use of a language for an interpreter if and *only if* it is *largely true* from the interpreter's perspective. Davidson finds that, similar to this truth-value criterion of languagehood, a criterion of alternative conceptual schemes lurks behind conceptual relativists' rejection of a sharp analytic–synthetic distinction. For many philosophers of science, Kuhn and Feyerabend in particular, to give up a sharp analytic–synthetic distinction as basic to the understanding of language is to give up the idea that we can clearly distinguish between theory and language/meaning. When switching from one theory to another, with the change of meanings of the terms that are determined by some principles or meaning postulates of the corresponding language, a sentence containing those terms, which were accepted as true within an old theory ('motion' in Aristotelian physics), could come to be accepted as false within a new theory (Newtonian dynamics). Davidson concludes:

> We may now seem to have a formula for generating distinct conceptual schemes. We
> get a new out of an old scheme when the speakers of a language come to accept as true

an important range of sentences they previously took to be false (and, of course, vice versa). (1984, p. 188)

Quine clearly thinks along the same line in terms of his holistic picture of scientific development. For Quine, any scientific theory is like a field of force with experience as its boundary conditions, and the total field is underdetermined by the totality of all possible experience. A new theory could emerge from an old one by making sufficient adjustments in the interior of the field. After readjustments 'truth values have to be redistributed over some of our statements' (Quine, 1980, p. 42). It is conceivable that if an old theory is undergoing a radical readjustment to the extent that truth-values over sentences have to be redistributed in a systematic way, then the new theory that emerges is radically distinct from the old one. In this case, we can say that two theories embody two different conceptual schemes. Thus we have reached a truth-value criterion of identity of conceptual schemes independent of the translatability criterion, implicit in the Quinean conceptual relativism:

(TV) Two conceptual schemes or languages differ when some substantial sentences of one language are not held to be true in the other in a systematic manner.

In other words, the difference between two conceptual schemes is semantically signified by the redistribution of *truth-values* over the sentences of two languages that embody the conceptual schemes.

The above truth-value criterion of conceptual schemes is not just different and independent from, but is actually more fundamental than, the translatability criterion. The latter is in fact a logical consequence of the former. Because of a systematic redistribution of truth-values from one language to another, the truth-preserving translation between them can hardly be carried out, and is even impossible in principle. This is why Davidson insists that we cannot get around the translatability criterion and that untranslatability is a necessary condition for distinguishing one conceptual scheme from another.

We can dig still deeper into a tacit assumption behind the Quinean truth-value criterion of conceptual schemes. Redistribution of truth-values over the sentences between two competing languages presupposes that although the speaker and the interpreter in discourse may assign opposite truth-values to some sentences in the other's language, they agree that all sentences in the alien language are either true or false. That means that they agree on the truth-value status of these sentences. Thus, there is no truth-value gap between two languages:

(TF) Most sentences in the speaker's language are either true or false from the point of view of the interpreter's language no matter how disparate the two languages are.

Davidson strengthens the grip of the truth-conditional claim of conceptual schemes further to require that, according to his truth-functional interpretation of the principle of charity, most sentences of an alternative conceptual scheme not only have truth-values, but also have to be true, from both the viewpoints of the speaker and the interpreter. That means the speaker and the interpreter must and *actually do* share the same notion of truth and the same truth conditions if understanding is possible. This in turn presupposes the above assumption that the speaker and the interpreter have a shared belief of the truth-value status of the sentences in each other's language. All the sentences in an alien language are either true or false from the interpreter's point of view. Consequently, the possibility of the occurrence of a truth-value gap between two alien languages is excluded *a priori* as long as we stick to the principle of charity in interpretation. 'If we cannot find a way to interpret the utterances and other behavior of a creature as revealing a set of beliefs largely consistent and true by our own standard, we have no reason to count that creature as rational, as having beliefs, or as saying anything' (Davidson, 1984, p. 137).

As I will argue extensively in the rest of the book, my major reservation with the Quinean notion of conceptual schemes is not just the many theoretical difficulties it faces, but rather with its basic assumption (TF); for it does not square with observations of many celebrated conceptual confrontations between opposing conceptual schemes revealed in the history of natural sciences and cultural studies, especially those under the name of incommensurability. These familiar conceptual confrontations are, to me, not confrontations between two conceptual schemes with different distributions of truth-values over their assertions, but rather confrontations between two scientific languages with different *distributions of truth-value status* over their sentences due to incompatible metaphysical presuppositions. The advocate of an alien conceptual scheme not only does not hold the same notion of truth as ours, but also does not agree with us on the truth-value status of the sentences in question. These scheme innovations, at bottom, turn not on differences in truth-values, but on whether or not the sentences in the alternative conceptual scheme *have* truth-values. We simply cannot find the agreement taking 'the form of widespread sharing of sentences held true' between two incommensurable languages.

3. The Charge of the Third Dogma of Empiricism

Conceptual schemes and contents come in a pair in scheme–content dualism but could be treated separately. In his arguments of inverifiability, Davidson focuses on the scheme side of the dualism. Such a strategy turns out not to be successful for at least two reasons. First, it is verificationist in nature. The conclusion Davidson is trying to reach is not that alternative conceptual schemes absolutely do not exist, but rather that it is meaningless or unintelligible to assert that they do in that we cannot semantically verify them. The critics, of course, could challenge such verificationist tactics: From the fact that there is no reliable way of telling whether something is the case, it does not follow that it is not the case, nor it is unreasonable to believe that it is the case. In any case, why do we have to hang our hat on the post of verificationism? Second, it only applies to one kind of conceptual scheme, namely, Quinean sentential language, but leaves other possible kinds such as S1 untouched.

Fortunately, both sides do not need to remain deadlocked on the issue of verificationism since there is another way in which conceptual relativism can be made intelligible without appealing to any verifiable identity criteria of conceptual schemes, that is, scheme–content dualism. As Davidson himself acknowledges, 'If we could conceive of the function of conceptual schemes in this way, relativism would appear to be an abstract possibility despite doubts about how an alien scheme might be deciphered' (2001a, p. 41). Besides, a conceptual scheme, no matter whether it is S1, S2, or any other possible bearer, can make sense only if there is some content to schematize. A dismantlement of scheme–content dualism would render any form of conceptual scheme baseless.

Kantian Scheme-Content Dualism

Even before Kant, we have been accustomed in empiricist tradition to 'split the organism up into a receptive wax tablet on the one hand and an "active" interpreter of what nature has there imprinted on the other' (Rorty, 1982, p. 4). 'Since Kant, we find it almost impossible not to think of the mind as divided into active and passive faculty, the former using concepts to "interpret" what "the world" imposes on the latter' (Rorty, 1982, p. 3). In reference to scheme–content dualism, the empirical contents schematized by our mental conceptual schemes in Kant's model would be either the world-as-it-is or what the world imposes upon us as 'given', i.e., unconceptualized, ineffable raw physical thrust of stimuli upon our organs, passively received through our two sensible intuitions. Both Davidson and Rorty are convinced that such a sharp Kantian 'given–interpretation' distinction must be the foundation of any workable scheme–content dualism. That means that the empirical contents of a conceptual scheme have to be either uninterpreted 'reality (the universe, the world, nature)' outside of the mind, or 'experience (the passing show, surface irritations, sensory promptings, sense data, the given)' within.[6] For Davidson, the notion of 'experience' E in the scheme–experience distinction has to be totally and purely experiential, scheme-free, and universal for all human beings, as we have defined as E1 in Quinean scheme–content dualism. The notion of 'the world' W used in the scheme–world distinction cannot be, as Rorty tries to convince us, the world as we have been experiencing it, full of familiar objects. Instead, it 'must be the notion of something completely unspecified and unspecifiable—the thing-in-itself' (1982, pp. 14-17). The world or experience (as empirical contents of schemes) is, putting it metaphorically, organized or interpreted by different conceptual schemes to produce different ideologies, theories, or beliefs about the world (as cognitive contents of schemes). To describe them fully here:

(E1) thin experience (experience narrowly construed): sensory experience independent of all conceptual or linguistic schemes, theories, and interpretations, or 'the sensuous given' imposed from the world upon our receptive minds, i.e., sense data or 'uninterpreted sensations', including Kant's 'passively received intuitions', Locke's 'simple ideas', Hume's 'impressions', and Quine's 'surface irritations' or 'sensory promptings'.

(W1) the world-as-it-is: the world outside of any possible conceptualization, like the Kantian 'thing in itself', which is unnamable, unspecifiable, and thereby an ungraspable physical thrust of stimuli or neutral materials.

These two possible empirical contents constitute the core of Kantian scheme–content dualism. Kantian scheme–content dualism, Davidson and Rorty argue, is presupposed by any notion of conceptual schemes—or more precisely, what Rorty calls 'the Kantian notion of "conceptual framework"—the notion of "concepts necessary for the constitution of experience, as opposed to concepts whose application is necessary to control or predict experience"' (Rorty, 1982, p. 5). Clearly, Quinean scheme–content dualism is a version of Kantian scheme–content dualism.

The Third Dogma of Empiricism

Davidson urges us that the above Kantian scheme–content dualism cannot be made intelligible and defensible, which is actually one more instance of the various harmful dualisms that are part of the Cartesian and empirical philosophical legacy, i.e., the third dogma of empiricism.

   In what sense is scheme–content dualism the third dogma of empiricism? To answer this, we need to identify what Davidson has in mind in referring to 'empiricism'.[7] Empiricism can be construed as a theory about the formation of beliefs/knowledge, and especially, how the mind acquires its cognitive content about the world. This version of empiricism, as a theory of the contents of the mind, has been traditionally associated with the representational theory of mind/knowledge that claims that our mental states, including ideas and beliefs, represent aspects of the world. On the other hand, empiricism can be understood as a theory about the justification of belief/knowledge. Empirical foundationalism, which appeals to some sort of universal sensory experience (usually sense data or the given) as the ultimate evidence or final foundation of our knowledge or warranted beliefs, has been a dominant theory of epistemic justification. The dominant empiricist theory of truth, namely, the correspondence theory, is a hybrid of the representational theory and foundationalism.

   Kantian scheme–content dualism is seen as the third dogma of empiricism by Davidson because it is either a logical consequence or a presupposition of the two versions of empiricism. The scheme–world distinction, usually associated with the fitting model R3, is nothing but another version of the representational theory of mind/knowledge; for both presuppose 'a theory-neutral reality' on the one hand and a mental entity (ideas or conceptual schemes), which represents or corresponds to the reality, on the other. For many conceptual relativists, a conceptual scheme is not just a convenient way of 'saving the phenomena', but rather a mental framework for facing, fitting, or representing reality. J. Searle, for example, regards conceptual schemes as a subspecies or central conceptual core of 'system of representation' of a language-independent reality. By comparison, sense data or the given in the scheme–given distinction, usually associated with the organizer model R1, is commonly thought of as the neutral raw materials from which our beliefs about the world are formed and justified. Such an ultimate source of beliefs is carved out and their foundation is secured by the scheme–given distinction. In this sense, both empiricism as a theory of justification and that as a theory about mental contents presuppose the scheme–given distinction. Furthermore, as Davidson notes, a common feature shared by both empiricism and Kantian scheme–content dualism is that they impose epistemological intermediaries between the mind and the world, that is, our beliefs and the world consisting of familiar objects our beliefs are about. More precisely, the scheme–world distinction turns schemes into intermediaries between the mind and the world, while the scheme–given distinction introduces two layers of intermediaries, sense data/the given from the content side as one layer and schemes from the scheme side of the distinction as the other. If one thinks these intermediaries are necessary or inevitable for us to understand the mind–world relationship, as Davidson believes that conceptual relativists do, one falls prey to the third dogma.[8]

Why is such a third dogma so harmful and in need of rejection? A simple answer from Davidson is, those imposed intermediaries between the mind and the world, no matter whether it is sense data/the given to be organized or schemes doing the organizing, simply do not play *any epistemological role* in either determining or justifying the contents of our beliefs about the world.

Let us start with the scheme–given distinction presupposed by both empiricism of belief formation and that of belief justification. Both Davidson and Rorty discover that the scheme–sense-data distinction is deeply rooted in the Kantian 'picture of the mind as a passive but critical spectator of inner show', the active faculty of the mind using schemes to 'interpret' what the world imposes on the passive faculty, i.e., the given. What are imposed onto the mind via its receptive faculty, i.e., sense data/the given, are purely private, subjective 'objects of the mind', which not only supply neutral raw materials in formation of our mental contents, but also constitute the ultimate evidence for our beliefs about the world.[9] Such a Kantian model of a picture of the mind, according to which the subjective ('experience') is the foundation of objective empirical knowledge, sanctions the search for an empirical foundation for justifying our beliefs outside the scope of the totality of beliefs, by grounding them in one way or another on the testimony of the senses, i.e., E1. Davidson believes that such a foundationalist view of justification is essentially incoherent. Confronting the totality of one's beliefs with the tribunal of experience simply does not make sense, 'for of course we cannot get outside our skins to find out what is causing the internal happening of which we are aware' (Davidson, 2001b, p. 144). 'Our beliefs purport to represent something objective, but the character of their subjectivity prevents us from taking the first step in determining whether they correspond to what they pretend to represent' (Davidson, 2001a, p. 43). In fact, 'nothing can count as a reason for holding a belief except another belief'. Davidson argues that we should abandon the search for a basis for knowledge outside the scope of our beliefs, because 'empirical knowledge has no epistemological foundation, and needs none'.

On the other hand, Davidson reminds us, the existence of the given and its role in belief formation has been challenged by a revised view of the relation of mind and the world. For instance, according to the causal theory of meaning, meanings cannot be purely subjective or mental; instead, words derive their meanings from the objects and circumstances in whose presence they were learned. Similarly, the truth-conditional theory of meaning tells us that our sentences are given their meanings by the situations that generally cause us to hold them as true or false. In general, all states of mind, like beliefs, thoughts, wishes, and desires, are identified in part by the social and historical contexts in which they are acquired. The fact that the states of mind are identified by causal relations with external objects and events convinces us that sense data or the given plays no *epistemological role*[10] in determining the contents of the mind, no matter whether it is belief, meaning, or knowledge.

Therefore, Davidson concludes, the scheme–given distinction is a deep mistake because 'an adequate account of knowledge makes no appeal to such epistemological intermediaries as sense data, qualia, or raw feels'. 'Epistemology has no need for purely private, subjective "objects of the mind"'. Since scheme and content come as a pair, the notion of scheme interpreting and organizing the given goes by the wayside with the given (Davidson, 2001a; 2001b, pp. 141-6).

What about the scheme–world distinction associated with representationalism and the correspondence theory of truth? Could it remain intact? Do we still need schemes to form our beliefs about the world in order to have true beliefs? Davidson's answer is negative. If sense data or the given (E1) cannot justify whether a belief is true, neither can the world (W1) since only beliefs can justify beliefs. Fortunately, although no epistemological justification can be given, nor is it needed for truth, we have a good reason, not a form of evidence, for supposing that a vast majority of our beliefs are true according to Davidson's theory of radical interpretation based on the principle of charity. In this sense, 'beliefs are in nature veridical' (Davidson, 2001b, pp. 146-53). Davidson is convinced that beliefs are true or false, but they not only do not need any epistemological justification, they also do not represent anything. 'Nothing, however, no *thing*, makes sentences and theories true: not experience, not surface irritations, not the world, can make a sentence true' (Davidson, 1984, p. 194). There is no need at all to introduce the world (W1) and schemes that face or fit the world since truth does not consist in a confrontation between what we believe and reality. Without the scheme–world distinction, the representational theory of perception and its twin sister, the correspondence theory of truth, are undermined.

In addition, Davidson warns us that the epistemological intermediaries imposed by the third dogma are not just impotent, without any significant epistemological role to play in the formation and justification of beliefs and knowledge, but are harmful. They cause more problems than they provide the answers that they are supposed to supply. The major motivation behind introducing sense data or the given into epistemology starts with a genuine concern about the relationship between the mind and the world: How can we be sure that our beliefs faithfully represent the world or the reality out there? Kantian scheme–content dualism has been frequently presented as a response to such an epistemological quandary by securing our grip on the world through introducing some unassailable 'given' as ultimate evidence of justification. However, Davidson argues that instead of finding any solace for our epistemological anxieties, 'introducing intermediate steps or entities into the causal chain, like sensations or observations, serves only to make the epistemological problem more obvious ... since we cannot swear intermediaries to truthfulness' (Davidson, 2001b, p. 144). 'The disconnection creates a gap no reasoning or construction can plausibly bridge', which prevents us from holding directly unto the world. Consequently, 'idealism, reductionist forms of empiricism, and skepticism loom' (Davidson, 2001a, p. 43).

The conclusion is obvious: 'We should allow no intermediaries between our beliefs and their objects in the world' (Davidson, 2001b, p. 144) by getting rid of Kantian scheme–content dualism that imposes them.

4. A Non-Dogmatic Scheme–Content Dualism

I do not intend to defend Kantian scheme–content dualism. It is only too clear today that a pre-schematic raw material, the given or sense data, is indeed objectionable. It is a well-worn topic argued repeatedly by many, including many conceptual relativists criticized by Davidson, such as Kuhn and Feyerabend. What Davidson said above is nothing new except that he unfairly attributes such a myth to conceptual relativism. I doubt that any contemporary conceptual relativists would continue to back up the scheme–E1 distinction. Kantian thing-in-itself W1 is not a viable option either. Let all this be granted, as it should be.

Of course, Davidson and Rorty have much bigger fish to fry than simply rejecting Kantian scheme–content dualism. They want to undermine conceptual relativism by removing its foundation, namely, *any form* of scheme–content dualism. To claim a victory, Davidson and Rorty have to show us that *all attempts* to disjoin scheme and reality are doomed to failure. In other words, they have to convince conceptual relativists that no scheme–content distinction of a certain sort can be drawn, which is innocent enough to be immune from the charge of the third dogma of empiricism, but solid enough to supply a foundation for conceptual relativism. Obviously, Davidson and Rorty have not done so. What they have done is to identify and criticize scheme–content dualism of a certain sort, i.e., Kantian scheme–content dualism, and jump to a wholesale dismissal of scheme–content dualism in general.

Following suggestions from J. McDowell (1994) and others (M. Baghramian 1998), I believe there is at least one kind of non-Kantian scheme–content dualism that can sustain conceptual relativism. The central question is: Are there other kinds of empirical contents, neither E1 nor W1, available for scheme–content dualism? Truth is always nearby if you know where to look. How about our common-sense experience and the world as it is experienced by us?

Let us begin with experience. Sense data such as a patch of color, an indescribable sound, or a fleeting sensation are not our lived-experience like the perception of a yellow ball, loud music, or a feeling of love. Sense data is what W. James (1909) and C.I. Lewis (1929) called 'thin experience of immediate sensation', which should be distinguished from the so-called 'thick-experience of everyday life', such as our reflection of happiness and sadness, our perceptions of trees, rivers, and other people, as well as our experience of the world full of common objects, horses, buildings, etc. Davidson has never seriously entertained the possibility of thick-experience as the content of schemes. He always speaks of 'experience' in the sense of 'thin experience'. Even when he does mention experience with plurality, which is thick-experience in nature—'events like losing a button or stubbing a toe'—he dismisses it right away by reducing those events into thin experience (Davidson, 1984, p. 92). However, Davidson does touch, I think, the notion of thick-experience under a different name, i.e., the world as it is experienced by us. Remember that Davidson's mission is to 'restore the unmediated touch' with the world. 'In giving up the dualism of scheme and world, we do not give up the world, but re-establish unmediated touch with the familiar objects whose antics make our sentences and opinions true or false' (Davidson, 1984, p. 198). Clearly, the world that Davidson wants to keep is not the Kantian world-as-it-is, i.e., the world Rorty (1982) commends Davidson for bidding farewell. Instead, it is our common-sense phenomenal world with familiar objects like trees, tables, people, rocks, and 'knives and forks, railroads and mountains, cabbages and kingdoms'; that is, the robust common-sense world defended by many common-sense realists. In this sense, Davidson's notion of the world is much like the pragmatist notion of 'funded experience'—'those beliefs which are not at the moment being challenged, because they present no problems and no one has bothered to think of alternatives to them.' As Rorty notes, Davidson seems to perform the conjuring trick of substituting the notion of 'the unquestioned vast majority of our beliefs' for 'the notion of the world'. From this view, since the vast majority of our common beliefs must be true, the vast majority of the objects that our common beliefs are about must also exist. It turns out that we can start with 'the unquestioned vast majority of our beliefs', which is just 'our funded experience' or 'thick-experience', but end with the world as it is experienced by us. 'The thick-experience of everyday life' or 'the thick-experience of the world of things' is in fact the world as it is experienced by us. For this reason, I will place both 'thick-experience' and 'the world as it is experienced by us' under one roof:

(E2/W2) either our thick-experience or our common-sense world, namely, the world which the unquestioned vast majority of our beliefs are thought to be about, or the world full of familiar objects, the stars, the trees, the grass, the animals, the rivers, the people, and so on.

Could E2/W2 be the proper content of a scheme–content dualism that can void the pernicious Kantian scheme–content dualism under attack? It is certainly innocent enough to rebut the charge of the third dogma of empiricism. Obviously, such a dualism does not prevent us from having direct contact with the world. The remaining question is whether E2/W2 is so innocent as to not be pure or solid enough to qualify as a proper empirical content of scheme–content dualism. Based on his reading of the upshot of Davidson's arguments against the third dogma of empiricism, Rorty thinks so:

> The notion of 'the world' as used in a phrase like 'different conceptual schemes carve up the world differently' must be the notion of something completely unspecified and unspecifiable—the thing-in-itself, in fact. As soon as we start thinking of 'the world' as atoms and the void, or sense data and awareness of them, or 'stimuli' of a certain sort brought to bear upon organs of a certain sort, we have changed the name of the game. For we are now well within some particular theory about how the world is. (1982, p. 14)

Davidson certainly agrees with Rorty on his dismissal of W2.

To understand fully Davidson's and Rorty's rejection of E2/W2 as a possible content of scheme–content dualism, we need to dig deeper into some basic assumptions which they associate with the dualism. Davidson believes that conceptual relativism apparently presupposes a commonality underlying alternative schemes: If there are alternative conceptual schemes, there must be one common element for them to conceptualize (Davidson, 2001a, p. 39; 1984, p. 195). I do not think this commonality requirement of conceptual relativism is objectionable as long as it is modest. What is controversial is how to explain it. Does 'to be common' mean 'to be neutral to any schemes or be free of any interpretations or conceptualizations'? Davidson and Rorty think so. For Davidson, for the notion of alternative conceptual schemes to make sense at all, there has to be some common element shared by *all* possible conceptual schemes, and such a common element has to be empirically pure or theory/concept neutral in the sense that 'there be something neutral and common that lies outside *all schemes … The neutral content* waiting to be organized is supplied by nature …' either as E1 or as W1 (Davidson, 1984, pp. 190-91; my *italics*). Similarly, Rorty believes that conceptual relativism commits to the existence of some 'neutral material' shaped by our concepts. For Rorty, the dismissal of the notion of 'neutral material' marks the downfall of the notion of the conceptual. We thus have reached Davidson and Rorty's first assumption of scheme–content dualism:

(D1)   The neutrality of the content: Any empirical content of scheme–content dualism must be neutral to and beyond all schemes, theories, languages, and ideologies.

If the content as 'the scheme-neutral input' or 'the theory-neutral reality' is 'untouched by conceptual interpretation', then it cannot be contaminated by any concepts or interpretation. Then the scheme–content distinction has to be fixed and sharp; not any overlapping or intertwine between scheme and content is possible. This is Davidson's and Rorty's second assumption of scheme–content dualism:

(D2)   The scheme–content distinction has to be *rigid and sharp*, without any overlapping, intertwining, or shifting between scheme and content.

Now we are able to understand fully why E2/W2, according to Davidson and Rorty, could not qualify as the empirical content of scheme–content dualism of the sort we are looking for. It has become common wisdom since Kant that there cannot be a purely unconceptualized content to our experience. Despite the decline of Kant's transcendental philosophy, the Kantian doctrine of concept-ladenness of experience is still very much alive today. During the last century, we have witnessed various attacks on the contrast between the observed and the theoretical, which leads to the thesis of the theory-ladenness of observation, and the dismantlement of the very notion of the given (in e.g., Kuhn, Feyerabend, and Sellars). Of course, our thick-experience or the common-sense notion of the world as we experience it is not exempt from the invasion of concepts either. As McDowell argues, those experiences are already equipped with conceptual content. This is why Rorty insists that E2/W2 cannot be the content of scheme–content dualism since it is not neutral and universal to concepts and theories. When conceptual relativists introduce E2/W2 into the scene, the rules of 'the game' set up by D1 and D2 are violated.

Davidson and Rorty present a dilemma to conceptual relativism: To make sense of scheme–content dualism, we need to clarify its empirical content, which is either concept-neutral (as with W1 or E1) or concept-laden (as with E2/W2). If the former, it has been proved to be a metaphysical myth—it simply does not exist—or a third dogma of empiricism; if the latter, it does not qualify as the content required by scheme–content dualism. Conceptual relativism's response is simply this: Conceptual relativism faces such an inescapable dilemma only if we accept D1 and D2 as Davidson and Rorty try to seduce us to do. However, clearly conceptual relativists do not have to take this route, which would only lead to its own peril. Although D1 and D2 are indeed essential for Kantian scheme–content dualism, they are by no means essential for the conceptual relativist notion of a conceptual scheme.

In fact, many conceptual relativists reject the requirement of neutrality out-of-hand, or they even build their relativism upon such rejections. As Rorty has noticed, it is conceptual relativists, such as Quine, Kuhn, and Feyerabend, who take the lead in the battles against 'givenness' and 'analyticity'. Rescher has put his finger right on the issue:

> The idea of a pre-existing 'thought-independent' and scheme-invariant reality that is seen differently from different perceptual perspectives *just is not* a presupposition of the idea of different conceptual schemes. The Kantian model of a potentially differential schematic processing of uniform preschematic epistemic raw material is nowise essential to the idea of different conceptual schemes. (1980, p. 337)

One can abandon entirely the myth of neutral content without giving up the idea of alternative conceptual schemes. On the contrary, as I have argued earlier, it is exactly the denial of a fixed and absolute analytic–synthetic distinction and the abandonment of the given and *a priori* that turns Kantian conceptual absolutism upside down and thus makes conceptual relativism possible.

However, if there is no preschematic 'given' for all conceptual schemes to process, what is the empirical content of our schemes? My answer is: the thick-experience. The answer may face some quick rebuttals such as this: Since our experience is already scheme-laden, how can we separate a scheme from its content if we still want to make sense of the scheme–content distinction at all? Especially, how can two competing schemes share a common content if the content itself could be the very making of the schemes involved?

To address those questions, we can think of a conceptual scheme as a central point along a rope—or fishing net if you wish—with one side connected to our experience of the world and the other side to our beliefs about the world. 'Our conceptual mechanisms evolve in a historical dialectic of feedback dialectic between cognitive projection on the one hand and experiential interaction with nature on the other' (Rescher, 1980, p 340). On the one hand, with the abandonment of the sharp, rigid analytic–synthetic distinction, our concepts are no longer thought of as being *a priori*, but rather being *a posteriori* in the sense that meaning and concepts are both the bearers and the products of our factual beliefs. Concepts are not only the tools of inquiry but also its products. On the other hand, a similar dialectic interaction exists between scheme and experience. Our concepts are products of our past experience in the form of an unquestioned vast majority of our beliefs about nature. But our experience itself is richly endowed with conceptual inputs. We will never encounter situations with a conceptual *tabula rasa*. Clearly, there can be no rigid, sharp distinction between a scheme and its empirical contents.

However, it does not mean for one moment that no meaningful distinction between certain schemes and experience can be drawn at all. To make the point vivid, imagine that our conceptual framework consists of multiple layers of schemes. At the bottom is our most fundamental set of concepts, dispositions, pre-judgments, or absolute presuppositions, which I will call basic experiential concepts. Like Kantian concepts, these concepts are highly general and pervasive, permeate every facet of our sensory experience, and are presupposed by our experience in general. In this sense, the basic experiential concepts can be plausibly said to 'structure' or 'schematize' our everyday experience. Unlike Kantian concepts, which are *a priori*, independent of any experience, the scheme of our basic experiential concepts is *globally a posteriori* as a product of past experiences. They evolve through human interaction with the natural and social environment during millions of years of human evolution. Unlike Kantian concepts, which are absolutely basic, i.e., having a fixed and invariant structure, our basic experiential concepts are not absolutely basic in a Kantian sense. Instead, they are *hypothetically* or *historically* basic in the sense that on the basis of our past evolutionary history and current structure of environment, those concepts are foundational—that is, universally presupposed by our experience—but we acknowledge that changes in the nature of those concepts could occur over time—for example, if our evolutionary path is altered in the future due to some unforeseeable dramatic environment change. Although we cannot give an exhaustive list of all basic experiential concepts here (I doubt we could ever do so), they are concepts mostly related to our sense perception, individuation, duration, and identity of objects within space and time. For example, to enjoy and experience the lovely, beautiful flower in front of me on the table, I need to have a concept of differentiation (such that I can distinguish the flower from its background), a concept of relative stability of the object (I know that it will not melt into thin air in the next moment), a concept of identity (I know it is the same flower sent to me by my lover yesterday), a concept of myself (I am the subject who is enjoying the flower), a concept of space and time, and so on. We can safely assume, based on Darwinian evolution theory, that there are some basic experiential concepts shared by human cultures and societies.[11] In this sense, they are global or universal.

Accordingly, our thick-experience is the product of the dialectic interaction between our basic experiential concepts and experiential input from nature, whatever it may be, such that there is no way to separate which is form and which is content. Some may notice that I continue to use a suspicious notion, 'experiential input from nature, whatever it may be'. But the notion used here is, borrowing Rescher's comment on the notion of 'scheme-independent reality', 'not *constitutive,* not as a substantial constituent of the world—in contrast with "mere appearance"—but a purely regulative idea whose function is to block the pretensions of any one single scheme to a monopoly on correctness or finality' (1980, p. 337). In other words, I intend to use the notion to emphasize the empirical root of our basic experiential concepts. Our thick-experiences thus can be thought to be a common content shared by other higher-level schemes. Therefore, we can have commonality without neutrality.

Besides our basic experiential concepts as the foundation of our conceptual and experiential life, there are some more advanced sets of concepts or metaphysical presuppositions associated with cultures, intellectual traditions, or languages. Some of those conceptual schemes are radically distinct and schematize our rudimentary thick-experience in different ways to form different worldviews, cosmologies, or ways of life. The presence of different conceptual schemes manifests itself most dramatically when we come across some different ways of categorizing, a way of conceptualization, what seems to be the same experience. The best-known case would be classification of color in different cultures. As the literature on color amply demonstrates, apparently individually identifiable color samples, which two cultural groups presumably experience in the same way (assuming they all have normal vision), are often categorized into totally different color systems using different concepts of color. Similar examples are plentiful in anthropological and historical literature. If not only what one experiences determines what one believes, but also what one believes shapes what one experiences (the thesis of the theory-ladenness of observation), communities that adopt two radically distinct conceptual schemes, and thus different worldviews, would have different 'more conceptually enriched experiences', so to speak. This is the reason why Kuhn makes a seemingly absurd claim that such communities 'live' in two different 'worlds'.

5. Davidson Misses Target

Despite Davidson's very influential criticism of the *very notion* of conceptual schemes, the notion continues to enjoy its popularity and remains ubiquitous not only in contemporary philosophy, but also in many other interpretative disciplines such as cultural studies, historical and classical studies, anthropology, comparative linguistics, cognitive science, and the history of science. Accordingly, conceptual relativism is still very much alive and remains in power. The sustained resistance to Davidson's criticism is, I think, not simply due to its limited scope and some internal flaws of Davidson's arguments, but mainly due to the explanatory power and intuitive appeal of the notion itself. For many philosophers and theorists in related fields, the notion serves as a conceptual foundation for their pet theories, such as social constructionism, multiculturalism, feminism, and postmodernism. For others, the notion plays a crucial role in many significant philosophical investigations, such as the issues of realism vs. anti-realism, incommensurability, inter-language translation, cross-language and cross-cultural understanding/communication, etc. The wholesale dismissal of the notion would leave us with the problems of how to explain different ways in which the mind can mediate reality in general, and how to explain the variability of the ways the world is understood and conceptualized by different cultures, traditions, and languages in particular.

However, there is still another major reason responsible for Davidson's failure that has not been widely recognized:[12] Davidson's attacks miss the very target that he wanted to hit. What Davidson attacks fiercely is not the *very notion* of conceptual schemes as it identifies in his 'war declaration', but *a notion* of conceptual schemes or, more precisely, the Quinean notion of conceptual schemes and its underlying Kantian scheme–content dualism. This is because, as we have argued above, Davidson's two lines of arguments against conceptual relativism have a very limited scope: His arguments of inverifiability can only apply to the Quinean notion of conceptual schemes, which mistakenly construes conceptual schemes as sentential languages; his arguments against scheme–content dualism are targeted at Kantian scheme–content dualism only. However, both Quinean notion of conceptual schemes and Kantian scheme–content dualism are unsatisfactory for many conceptual relativists also, not just for the reason that they indeed face many conceptual difficulties as Davidson points out, but mainly because they simply cannot carry the weight of conceptual relativism. They cannot catch the essence of conceptual relativism. As such, Davidson is successful insofar as the Quinean notion of conceptual schemes and Kantian scheme–content dualism are concerned, but 'the victory' cannot be validly extended to a more robust notion of conceptual schemes. The three essential components of conceptual relativism are undamaged by Davidson's attacks.

I have argued that conceptual schemes are not identical to sentential languages, and that scheme–content dualism neither entails nor presupposes Kantian scheme–content dualism—in giving up Kantian scheme–content dualism, we do not give up scheme–content dualism. Conceptual relativism can sustain a viable version of scheme–content dualism that is not subject to the charge of the third dogma.

Conceptual relativism does not need to detach itself from the world as it is experienced by us through introducing any epistemological intermediary between the world and us. We stand together with Davidson and Rorty to applaud the loss of the Kantian world-as-it-is and the dissolution of the given, but we do not lose our world as it is experienced by us on the way. On the contrary, in terms of our conceptual schemes, we are connected to the world as close as it could be. And only through conceptual schemes can we be connected to the world. As a non-dogmatic empiricism, conceptual relativism, as it is properly construed, acknowledges our common experiential root and celebrates our conceptual diversity at the same time.

This, I think, is a better reply to Davidson.

Notes

1   Davidson, 1984, pp. 184, 185f; 2001a, pp. 39-40.
2   It is not controversial as to whether translatability is a sufficient condition of languagehood.
3   For example, see Rescher, 1980, pp. 326-9.
4   For a few good comprehensive criticisms of Davidson's arguments from inverifiability, please refer to N. Rescher, 1980, M. Forster, 1998, S. Hacker, 1996, and D. Henderson, 1994.
5   Please refer to the list of readings given in the previous note.
6   Davidson, 1984, p. 192; 2001a, p. 41; 2001 b, pp. 140-44.
7   In response to Davidson's early criticism, Quine (1981) identifies two versions of empiricism, one as a theory of truth and the other as a theory of evidence. Based on her reading of Quine's distinction, M. Baghramian (1998) explores the connection of these two versions of empiricism with the scheme-content in detail. Many of my comments on the topic draw inspiration from Baghramian (1998).
8   Davidson, 1984, p. 198; 2001a, p. 52; 2001b, pp.143-4.
9   Rorty, 1982, pp. 3-40; Davidson, 2001a, p. 52.
10 Davidson does not deny that senses do play a causal role in knowledge and acquisition of language. He admits that senses are crucial in the causal process that connects beliefs with the world (2001a, pp. 45-6).
11 Many wild thought-experiments about possible creatures endowed with very different thick-experience from humans' (such as the visitors from Mars with different ranges of sensory perceptions from humans) assume a radically different environment from ours.
12 M. Lynch (1998) is a notable exception. I have drawn many inspirations from his essay.

Chapter 5

Case Studies:
The Emergence of Truth-Value Gaps

My dissatisfaction with the translation-failure interpretation of incommensurability and the Quinean notion of conceptual schemes is not only caused by the many theoretical difficulties they face, but is also due to the observation that they are unable to elucidate many classical confrontations between alleged incommensurable theories or languages. Some of classical incommensurable cases that I have in mind are: Aristotelian physics versus Newtonian physics; Newtonian mechanics versus Einsteinian relativistic mechanics; Lavoisier's oxygen theory versus Priestley's phlogiston theory of combustion; Galenic medical theory versus Pasteurian medical theory; Newton's view versus Leibniz's view of space-time; Ptolemaic astronomy versus Copernican astronomy; quantum mechanics versus pre-quantum mechanics, and so on. My presuppositional interpretation emerges in a close reading of those celebrated cases of incommensurability.

How should we understand and explain these classical cases of incommensurability? Specifically, how can we identify the conceptual disparity between the incommensurables? What is the hallmark of these conceptual confrontations, and what are their genuine sources? These familiar conceptual confrontations are, to me, not confrontations between two scientific languages with different *distributions of truth-values* over their assertions due to the radical variance of the meanings (sense and reference) of the terms involved, but rather are confrontations between two languages with different *distributions of truth-value status* over their sentences due to two incompatible sets of metaphysical presuppositions underlying them. Consequently, a communication breakdown between the proponents of two incommensurable theories is not signified by the untranslatability between the languages of the theories, but is rather indicated by the occurrence of a truth-value gap between the languages.

The above insights are derived from two case studies. One is a not-yet-well-known study comparing contemporary Western medical theory and traditional Chinese medical theory; the other is the well-known debate on the absoluteness of space between Newton and Leibniz. I choose the first case because both theories arise from two highly developed and sophisticated cultural and intellectual traditions so disparate that cross-language communication between them often breaks down. I will show in the later chapters that this case represents an extreme in which two scientific languages are radically incommensurable. By comparison, the second case represents a moderate version of incommensurability.

1. Traditional Chinese Medical Theory[1]

Traditional Chinese medical theory (hereafter CMT) emerged more than two thousands years ago. It is still practiced widely inside and outside of Pan-Chinese culture and is very successful today. During its more than 2000-year development, CMT has established a complete conceptual system including its own physiological theory, pathological theory, diagnosis, and treatments. Its physiological and pathological basis consists of the yin-yang doctrine, the five-elements doctrine, the viscera doctrine, and the jingluo doctrine. Z. Lan has a clear and concise summary of these doctrines worthy of full quotation:

> The yin-yang doctrine takes the human body to be a unity of opposites. It is composed of two parts, the yin part and the yang part. The yang part includes the upper, exterior, back, outer side, and the six hollow organs, while the interior, the abdomen, the inner side, and the five solid organs belong to the yin part. There are further divisions within individual organs such as the heart and the liver.
>
> The theory of five elements holds that everything is made of five different elements: metal, wood, water, fire, and earth. One of these elements is found in each of the five solid organs of the human body: wood in the liver, fire in the heart, metal in the lung, earth in the spleen, water in the kidney. These organs are inter-related. The kidney essence nourishes the liver, the liver stores blood which is supplied to the heart, the heat generated by the heart warms the spleen, the spleen extracts vital substances from water and cereals to feed the lung, and the lung, in turn, assists the kidney by keeping the kidney fluid pure.
>
> Pathological changes in one organ affect the other solid organs, the limbs, the bones, the five sensory organs, the nine orifices, the tendons, and the blood vessels. These interactions are governed by the viscera-state doctrine. According to this doctrine, the eye is the orifice leading to the liver, the tongue leads to the heart, the ear to the kidney, the nose to the lung, and the mouth to the spleen. As a result, eye disease is treated by clearing away the liver fire, kidney stone by compressing the ear, and so on. (Lan, 1988, pp. 229-30)

Applying the above theories or doctrines to medical practice and health care, CMT has its own unique way of diagnosis and treatment. According to the yin-yang doctrine, it is the balance between the yin and yang parts of the human body that ensures its normal working and health. Loss of the yin-yang balance of the human body invites evils that lead to diseases. All symptoms related to diseases are classified as eight principal syndromes, which can be grouped further into four matched pairs: the yin versus the yang syndrome, the superficial versus the interior syndrome, the cold versus the heat syndrome, and the asthenia versus the sthenia syndrome. Of the eight principal syndromes, the yin and the yang are the leading ones. The yin syndrome governs the superficial, the asthenia, and the cold syndromes, while the yang syndrome controls the interior, the sthenia, and the heat syndromes. Therefore, all diseases arising from loss of the yin-yang balance can be diagnosed as the result of either a yin syndrome or a yang syndrome.

Treatment is a matter of restoring the balance between the yin and the yang inside of the human body. Different principles of treatment are applied to the different syndromes, such as the cold-heat principle, the asthenia-sthenia principle, and the yin-yang principle. For example, when an excess of the yin leads to a weakness of the yang, which is diagnosed as the yin syndrome, herbs and other healing techniques (such as acupuncture and moxibusion, cupping, Qigong[2]) are administered to restore the yin-yang balance by making up the deficiency of the yang. As another example, CMT holds that the spleen is responsible for transport and conversion. The spleen affects the upward movement of vital substances and controls blood. Many spleen diseases (the important symptoms of which are abdominal distension, loose stools, inappetence, phlegm-retention, oedema, diarrhea, blood in stools, and so on) are caused by the imbalance between the yin and the yang within the spleen, which is manifested as either an asthenic or a sthenic spleen. The cure for these diseases lies in the nourishment of the spleen to restore the yin-yang balance.

In addition, traditional CMT regards the human body as one part of an organic universe in a process of constant transformation in which everything is interrelated and interacted. There is mutual influence between the human body and Heaven (as Nature or the Universe). Therefore, many symptoms can be attributed to the associations between natural forces and changes—which represent the yin or the yang principles of the universe—and the yin-yang parts of the human body that are supposed to correspond to the former. For example, according to Han Confucians, when Heaven is about to make rain (representing the yin) fall, people feel sleepy. This is because when the yin force in Heaven and Earth begins to dominate, the yin in the human body responds. In this way, the association between the yin and rain causes increased sleepiness in people.

From the above brief summary of traditional CMT, we can see that the physiological and pathological theories and the medical concepts of CMT are systematically different from those of contemporary Western medical theory. The former is holistic while the latter is analytic. In addition, both medical theories have very different systems of medical categories. Although CMT has some equivalents (as far as the referents are concerned) for certain Western anatomical terms—such as 'heart', 'liver', 'spleen', 'lung', 'kidney' (the so-called five solid organs), and 'gallbladder', 'stomach', 'large intestine', 'small intestine', 'bladder', and 'triple warmer' (the so-called six hollow organs)—the physiological and pathological connotations of these terms are not the same. The functions of corresponding referents are believed to be totally different. This is because the organs in CMT are as much physiological and pathological entities as they are anatomical entities. Take the heart as an example:

> In Chinese theory, the heart is not only an anatomical entity; it is part of the nervous system and can perform some of the functions which Western theory attributes to the cerebral cortex. The heart not only gives force to the circulation of the blood but also controls the mental and emotional faculties. The Chinese heart is similar to the Western heart in respect of its cardiovascular function. But it differs in its relationship to the

cerebral cortex. One consequence of this difference is that Chinese pathology tends to be holistic whereas Western theory tackles medical problems at the molecular and cellular level. (Lan, 1988, p. 231)

2. Is Untranslatability a Barrier of Cross-Language Understanding?

CMT is hardly intelligible to most Western physicians. They are very skeptical of Chinese medicine and even regard Chinese physicians as medicasters. Many Western physicians claim that Chinese medicine sounds strange and alien to them. As one Western physician complains, the sentence, 'The loss of balance between the yin and the yang in the human body invites evils which lead to diseases', sounds as nonsensical to him as the utterance, 'ooh ee ooh ah ah' does. What causes such an apparent communication breakdown between the two medical language communities? One might say that the failure of understanding is caused by the fact that Western physicians are unable to understand many terms used by CMT, such as 'the yin' and 'the yang'. Because of this, they cannot translate the sentences of CMT into their own language. Such untranslatability sets obstacles for understanding and thereby leads to a communication breakdown between the two communities.

However, untranslatability is not the issue at hand here. The Westerner does not lack expressions to convey many central concepts of CMT. For example, the two central notions of CMT, the yin and the yang, could be formulated in English as follows: The yin and the yang are two fundamental elements, forces, or principles in the universe. The yin, which represents the negative, passive, weak, and destructive side of the universe, is associated with cold, cloud, rain, winter, femaleness, and that which is inside and dark. The yang, which represents the positive, active, strong, and constructive side of the universe, is associated with heat, sunshine, spring and summer, maleness, and that which is outside and bright. Of course, the terms 'yin' and 'yang' have different connotations in different schools of Chinese philosophy. To Tung Chung-Shu, for example, the yin and yang are two kinds of ch'i, which has an exceptionally varied number of meanings in the Chinese language and philosophy. But this is not the problem of translation, but rather the issue of different philosophical interpretations of the concepts. Furthermore, the meanings of the terms 'yin' and 'yang' in Chinese medical theory are relatively limited. We can use the above plain definitions to convey their basic meanings in English. Then, the English sentence,

(4)   The association of the yin and rain makes people sleepy,

is a close translation of the original Chinese sentence.

However, the defenders of the untranslatability thesis might quickly point out that the above argument confuses two different kinds of languages—the theoretical language used to formulate Western medical theory and the natural language used to code the theory. To say that CMT (which is written in ancient Chinese) can be translated into English is not the same thing as to say that the theory can be translated into the language of Western medical theory. The thesis of untranslatability is about the impossibility of translation between two scientific languages. It cannot be applied to, without further qualification, the impossibility of translation from one scientific language into one natural language.

I agree that the literal translation between the two medical languages cannot be done, for many reasons. Let me mention only one of them: because the two languages have sufficiently disparate medical category systems. To have an exact translation from Chinese medical language into Western medical language, we have to alter the whole taxonomic structure of the target language (the Western medical language) and extend dramatically its semantic resources by semantic enrichment. That means that we have to change the whole target language, which is not permissible for a literal translation.

The real issue is, however, that such untranslatability between the two languages does not ensure the failure of understanding between the two language communities; for translatability is not necessary for understanding. Without appealing to translation, the Western physician can still manage to understand Chinese medical theory by language-learning. In fact, there are many people from the West studying Chinese medicine in China today. Many Western medical practitioners successfully adopt ancient Chinese healing techniques in their clinical practice in the US. As one Western physician said, after he learnt the theory, there is no longer 'ooh ee ooh ah ah' about it. Sentences that used to be senseless suddenly turned to being perfectly understandable.

In either case, the difficulty of the intertranslation between the two languages is not exactly a barrier to the achievement of cross-language understanding.

3. What Causes Cross-Language Communication Breakdown?

If untranslatability between Chinese medical language and Western medical language cannot account for the failure of understanding and communication breakdown between the two language communities, then two questions remain. First, what does the real difficulty of mutual understanding and communication between the two language communities derive from? Second, could we identify a strong linguistic correlate, if any, of such problematic understanding?

The answer to the first question lies in the following observations. Let us first consider the case in which Chinese medical language can be translated into modern English (not into Western medical language). Even so, Westerners who are not familiar with the traditional Chinese mind and the pre-modern Chinese way of thinking would still be left in a fog. Even supposing that they know the general meanings of the terms, 'the yin' and 'the yang', the question still remains, what is *the point* of what is expressed by a sentence such as (4) or (5)?

(5)   All diseases are due to the loss of the balance between the yin part and the yang part of the human body.

What is the (cognitive or empirical) content asserted by (5)? What is the point of what is being presented or argued by (4) or (5)? It is not the words of (5) that Westerners cannot make sense of, but the mode of reasoning or the way in which the assertion is made and defended is entirely alien to them. Besides, it is based on a whole system of categories that is hardly intelligible to Westerners. Without grasping the unique mode of reasoning and the categorical system underlying the sentence, one cannot understand it effectively. Due to the lack of an alternative— since Westerners can identify neither the underlying way of thinking nor the category system embedded in CMT—Westerners have to approach it from the mode of reasoning and the category system of their own time with which they are familiar. Westerners would naturally try to project their own way of thinking and categories onto what they try to understand. Such *a projective way of understanding* would distort the Chinese text and lead to the failure of genuine understanding, and thereby a communication breakdown. This is exactly what happens when Westerners feel lost when reading sentences like (4) and (5).

Let us turn to another case in which a literal translation between the two medical languages is not available, but one side can understand the other by language-learning. In general, language-learning is a process in which the learner is not just trying to understand some individual words of an alien language, but also trying to understand the language *as a whole*. In our case, the language as a whole that Westerners should learn is not only the theoretical language employed by CMT, but also the Chinese language used to code it. This is because ancient Chinese philosophy had such a strong influence on the formation and development of Chinese civilization that it left its print on almost all aspects of Chinese culture and institutions. Even today, the life of the Chinese people is permeated with Confucianism and Taoism. The Chinese language, as the linguistic representation of this great civilization, is no exception. Of course, one might argue the other way around: It is not ancient Chinese philosophy that shaped Chinese civilization and its language, but it is Chinese civilization, especially its language, that shaped ancient Chinese philosophy. I do not know how to cut into this circle, but one thing is clear: There is an intrinsic connection between the Chinese language and Chinese philosophy. For example, suggestiveness, instead of articulateness, is the particular way in which Chinese philosophers expressed themselves. Ancient Chinese philosophers were accustomed to express themselves in the form of aphorisms, apothegms, allusions, and illustrations. Chinese, as a pictographic and ideographic language full of imagination, is the ideal vehicle to convey the suggestiveness of the sayings and writings of Chinese philosophers. Strictly speaking, the rich content and implications of Chinese philosophy can be formulated precisely (without loss and distortion) in Chinese only. On the other hand, the Chinese language is fully loaded with Chinese philosophical ideas. Many concepts of Chinese philosophy have become the common vocabulary of Chinese, such as the yin and the yang, Tao, etc. For example, in Chinese the penis is called the yang zhui (the positive tool); the vagina is called the yin tao (the negative way). Even the terms in modern science still adopt the yin-yang division. The positive electrode is called the yang ji; the negative electrode is called the yin ji. As I will argue in detail in chapter 10, the pre-modern Chinese way of thinking and its underlying cosmology, which are embedded within CMT, were rooted in and developed from ancient Chinese philosophy. Due to the internal connection between Chinese philosophy and the Chinese language, the pre-modern Chinese way of thinking is embedded in the Chinese language. For this reason, to learn CMT, one had better learn it in Chinese. (Of course you can learn it in English. But to do so is akin to tasting Chinese food in a Chinese restaurant in the US that does not serve authentic Chinese food.) I will call the language of CMT coded in Chinese the language of Chinese medical theory as a whole, or the Chinese Medical-Language, in brief, below.

The Chinese Medical-Language reflects a unique belief system and embodies a specific form of life. One effective way to understand this language (if one could do so) is to immerse oneself into its unique belief system and its form of life as adopted by the native speaker living in the pre-modern period of China. During its long historical development, this set of beliefs and the form of life associated with the language have been internalized into the language[3] and have formed their own specific mode of reasoning or rational justification (known as the associated way of thinking) as well as their own specific categorical framework.[4] Therefore, this language is fully intelligible to Westerners only if they are able to grasp its specific mode of reasoning and categorical framework embedded in the language.[5] For Westerners to understand effectively the Chinese Medical-Language, they have to get sufficiently into the pre-modern Chinese way of thinking and categorical framework, which are barely recognizable by and hardly intelligible to most contemporary Westerners.

After understanding the words and grasping the mode of reasoning as well as the categorical framework of the Chinese Medical-Language, Westerners can understand the language quite well. If Westerners were physicians, they would eventually become bilinguals, in our case, bi-medical practitioners. However, a *bi*-medical practitioner (who can speak both medical languages) is not necessarily a *meta*-medical practitioner (who can speak a metalanguage with the two medical languages as its sublanguages). Such bi-medical practitioners who live in the boundary between the two language communities often find themselves in an awkward situation. They can eventually understand the Chinese Medical-Language by getting into the pre-modern Chinese way of thinking, but, at best, they can start talking the pre-modern Chinese way only if they become *alienated or dissociated* from the modern Western way of thinking practiced in their own time. Westerners cannot hold both languages side-by-side in mind to make a point-to-point comparison. They cannot think in both languages at the same time. They cannot use one language to understand the other or incorporate one language into the other.

There has been an interesting initiative, suggested by the Chinese government and medical professional organizations, aimed at incorporating Western medical theory into CMT under the goal of developing traditional CMT. Many Chinese physicians and researchers thought that Chinese and Western medicine might (or should) complement one another so that one makes up for the disadvantages of the other. They called this initiative the Combination of Chinese and Western Medicine (CCWM). Encouraged (actually, required) and funded by the Chinese government, many hospitals and research institutions established research branches of CCWM. In the 1970s, it was even suggested that the yin-yang doctrine could be explained in terms of the Western theory of regulation (C-AMP and C-GMP), but it soon became obvious that the Western theory of regulation cannot be used to substitute for the yin-yang doctrine. The holistic framework of CMT cannot be assimilated into the atomistic framework of Western medical theory. For example, the physiological basis of the viscera doctrine cannot be found in the workings of C-AMP and G-AMP, or nucleic acid. After about thirty years of experiments, it became clear that CMT could not be incorporated into Western medical theory. The same conclusion applies to the opposite incorporation from Western medical theory into its Chinese counterpart. We cannot locate a common theoretical ground to combine the two into one coherent theoretical system.

Of course, the failure of the theoretical integration of Chinese and Western medicine does not mean that our bi-medical practitioners cannot apply both theories and ways of treatment in their clinical practice. Actually, what a bi-medical practitioner does, under the name of CCWM, is just that: either to apply different medical theories to treat different diseases, or to combine both healing techniques (not the two theories) to speed up healing of the same disease. What these bi-medical practitioners have done is not to adopt both theories at the same time, but to move back and forth between the two theories according to their needs in different situations.

In sum, full communication between Chinese and Western medical communities is hardly attainable. A communication breakdown between them is not due to the difficulty of translation, but rather is either due to one side failing to grasp the mode of reasoning and the categorical framework of the other language (hence lack of understanding) or is due to the lack of compatible modes of reasoning or matchable category systems between the two languages (even if understanding can be restored by language-learning). In this sense, we say that the two medical languages are incommensurable.

4. The Emergence of a Truth-Value Gap

How can we identify a failure of understanding and a communication breakdown between the two languages? To say that the underlying modes of reasoning or categorical frameworks of the two languages are incompatible does not help, since those attributes are not something that can be easily identified.

As an example, imagine that a Chinese physician Dr Wong, when asked why people tend to become sleepy on rainy days, claims:

(4)   The association of the yin and rain makes people sleepy.

What is a likely response of a practitioner of Western medicine, say, Dr Smith? To see his possible responses, we need to consider two different hypothetical situations.

First, let us suppose that Dr Smith knew nothing about CMT and had no background in Chinese culture, philosophy, and language. Suppose further that he was told the meaning of the term 'the yin' by a definition which we have given before. Under these suppositions, he would certainly not say, 'No, what you have said is simply *false*'. He could not understand (4) since he could not grasp the way in which the assertion was proposed and justified. The content of the sentence lies outside his conceptual reach. It is not even clear for him whether the sentence really asserts anything. It is, hence, very likely that he would say something like: 'I do not know what you are talking about' or 'What is the point of what you are saying?' The implication behind these responses is that the question of whether the assertion (4) was true or false simply did not arise. For Dr Smith to claim that it was true or false presupposes that he understood the sentence. This is excluded by our suppositions. (4) is nonsense to him and cannot be answered in the way the question is put to him.

Secondly, let us suppose that Dr Smith happened to know CMT and the yin-yang doctrine in ancient Chinese philosophy, and was also able to comprehend the pre-modern Chinese mode of reasoning. Suppose further that he did not adopt the yin-yang doctrine and the corresponding mode of reasoning. If we asked him whether he thought Dr Wong's assertion is true or false, whether he agreed or disagreed with it, would he directly deny the assertion by replying, 'No, (4) is false'? I do not think so for the following reasons. We can reformulate the suggested Dr Smith's answer as follows:

(6)   It is false that the association of the yin and rain makes people sleepy.

Then, both (4) and (6) somehow strongly 'suggest' or 'imply' (more precisely, 'presuppose') a third sentence:

(7)   There is a fundamental element, force or principle in the universe, namely, the yin, and there is a pre-established connection between the human body and natural forces.

Therefore, Dr Smith (suppose that he is a critical thinker and is able to sense the implication behind such an answer) cannot simply deny (4) on its face; for such a denial seems to set a trap for him (recall the infamous question, 'Do you still beat your wife?'). Therefore, he will hesitate to deny (4) directly.

Dr Smith's uneasiness about how to answer (4) could have been strengthened further if he had realized the difference between the truth-value status of sentence (4) and a syntactically similar sentence:

(8)   The association of the falling of the American stock market and rain makes people sleepy.

It is intuitively recognizable that (8) is untrue in a way which is different from the non-truth of (4). The non-truth of (4) is due to the failure of its 'felt implication', namely, (7). But the non-truth of (8) is caused by a false connection between unrelated events. Therefore, it is possible for the negation of (8) (It is false that the association of the falling stock market makes people sleepy) to be true while the negation of (4), namely, (6), is untrue.

In addition, due to his uneasiness about answer (6), Dr Smith will very likely respond to the question by saying, 'I do not think I can answer the question since there is no such thing as the yin'. To assert that there was no such thing as the yin is certainly not to contradict (4). He is rather giving a reason for saying that the question of whether (4) is true or false did not arise.

In both situations, we reach the same conclusion: For the Western physician, the question of whether (4) is true or false simply does not arise. A similar analysis can be extended to other core sentences of CMT. Either Western physicians fail to understand these sentences due to a failure to grasp the mode of reasoning and thereby cannot assign truth-values to them; or even if they are able to understand these sentences they are still not willing to assign truth or falsity to them because they do not adopt the underlying mode of reasoning and related doctrines. There is no way to match what the Chinese physician wants to say to anything the Western physician wants to say at the theoretical level. The confrontation between the two medical theories does not lie in the sphere of *disagreement or conflict* of the sort arising when one theory holds something to be true that the other holds to be false (logical inconsistency). The difference between them is not that Western medical theory has a different theory of the operation of the yin and the yang from that of its Chinese counterpart, or that the Chinese physician says different things about, say, bacteria and viruses. Rather, the difference lies in the fact that one side has nothing to say about the other. When Dr Wong diagnoses a disease as an excess of the yin within the spleen, it is not that Dr Smith thinks that Dr Wong is making a mistake in diagnosis or a false assertion, but that he simply cannot assign any truth-value to it at all. He has nothing to say about it. It is not that they say *the same thing* differently, but rather that they say totally *different things*. The key contrast here is between saying something (asserting or denying) and saying nothing. The Western physician can neither assert nor deny what the Chinese physician claims.

Consequently, Western physicians do not regard as false many core sentences of the language of CMT; they simply cannot assign truth-values to those sentences because their contents lie outside their conceptual reach. They had better keep silent about what they cannot grasp. Thus, a truth-value gap occurs between the two medical languages. It is this truth-value gap that indicates semantically that we encounter two alternative medical conceptual schemes embodied by the theories and that the communication breaks down between the two medical communities.

5. The Newton–Leibniz Debate on the Absoluteness of Space[6]

For a comparison, let us turn our attention to a classical debate within the same cultural and intellectual tradition—the Newton–Leibniz debate on the absoluteness of space.

Newton thinks that classical three-dimensional Euclidean space describes an independently self-existing physical space. The world of physical bodies and events, for a Newtonian, can be pictured as if it moves through a self-existing empty spatial continuum, as a ship moves through a pre-existing absolutely static ocean. All the physical and geometrical properties of Newtonian space exist independently of configurations of the physical bodies within it. To visualize Newtonian absolute space,

> Imagine an empty space without any physical bodies, where each point has an identity of its own. You may stick imaginary labels on these points, naming each. After putting bodies into the space, each will occupy at any given time a certain region, i.e.—a certain set of space-points. (Gaifman, 1984, p. 324)

On the contrary, Leibniz rejects the absoluteness of space. For a Leibnizian, Euclidean space does not describe an independently existing physical space, although it might correspond to the structure of spatial relations in our world. There is no such self-existing empty space. Instead, space is determined by physical body configurations. Leibniz holds that 'space is nothing else than the order of existing things', 'the order of bodies among themselves', or 'an order of things which exist at the same time' and thus, it is not 'something in itself'; it 'has no absolute reality', but it is 'something merely relative' to body configurations within it.[7]

For comparison, imagine that we are able to move the whole configuration of bodies in a space, if any, keeping all distances between bodies and body parts constant. For Newtonians, the worlds before and after the movement will be different since each body in the configuration, after the movement, occupies a different region in absolute space. For Leibnizians, it will be the same world, since the configuration of bodies remains unchanged before and after the movement. Actually, it is not even possible, for Leibnizians, to move the same body configuration in a space because this presupposes the existence of absolute space.

The controversy about the absoluteness of space between Newton and Leibniz has become common knowledge among historians and philosophers of science. But as H. Gaifman points out, the question still remains concerning the real meaning of the debate. What is the real difference between Newton's framework of space and that of Leibniz's? To say that Newton believes in absolute space but Leibniz does not is of no big help, unless the notion of absolute space can be given a precise definition. But what we have right now is at most the illustration of it. Furthermore, people have many different ways of visualizing a given state of affairs, or can picture the same world in different ways. For example, what is the difference between saying 'the car is moving toward me and saying 'I am moving toward the car'? What difference does it make if you visualize natural numbers as a horizontal series going from left to right, but I visualize them as going up vertically? Perhaps the debate about the absoluteness of space is only restricted at the metaphorical or even psychological level? Therefore, it is Leibniz's burden to justify his rejection of Newtonian absolute space by showing that the difference between Newton's and his own picture of the spatial world is real and not just a difference involving favored ways of visualizing the same spatial world.

Leibniz's way of justifying his framework of space has proven to be effective. Leibniz does not attack the Newtonian notion of absolute space head-on by directly addressing the question, 'Is there a self-existing empty space?' Instead, Leibniz questions the truth-value status of some core Newtonian sentences about comparison of positions of a given body at different times under the supposition that the whole body configuration remains the same so that the body's position with respect to other bodies is unchanged (supposition C). The following sentence in modal form represents one such core Newtonian sentence.

(9)   The body b at time t could have located in a different place.[8]

Put (9) in a non-modal form:

(10)   The spatial location of the body b at time t_1 is different from its location at time t_2.[9]

(9) and (10) are two core Newtonian sentences because they presuppose the existence of Newtonian absolute space. In other words, (9) and (10) are true or false only if (11) is true.

(11)   Newtonian absolute space exists (a self-existing empty spatial continuum, all the physical and geometrical properties of which exist independently of different configurations of the physical bodies within it.)

To admit that (9) and (10) are true or false is actually to claim the existence of Newtonian absolute space. Therefore, Newton's belief in absolute space can be expressed by admitting comparisons of positions of a given body at different times as factually meaningful. In this way, Newton's position of absolute space can be redefined precisely as an affirmation of the *truth-or-falsity* of core Newtonian sentences like (9) and (10).

Leibniz challenges the Newtonian notion of absolute space by questioning the truth-value status of these core Newtonian sentences like (9) and (10). In Leibniz's view, (10) is truth-valueless under supposition C. For a Leibnizian, these Newtonian sentences would be true or false if relative to a frame of reference. It is meaningful, for Leibniz, to ask whether the distance between two bodies changes at different times; or, more generally, whether the spatial locations of a given body relative to a given system of bodies (used as a frame of reference) are different at two different times. However, one cannot speak of the absolute change of locations independently of a body configuration. Here, Leibniz presents his own position as denying that the above Newtonian sentences have truth-values and thereby rejects the existence of Newtonian absolute space.

In short, here is the real issue of the Newton–Leibniz debate on the absoluteness of space: Certain sentences of the Newtonian language, which have truth-values within Newton's framework of space, lose their truth-values when considered from Leibniz's point of view.

To reconstruct the Newton–Leibniz debate on a more exact level, we need to identify precisely the set of sentences that have truth-values in the Newtonian language and the set of sentences that have truth-values in the Leibnizian language. To do so we need to formalize the two languages in question. Gaifman provides us with the following formal language that can be modified to serve our purpose.

Euclidean language L_E Let us first set up a formal language L_E, which treats space and time directly as primitives, to describe the 'pure' spatio-temporal structure. L_E is a many-sorted language. It is composed of the following factors:[10]

(a) A pure mathematical part, in which we can speak about real numbers, numerical functions, and constants, and whatever mathematics we wish.

(b) Time terms (constants and variables) u, u_1 ... v, v_1 ... ranging over time points.

(c) Space terms (constants and variables) x, x_1 ... y, y_1 ... ranging over space points in a three-dimensional Euclidean space.

(d) (i) Temporal predicate $<_t$ for temporal precedence (' $t_1 <_t t_2$ ' asserts that the time point t_1 precedes the time points t_2); (ii) equality symbols ' $=_t$ ', ' $=_s$ ' for time and space, respectively (for example, ' $u =_t v$ ' states that the time-point u is equal to the time-point v).

(e) Quantifiers and the usual sentential connectives.

Holding these factors, all the concepts and axioms of elementary geometry can be formulated in L_E, including the axioms of elementary real number theory, the axioms that space is a three-dimensional Euclidean space, and that the time line is a Euclidean line. These are classical Euclidean axioms of space and time.

Both Newtonians and Leibnizians would regard these classical Euclidean axioms as general truths. Newtonians and Leibnizians should have no disagreement about the truth-value status of the sentences of L_E. Therefore, all the sentences of L_E, which we call the class of sentences S_E, are true or false for both Newtonians and Leibnizians. As long as both work within L_E, no incongruity arises between them.

Newton's language L_N Extend L_E to L_N by introducing physical bodies and events.[11] This means adding two more types of names and two more function symbols to the five factors of L_E:

(f)   Proper names of particular material bodies b_1, b_2 ... and proper names of particular events e_1, e_2 ...

(g)   Two function symbols, a unary 'τ' and a binary 'σ' Here, τ(e) denotes the time point of the event e while σ (b, t) denotes the center of gravity of the body b at the time t.

Based on the above conventions, the Newtonian framework of space and time can be expressed as the following three claims:[12]

(N1)   The usual rules of first-order logic apply in L_N.

(N2)   Classical Euclidean axioms for space and time are valid.

(N3)   All the sentences of L_N, which we call Newton's class of sentences S_N, are candidates of truth-or-falsity.

Of the three claims, N3 is the most significant for Newton's notion of absolute space. Actually, N3 is the precise expression of Newton's belief in absolute space. This is because many core sentences of L_N, more specifically, a set of sentences within L_N but outside L_E, presuppose the existence of absolute space. Take sentence (10) as an example again. If *b* is a physical body and *e* is an event, then for Newton, τ(e) is an absolute time-point and σ (b, τ(e)) is an absolute space-point. Then, in Newton's conception of space, the question, 'Is the location of the body *b* at time t_1 different from its location at time t_2 (under supposition C)?' has a determinate yes-or-no answer since the corresponding sentence (10) is true or false. Reformulating (10) in our formal language into (12):

(12) $\sigma (b, \tau (e_1)) \neq \sigma (b, \tau (e_2))$

(12) asserts that the space-points of *b* at the time of event e_1 and at the time of event e_2 are different. According to Newton, (12), like (10), is a candidate for truth-or-falsity. In contrast, for Leibniz, space-points indicate only relative positions at a given time. Hence, (12) is true or false only if the configuration of physical bodies, including the body *b*, has changed. In other words, (12) is truth-valueless unless this precondition is satisfied.

From the examples we can see that certain atomic sentences, such as (12), that are true or false in relation to Newton's conception of space will be regarded as truth-valueless in relation to Leibniz's conception of space. That means that Leibniz would reject N3. According to Leibniz, some sentences of L_N do not have truth-values. In this way, the disagreement between Newton's and Leibniz's conceptions of absolute space can be reduced to the disagreement about the truth-value status of certain core Newtonian sentences about the positions of physical bodies.

Leibniz's language L_Z   We can further identify the language which determines the class of sentences S_L, which are true or false based on Leibniz's conception of space. First, it is technically convenient to eliminate the variables of space terms. This is because the Newtonian conception of space is fully expressible in a sublanguage of L_N in which no variables of space terms exist. Let K_N be the sublanguage of L_N obtained by omitting from L_N the variables of space terms. Thus, every possible fact expressible in L_N is expressible in K_N. Accepting Newton's framework of space amounts to accepting all the sentences of K_N as true or false. Second, let L_Z be the sublanguage of K_N obtained by imposing the following restriction: An inequality whose sides are occupied by the terms of space terms cannot figure in a formula unless the body configuration has changed. Thus Leibniz's space-time ontology can now be expressed precisely as an affirmation of the truth-or-falsity of all the sentences of L_Z.

In sum, within the Newtonian framework of space, every sentence of L_N and K_N (a sublanguage of L_N) is true or false. In contrast, within Leibniz's framework of space, all the sentences of L_Z, which we call Leibniz's class of sentences S_L, are candidates for truth-or-falsity. As to the sentences of L_N and K_N that are not in L_Z, they are neither true nor false from Leibniz's point of view unless they, by some agreed conventions, can be translated and reread as sentences of L_Z. Thus, the substantial difference between the Newtonian and the Leibnizian languages of space consists in a different assignment of the truth-value status of the sentences of L_N that are not in L_Z, namely, the class of sentences S_N-S_L. In other words, there is a truth-value gap between the Newtonian and the Leibnizian languages of space as to the class of sentences S_N-S_L.

The confrontation between the Newtonian and the Leibnizian languages of space seems to be very different from the confrontation between Chinese and Western medical languages. The Leibnizian language L_Z is a sublanguage of the Newtonian language L_N. The Newtonian and the Leibnizian languages have a great deal of overlap while the languages of Chinese medical and Western medical theories are radically disparate. Both the Newtonian and the Leibnizian languages have much common ground: both contain the Euclidean language L_E as sublanguage; both share the same category system; both are embedded in the same intellectual tradition with the same mode of reasoning; and so on. Because of these commonalties, the Leibnizians can identify and understand the underlying presupposition of the Newtonian language, namely, the existence of Newtonian absolute space. Consequently, each language is perfectly intelligible to the other. In addition, the Leibnizians have some translation that gives the 'true meaning' of the claim for every testable Newtonian claim about the actual world. A major source of the divergence between the two languages consists in whether two ostensibly different descriptions in terms of relations to space-time points correspond to two possible worlds or just one.

Irrespective of these commonalities, the confrontation between the two languages is still ontologically significant. In general, one can understand a language without affirming it just as one can understand a fiction story without believing in the reality of the events described. Indeed, one's rational rejection of a language is often based on one's comprehension of the language. Modern physicists understand very well what the sentences in the language of Newtonian mechanics pretend to express, yet they deny that they describe objective reality. In this sense, a rational confutation seeks understanding. In our case, although the Leibnizians are able to identify and comprehend the underlying presupposition of the Newtonian language, i.e., the existence of absolute space, they categorically deny its truth. This is the reason why the Leibnizians regard the set of sentences S_N-S_L (actually, but not potentially) truth-valueless.

However, cross-language understanding does not guarantee successful cross-language communication. The existence of a truth-value gap between the Newtonian and the Leibnizian languages due to the denial of the underlying presupposition of the Newtonian language signifies that the full communication between the two language communities is at risk. Both sides cannot engage in a full communication although they appear to 'talk' with one another. A Leibnizian who can understand the Newtonian language is actually bilingual, not metalingual. Such a bilingual faces many difficulties during communication. The communication between the two language communities is incomplete. Of course, compared with the communication breakdown between Chinese and Western medical communities, the communication breakdown between the Newtonians and the Leibnizians is moderate in nature.

6. The Newton–Einstein Debate on the Absoluteness of Time

For a further illustration, let us take a look at another similar case study from Gaifman (1984), the Newton–Einstein debate on the absoluteness of time. Consider the following two sentences about simultaneity and precedence:

(13)   Event e_1 and event e_2 are simultaneous: $\tau(e_1) = \tau(e_2)$.
(14)   Event e_1 precedes event e_2: $\tau(e_1) < \tau(e_2)$.

These two sentences make perfect sense and are true or false in Newtonian physics. Newton pictured the world of events and physical bodies as if it moved through a self-existing empty time consisting of an ordered line of time points that are independent of any event. Thus, we can redefine Newton's position as an affirmation of the truth of sentences (13) and (14).

However, according to relativity theory, precedence may depend on the coordinate system from which the events are viewed. More precisely, if in some coordinate system the events are separated by distance d and time Δt and $d > c \times \Delta t$ (c = light velocity)[13] then their temporal order depends on the coordinate system. Therefore, to ask the questions, 'Does event e_1 precede e_2?' or 'Are e_1 and e_2 simultaneous?' without specifying a coordinate system is to ask factually meaningless questions. Thus, (13) and (14) have no truth-values from the relativistic point of view.

The difference between Newtonian and relativistic physics 'is often described by saying that in Newtonian physics we have an absolute time ordering and in relativistic physics we do not. But the more precise way of expressing the difference is by pointing to the sentences whose true-or-false status differs from one theory to another' (Gaifman, 1984, p. 321). Actually, this is the way Einstein treats the concept of time. Einstein does not directly address the question, 'Does there exist an absolute time ordering?' Instead, his analysis of the concept of time focuses on the factual meaning of the assertions about the simultaneity of spatially separated events. Einstein told the physicists and philosophers: You must first say what you mean by simultaneity. You can answer it by showing how N_5 is tested. But in so doing you will find that you have to specify a coordinate system in which these events happen. It will soon become clear that to ask the question, 'Are event e_1 and event e_2 simultaneous?' without specifying a coordinate system is to ask a factually meaningless question. Only relative simultaneity can be admittedly established; there is no such thing as absolute simultaneity. The concept of absolute simultaneity presupposes the existence of an absolute time ordering. Therefore, the disagreement on the truth-value status of some substantial sentences, for example, (13) and (14), can be used to represent the real difference between Newton's and Einstein's concepts of time.

Notes

1   The current case study mainly refers to Z. Lan (1988) which gives a brief summary of the main characteristics of Chinese medical theory. Also see Beijing College of Traditional Chinese Medicine (1978) for a complete introduction to traditional Chinese medicine.
2   Qigong (pronounced as Ch'i kung) is a combination of breathing and mental exercises performed to create vital energy within the body.
3   By 'the language' here I mean the Chinese Medical-Language. It might be doubtful whether ancient beliefs and forms of life really survive 'internalization into' a natural language, in our case, contemporary Chinese language.
4   I will illustrate the formation of the pre-modern Chinese mode of reasoning and its content in detail in chapter 10.
5   For further arguments for the necessary role of modes of reasoning and categorical framework in effective understanding, please see chapter 12.
6   This case study is adapted from H. Gaifman, 1975 & 1976, 1984. Gaifman uses the case to illustrate his Wittgensteinian explication of the notion of ontology.
7   Gaifman's discussion of the debate between Leibniz and Newton about the absoluteness of space is based on the Clarke–Leibniz letter exchange. Clarke did his debating for Newton while Newton supplied the arguments.
8   Another similar formulation of (9) without appealing to supposition C would be: An object in an otherwise empty space could have been located at any number of different spatial points.
9   We could express a similar Newtonian sentence without appealing to supposition C as follows: An object in an otherwise empty space is at rest.
10 Gaifman, 1975 and 1976, pp. 54-5; 1984, pp. 324-5.
11 The Leibnizian language is a sublanguage of the Newtonian language. The Leibnizians can extend their language in the same way. In essence, we introduce a coordinate

system onto the set of actual events (or possible events for some relationalists) that allows us to refer to specific events and objects and to describe temporal and spatial relations and distances of them.

12   Gaifman, 1975 and 1976, p. 56.
13   In this case, events are 'spacelike' separated or they could be connected only by signals or objects going faster than light.

Chapter 6

Toward the Presuppositional Interpretation

We have found, from our cases studies, that the confrontation between two incommensurable theories is actually a conceptual confrontation between two scientific languages with different *distributions of truth-value status* over their sentences. The communication breakdown between them is often semantically correlated with the occurrence of a truth-value gap between the languages. Hence, what we should focus on, in the cases of incommensurability and conceptual schemes, is not truth or truth-functional meaning and translation, but rather truth-value status. In fact, I. Hacking put his finger on the problem about 20 years ago:

> Many of the recent but already classical philosophical discussions of such topics as incommensurability, indeterminacy of translation, and conceptual schemes seem to me to discuss truth, where they ought to be considering truth-or-falsity. (1982, p. 49)

Hacking here clearly identifies the very root of both the received interpretation of incommensurability as untranslatability and the Quinean notion of conceptual schemes: neglect of truth-value status and denial of possible truth-value gaps between two incommensurable languages (or two alternative conceptual schemes). This is because both notions are rooted in bivalent semantics in which every declarative sentence is either true or false. As we have discussed in chapter 2, the translation used in the thesis of incommensurability as untranslatability is truth-preserving in nature, which requires the translation to preserve the truth-values of the sentences translated and to match the truth-values of the corresponding sentences between the target and the source languages. The alleged translation-failure between two incommensurable languages is believed to be caused by redistribution of truth-values between the sentences of the two languages due to meaning (sense and/or reference) variance of the terms involved. Therefore, the meaning relation becomes the determinate semantic relation between two incommensurable languages. The exact same line of reasoning is behind the Quinean notion of conceptual schemes and Davidson's rejection of it. To see the link, one only needs to identify a conceptual scheme with a sentential language, construed as the totality of sentences held to be true by its speaker. Thus, redistribution of truth-values, and consequently the failure of intertranslation, would become the criterion of distinguishing two alternative conceptual schemes. If one thinks along this line of bivalent semantics, it would seem natural to deal with the notions of incommensurability and conceptual schemes in terms of truth, meaning, and translation in Tarski's style, as Davidson does.

The furor over the translation-failure interpretation of incommensurability and conceptual schemes as well as the obsession with bivalent semantics have, I believe, obscured one of the most significant aspects of incommensurability: the obstruction of *the truth-value functional relation* (concerning the semantic relation between *truth-value status*, instead of *truth-values*, of the sentences between two competing languages) between the incommensurables. Too much attention paid to the meaning relation and lack of attention to the truth-value functional relation between two rival scientific languages is, to a large extent, responsible for the slow progress that has been made toward establishing the integrity and tenability of the notion of incommensurability. It is time to switch our attention from the meaning relation to the truth-value functional relation between rival scientific languages. As will become clear in the remaining chapters, it is the truth-value functional relation, instead of the meaning relation, that is the dominant semantic relation in the case of incommensurability.

Our new orientation not only emerges from the above cases studies, but also is inspired by I. Hacking's (1982, 1983) and N. Rescher's (1980) related works on the issue of incommensurability and the notion of conceptual schemes.

1. I. Hacking's Styles of Scientific Reasoning

Hacking fundamentally opposes the received interpretation of incommensurability because 'the idea of incommensurability has been so closely tied to translation rather than reasoning' (Hacking, 1982, p. 60). He also rejects the Quinean notion of conceptual schemes for a similar reason: It exclusively focuses on truth and translation. For Hacking, even Davidson's criticism of the Quinean notion does not fare well since 'like Quine, he [referring to Davidson] assumes that a conceptual scheme is defined in terms of what counts as true, rather than what counts as true-or-false' (Hacking, 1982, p. 62). Hacking believes that to think of conceptual schemes in terms of truth within the framework of bivalent semantics is misleading. 'Bivalence is not the right concept for science' (Hacking, 1982, p. 55). In fact, 'once you focus on truth rather than truth-or-falsehood, you begin a chain of considerations that call in question the very idea of a conceptual scheme' as well as the notion of incommensurability (Hacking, 1982, p. 59). Hacking thus urges us— in discussions of incommensurability, conceptual schemes, and other related issues—to switch our attention from truth and translation to truth-or-falsehood. Through the introduction of the notion of styles of scientific reasoning, Hacking starts to explore an alternative interpretation that focuses on truth-value status along the line of trivalent semantics.

Incommensurability as Dissociation

Like Kuhn, Hacking (1983) found that interpreters reading an out-of-date text characteristically encounter passages that are not fully meaningful to their contemporaries. However, not all old texts cause the same level of difficulty in understanding.

> An old theory may be forgotten, but still be intelligible to the modern reader who is willing to spend the time relearning it. On the other hand some theories indicate so radical a change that one requires something far harder than mere learning of a theory. (Hacking, 1982, p. 69)

When an old text involves an alien way of thinking and reasoning, even if it lies in the past of the interpreter's own culture, it may sound so strange that it is often incomprehensible. Hacking uses the medical theory of Paracelsus, a well-known sixteenth-century medical chemist, as an illustration of such a case. Paracelsus's works exemplify a Northern European Renaissance tradition of a bundle of Hermetic interests. He practiced many different disciplines, such as medicine, physiology, alchemy, herbals, astrology, and divination as a single art. For Paracelsus, the human body is a chemical system, similar in structure to the solar system. Each part of the human body corresponds to a celestial body: the heart to the sun, the brain to the moon, the liver to Jupiter, and so on. In this way, the human body mutually interacts with the celestial bodies and other chemical elements. This mutual interaction is exemplified by passages such as: 'Nature works through other things, such as pictures, stones, herbs, words, or when she makes comets, similitudes, halos and other unnatural products of the heavens'. Specifically, Paracelsan medicine prescribes associations between the chemical element mercury, the planet Mercury, the marketplace, and syphilis. Syphilis was treated by a salve of mercury and by internal administration of the metal, because the metal mercury is the sign of the planet Mercury, and that in turn signs the marketplace, and syphilis is contracted in the marketplace.

Paracelsan medical theory simply appears bizarre to modern Westerners. We might conclude that Paracelsus is just a witch doctor. But Hacking argues that, as Kuhn's experience with the reading of Aristotle's physics, things are not that simple. Paracelsan theory was very influential in the sixteenth and seventeenth centuries and was accepted by both the populace and intellectuals. It was recorded that the students in Paris and Heidelberg protested against the proscription of Paracelsan theory at the end of the sixteenth century. It is obvious that Paracelsan theory was perfectly intelligible to his contemporaries. His theory was believed either to be true by his proponents or to be false by his opponents. Either way, for his contemporaries, Paracelsus's assertions had definite truth-values. However, many Paracelsan sentences, such as (15), are hardly intelligible to us,

(15)   Mercury salve might be good for syphilis because of associations among the metal mercury, Mercury, the marketplace, and syphilis.

Why can we not understand Paracelsan sentences? Hacking contends that, contrary to the received interpretation of incommensurability as untranslatability, it is not because we are unable to understand the words of his sentences, nor because we cannot translate his sentences into our modern language (say, English). Although his works were written in dog Latin and Proto-German, they now can be translated into modern German or English. In the *Oxford English Dictionary*, we can find a definition of the Renaissance word 'Anatiferous': producing ducks or geese, that is producing barnacles, formerly supposed to grow on trees, and, dropping off into the water below, to turn into tree-geese.[1] Based on plain definitions like this, we can translate (liberal translation) Paracelsan sentences into English sentences. For instance, English sentence (15) cited above is a good translation of the original Paracelsan sentence, each word of which is plain enough for us.

The real trouble, Hacking finds, is that even when we are able to understand each word of Paracelsan sentences like (15), we are still left in a fog. What is the point of (15)? What does one argue or present by uttering (15)? Without grasping the point of (15), we cannot understand it effectively. In order to grasp the point of (15), we have to comprehend the forgotten mode of reasoning (the way of thinking and justification) that was central to Paracelsus's thought. However, the Paracelsan mode of reasoning is alien to us. The Renaissance medical, alchemical, and astrological doctrines of resemblance and similitude, which had internalized into the Renaissance mode of reasoning, are well nigh incomprehensible to us. The Renaissance mode of reasoning is especially hard to grasp if we try to approach it from our familiar way of thinking and justification (which we have to since there is no other alternative). In this sense, 'understanding is learning how to reason' (Hacking, 1982, p. 60). Paracelsan sentences would not be fully intelligible to us until we have learned how to reason in his way.

More significantly, Hacking notices that there is a strong semantic correlate of our failure of understanding of Paracelsus:

> The trouble is not that we think Paracelsus wrote *falsely*, but that we cannot attach *truth or falsehood* to a great many of his sentences. His style of reasoning is alien. (Hacking, 1983, p. 70; my italics)

Numerous Paracelsan sentences, when considered within the context of modern scientific theories, do not have any truth-values. We simply cannot *assert or deny* what was being said since there is no way to match what Paracelsus wanted to say against what we want to say. Let us put it into the terminology we have been using all along: There is a truth-value gap occurring between Paracelsus's discourse and ours. The occurrence of a truth-value gap between the Paracelsan language and ours strongly indicates that we have experienced a communication breakdown between Paracelsus and ourselves. Such a communication breakdown could not be restored easily through normal language learning due to the involvement of an alien style of reasoning. In this sense, Paracelsus's discourse is dissociated from or incommensurable with ours.

The Styles of Reasoning as Conceptual Schemes

To explain the phenomenon of incommensurability as dissociation, Hacking introduces the notion of the styles of scientific reasoning as a substitute for the Quinean notion of conceptual schemes. Hacking finds, following the lead of A.C. Crombie, that 'there have been different styles of scientific reasoning' within the Western scientific tradition, such as the Euclidean style of thought in ancient Greece, and the Galilean style of reasoning in modern time, each of which has 'specific beginnings and trajectories of development' (Hacking, 1982, pp. 48-51). A style of reasoning is characterized by introducing novelties into scientific inquiry, including new types of objects (such as abstract mathematical objects in Platonism in mathematics), evidence, sentences, laws or at any rate modalities, possibilities, and new types of classification and explanations. Among all these novelties, the most notable feature that distinguishes Hacking's styles of reasoning from the Quinean notion of conceptual schemes is that each new style brings with it new types of sentences, or a new way of being a candidate for truth-or-falsehood. Sentences that are meaningless and cannot be stated within one style of reasoning can be asserted to be either true or false within another style (Hacking, 1992, pp. 10-17). Thus, two scientific communities committed to different styles of scientific reasoning often find themselves experiencing a frustrated communication breakdown when one side tries an approach to the other—as with our experience with Paracelsus's medical theory.

In contrast with the Quinean notion of conceptual schemes, Hacking's styles of scientific reasoning have the following distinctive features.

(a) The Quinean notion of conceptual schemes is characterized by assignments of truth-values, that is, whether core sentences of a conceptual scheme are true. More precisely, 'a conceptual scheme is a set of sentences held to be true' and 'two schemes differ when some substantial number of core sentences of one scheme are not held to be true in another scheme' (Hacking, 1982, p. 58). A style of reasoning, in contrast, is concerned with truth-or-falsehood and is characterized by assignments of truth-value status. It is the styles of reasoning that 'create the possibility for truth and falsehood' and determine 'what is taken to be a legitimate candidate for truth or falsity' (Hacking, 1982, p. 57). 'The very candidates (of sentences) for truth or falsehood have no existence independent of the styles of reasoning that settle what it is to be true or false in their domain' (Hacking, 1982, p. 49).

(b) The styles of reasoning are not 'sets of sentences held to be true', but 'would be sets of sentences that are candidates for truth or falsehood' (Hacking, 1982, pp. 58, 64). Accordingly, the styles of reasoning are not sets of beliefs or propositions about the world, but rather the way beliefs or propositions are proposed and defended.

(c) The Quinean schemes are often characterized as a language confronting reality as in the fitting model R3. 'A style is not a scheme that confronts reality'. Therefore, such a notion would not fall into the dogma of scheme and reality that

Davidson resents (Hacking, 1982, p. 64).

(d) According to Davidson's interpretation, cross-language understanding is a matter of designing a truth-conditional translation in Tarski's style that preserves as much truths as possible as required by his charity principle. Since a style of reasoning is internal to any alien text, understanding is, for Hacking, not about truth-preserving translation, but 'is learning how to reason'. 'Understanding the sufficiently strange is a matter of recognizing new possibilities for truth-or-falsehood, and of learning how to conduct other styles of reasoning that bear on those new possibilities' (Hacking, 1982, p. 60).

(e) Much like Kuhn's paradigms, each style of reasoning is self-authenticating; there is no external justification as to which alternative styles of reasoning are better or worse (Hacking, 1982, p. 65; 1992, pp. 13-16).

If Hacking were forced to use the term 'conceptual schemes', he would define it as 'a network of possibilities, whose linguistic formulation is a class of sentences up for grabs as true or false' (1983, p. 71). Two schemes are distinct when the core sentences of one scheme are not held to be true or false in the other scheme.

2. N. Rescher's Factual Commitments

Rescher, like Hacking, is equally unsatisfied with the Quinean notion of conceptual schemes and Davidson's wholesale dismissal of the very notion of conceptual schemes and conceptual relativism within the framework of bivalent logic. Rescher therefore proposes a new version of conceptual schemes within 'a three-valued framework of truth-values, one that adds the neutral truth-value (I) of interdeterminacy or indefiniteness to the classical values of truth and falsity (T and F)' (Rescher, 1980, p. 332).

As the term suggests, a conceptual scheme is the mode of operation of concepts. 'Concepts' here for Rescher does not refer to mysterious Platonic entities, nor Kantian *a priori* mental schemes, but rather is a generic term used to denote all that is conceptual, such as 'categorical framework (descriptive and explanatory mechanisms)', 'taxonomic and explanatory mechanism', 'fundamental concepts', 'modes of classification, description, explanation'. Rescher believes that, as a product of temporal evolution, our concepts of things are moving rather than fixed targets for analyses. As Quine has convinced us that there is no clear distinction between the analytical/conceptual and the synthetic/empirical, our concepts are always correlative with and embedded in a substantive view of how things work in the world. In other words, concepts are themselves loaded with substantial empirical/factual commitments. Accordingly, since all our concepts are factually committal, our conceptual schemes for operation in the factual domain (in natural sciences in particular) come to be correlative with a set of factual commitments, which constitute the essence of a conceptual scheme. In this sense, the factual commitments embedded within a scientific theory/language are its conceptual scheme. 'Schemes differ in just this regard—in undertaking different sort of factual commitments', or in a different way of conceptualizing the purported facts (Rescher, 1980, pp. 329-31).

The issue of scheme differentiation can also be approached from the angle of conceptual innovation. A new conceptual scheme brings with it not only new phenomena never before conceived and described, new modes of classification, description, and explanation, but also new ways of looking at old phenomena. 'Such innovation makes it possible to say things that could not be said before—and so also to *do* new things' (Rescher, 1980, p. 330). Why? Because 'they lie beyond the reach of effective transportation exactly because they involve different factual commitments and *presuppositions*' (Rescher, 1980, p. 331; my italics). Like Hacking's styles of reasoning, factual commitments determine whether a sentence has a truth-value. In addition, only sentences that are either true or false can be used to describe presumptive facts, to make factual assertions or propositions, and are factually meaningful. Therefore,

> Innovation—that availability of assertions in one scheme that is simply unavailable in the other—is one important key to their difference. One scheme will envisage assertions that have no even remote equivalents in the other framework. (Rescher, 1980, p. 331)

Consequently, Rescher reaches the same conclusion about the scheme differentiation as Hacking does: 'The issue of scheme innovation at bottom turns out not on differences in determinate truth-values but on the having of no truth-value at all'. More precisely, 'the key schematic changes are those from a definite (classical) truth-status to *I* (i.e., from *T* or *F* to *I*) or those in the reverse direction (i.e., from *I* to *T* or *F*)' (Rescher, 1980, p. 332).

Now we can see, according to Rescher, why the Quinean model of conceptual schemes cannot reveal the genuine conceptual innovation.

> If the conceptual scheme C' is to be thought of as an alternative to C along the lines we have in view, then one cannot think of C' as involving a different assignment of truth-values to the (key) propositions of C. One must avoid any temptation to view different conceptual schemes as distributing truth-values differently across the same propositions. The fact-ladenness of our concepts precludes this and prevents us from taking the difference of schemes to lie in a disagreement as to the truth-falsity classification of one selfsame body of theses or doctrines. (Rescher, 1980, p. 331)

For this reason, the key contrast between competing conceptual schemes is not between affirmation (to be true) and counter-affirmation (to be false), but rather between saying something (to be either true or false) and saying nothing (to be neither true nor false). As such, two schemes do not dispute over the same things or facts; rather, they are about different things and facts. If so, Davidson's charge of the third dogma is off target since there is no common content for two conceptual schemes to process.

3. Introducing Presuppositional Languages

Hacking and Rescher have proposed a very promising version of conceptual schemes different from the Quinean model, and, accordingly, a new interpretation of incommensurability within trivalent semantics. But they have offered only scattered insights here and there. Many details need to be worked out in order to make it a full case. The major issue remaining is this: If the primary function of conceptual schemes—Hacking's styles of reasoning or Rescher's factual commitments—is to determine the truth-value status of the sentences involved, then we need a semantic mechanism to explain why this is so. In other words, we need a workable semantic theory of truth-value conditions (not truth conditions). However, Hacking and Rescher either fail to do so or are unable to give us a satisfactory one. This is part of the reason why their insights have not gained their deserved attention up to now.

Hacking claims that truth-value status is style-of-reasoning relative, but he fails to tell us how. For example, how can the introduction of new types of objects, evidence, classifications, and explanations change the way sentences stand as candidates for truth or falsehood? Hacking does not explain it except to claim that the new style of reasoning and the new type of sentences—or a new way of being a candidate for truth or falsehood—occur together. Worse still, Hacking specifies the introduction of a new type of sentences, which have no truth-values in some earlier language but are true or false with the new style in place, as a necessary condition of a new style of reasoning. Apparently, such a clarification of the notion of the styles of reasoning suffers from circularity. Which one is logically more primitive: the occurrence of the new type of sentences or the new style of reasoning? It cannot be the new type of sentences; otherwise, we need another more primitive notion to explain their occurrence. It has to be the new style of reasoning. If so, then we need a criterion independent of the new type of sentences to define it.

Presuppositions

Fortunately, there is one semantic theory available to explain the occurrence of truth-valuelessness, and hence truth-value status without circularity. I mean R. G. Collingwood's and P. Strawson's theory of semantic presupposition. To be fair to Rescher, he indeed uses the word 'presupposition' once in conjunction with 'factual commitments' as we have quoted earlier (Rescher, 1980, p.331). Nevertheless, Rescher never explores such a significant insight further except to imply that 'factual commitments and *presuppositions*' determine the truth-value status of the sentences involved (Rescher, 1980, p. 331; my italics). However, it will become clear that Rescher's factual commitments do function as metaphysical presuppositions underlying scientific theories.

As Collingwood observes:

> Whenever anybody states a thought in words, there are a great many more thoughts in
> his mind than are expressed in his statement. Among these there are some which stand

in a peculiar relation to the thought he has stated: they are not merely its context, they are its presuppositions. (1940, p. 21)

In fact, every statement is made potentially in answer to a question; and every question involves a presupposition, such as the notorious one, 'Do you still beat your wife?' Thus, every statement involves a presupposition. For example, the sentence,

(16)   The present king of France is bald,

could be used to answer a potential question:

(16q) Does the present King of France have hair?

The question in turn presupposes this sentence:

(16a) The present king of France exists.

Of course, a statement may have many different presuppositions. For example, sentence (16) has at least four different presuppositions, i.e., (16a), (16b), (16c), and (16d).

(16b) There is a country named France.
(16c) A person can have hair.
(16d) A subject (such as 'a person') can possess a property (such as 'of having hair').

Among those presuppositions, only one is the immediate presupposition, namely, the one from which the question immediately arises, such as (16a). The rest would be mediate presuppositions, which are indirectly presupposed by the original question.

More significantly for our later discussion is the distinction between relative and absolute presuppositions made by Collingwood (1940, p. 4). A relative presupposition is a proposition (a proposition could be true or false) that could be questioned or verified within a certain domain of inquiry, such as (16a), (16b), and (16c) for our common-sense non-metaphysical way of thinking. But some fundamental presuppositions are unquestionable and cannot be verified within a certain domain of inquiry: These are absolute presuppositions. For instance, (16d) is an absolute presupposition to be taken for granted in all of our non-metaphysical inquiry. Newtonian absolute space and time are presupposed absolutely within the Newtonian paradigm. In this sense, those absolute presuppositions are not propositions that could be false within a certain domain. Absolute presuppositions are not *asserted* (what is asserted could be either confirmed or negated), but made or *presupposed* (what is presupposed in a certain inquiry could not be denied). Any question challenging the legitimacy of an absolute presupposition, such as the question, 'Is it true?' or 'What evidence is there for it?' is a nonsense question to its believers. This is why when absolute presuppositions are challenged, their believers 'are apt to be ticklish'.

Strawson's notion of semantic presuppositions will be formally presented, clarified, and defended in chapter 8. Here I only want to mention Strawson's notion of basic concepts, which function like Collingwood's absolute presuppositions. Strawson approaches 'the absolute components' of our thoughts from the perspective of basic conceptual structures. For Strawson, our conceptual structure is a set of interlocking systems of concepts such that each concept, no matter which is simple or complex, could be properly understood only by grasping its connections with others, its place in the system. However, within a conceptual system, although no concepts are absolutely simple or fundamental, some concepts could be conceptually *a priori* to other concepts in the sense that 'the ability to operate with one set of concepts may *presuppose* the ability to operate with another set, and not vice versa'. Strawson calls those presupposed concepts, such as the concepts of body, time, change, truth, identity, knowledge, etc., philosophically basic concepts.

> A concept or concept-type is basic in the relevant sense if it is one of a set of general, pervasive, and ultimately irreducible concepts or concept-types which together form a structure—a structure which constitutes the framework of our ordinary thought and talk and which is presupposed by the various specialist or advanced disciplines that contribute, in their diverse ways, to our total picture of the world. (Strawson, 1992, p. 24)

Notice that for both Collingwood and Strawson, absolute presuppositions or basic concepts are not absolute or basic in the Kantian sense, i.e., to be *a priori*, ahistorical, and universal for all conscious beings. Rather, they are absolute or basic within certain contexts. For Collingwood,

> metaphysics is the attempt to find out what absolute presuppositions have been made by this or that person or group of persons, on this or that occasion or group of occasions, in the course of this or that piece of thinking. Arising out of this, it will consider ... whether different absolute presuppositions are made by different individuals or races or nations or classes. (1940, p. 47)

In this sense, 'all metaphysical propositions are historical propositions'. Strawson's basic concepts are basic sets of concepts that are pervasive within a particular historical, cultural, or linguistic context. What are presupposed (basic concepts) could change with what are presupposing (normal concepts and factual statements). Thus, alternative sets of absolute presuppositions or basic concepts are not just possible and desirable, but also are what really happened in the development and transformation of human intellectual history.

What happens if a presupposition of a statement does not hold? For Collingwood, if a presupposition (such as (16a)) of a question (such as (16q)) is not made, then the question simply does not arise. That implies that we cannot judge the truth-value of the corresponding initial statement that the question addresses (such as (16)) since it is not even a proposition. Strawson makes this conclusion bluntly clear. Based on his trivalent semantics (1950), a semantic presupposition of a sentence has to be held true in order for the sentence to be true or false. In other words, the truth of a presupposition of a sentence is necessary for the truth or falsity of the sentence. For example, (16) is true or false only when (16a) is true; otherwise, (16) is neither true nor false (has no classical truth-value).

Presuppositional Languages

In fact, a comprehensive scientific language is fully loaded with a set of absolute presuppositions (some of them are basic concepts). Its core sentences share one or more absolute presuppositions. For example, the existence of the yin and the yang as well as the five elements, as the absolute presuppositions, underlies core sentences of the language of traditional Chinese medical theory. Similarly, the existence of phlogiston is embedded within the very conceptual set-up of phlogiston theory and is presupposed absolutely by numerous core sentences of the language of phlogiston theory (say, 'Object *a* is richer in phlogiston than object *b*'). Likewise, the assumption that there exists absolute space and time underlies the core sentences of the Newtonian language of space and time, the denial of which is unimaginable within the conceptual framework of the Newtonian language.

For this reason, I call a scientific language a presuppositional language ('P-language' in brief hereafter). By a P-language I mean an interpreted language whose core sentences share one or more *absolute* presuppositions. Even languages we use in everyday discourse (not a natural language such as English *per se*) are P-languages to some extent. The fact that the sun exists, rises, and sets periodically may in many everyday discourses count as an inevitable presupposition. Denial of it would play chaos with everyday communicative activity. For instance, if I promise you that I will pay back your money tomorrow morning and you understand what I mean, then both of us take it for granted that the sun exists, rises and sets periodically.

The conceptual core of a P-language consists of a set of absolute presuppositions underlying the core sentences of the language, which I call metaphysical presuppositions ('M-presuppositions' hereafter) of the language. M-presuppositions of a P-language are contingent factual presumptions about the world perceived by the language community. As will become clear later, Hacking's styles of reasoning and Rescher's factual commitments are actually different kinds of M-presuppositions of scientific languages.

The factual presumptions of a P-language could manifest themselves in different ways. First, they could be basic existential presumptions about the entities existing in the world around a language community, such as phlogiston in the language of phlogiston theory. Second, they could be basic universal principles about the existential state of the world around a language community, such as

'Fermat's conjecture' in the language of classical arithmetic; the second law of motion in the Newtonian language of mechanics, etc. Third, they could function as basic categorical frameworks about the structure of the world perceived by a language community, such as the taxonomy of Copernican astronomy, or the taxonomy of the Aristotelian language of mechanics.

The traditional Chinese medical theory (CMT) is a typical P-language. All three types of M-presuppositions are identifiable within it. First, the existences of the yin and the yang as well as the five elements are existential presumptions of CMT. Both underlie numerous core sentences of the language of CMT. Second, CMT has its own unique medical category system. For example, all symptoms related to diseases are categorized as eight principal syndromes, which can be grouped further into four matched pairs: the yin versus the yang syndrome; the superficial versus the interior syndrome; the cold versus the heat syndrome; and the asthenia versus the sthenia syndrome. Third, CMT is richly embedded with a unique universal principle, the pre-established harmony and mutual influence between the human body and Heaven (as Nature or the Universe). I will analyze those and other aspects of CMT in detail in chapter 10.

As absolute presuppositions, M-presuppositions of a P-language are analytically true in the language. Denials of them signify a complete breakdown of the informative use of the language and a complete rejection of it. For example, the M-presuppositions that the sun exists and that it rises and sets periodically are presupposed in the very linguistic set-up of our natural language. Denial of them will play chaos in everyday conversation. The existence of phlogiston is analytically true for any conceptually possible interpretations of the language of phlogiston theory. Rejection of phlogiston means rejection of phlogiston theory.

It should become clear that the proper terms of incommensurability relationship are two distinct P-languages. To say that two theories, systems, or languages are incommensurable is to say that the two associated P-languages are incommensurable, or that the communication between the two language communities breaks down.

4. Conceptual Schemes Reconsidered

Metaphysical Presuppositions as Conceptual Schemes

A set of M-presuppositions of a P-language is what we normally call the conceptual scheme of the language. To construe a conceptual scheme as a set of M-presuppositions of a P-language is advantageous for several reasons. First, it can catch the essence of Kant's scheme–content dualism; that is, conceptual schemes are 'necessary for the constitution of experience, not just necessary to control and predict experience' (Rorty, 1982, p. 5). A conceptual scheme is not what we experience, what we believe consciously, but what makes our experience and beliefs possible. In other words, some conceptual structures are logically presupposed by all experiences and beliefs of a language community. Of course, what Kant tries to reconstruct are certain minimum conceptual structures that are essential to or universally presupposed by any conception of experience of all self-conscious beings. However, the possible existence of such a minimum limit of conceptual structure for all conscious beings does not exclude the possibility of a more localized conceptual structure embedded within a language, a tradition, or a culture that is presupposed by any specific experience and beliefs of its participants. Although those localized conceptual structures themselves, unlike Kant's *a priori* categorical concepts, are factually committal (using Rescher's terms) and could change with contexts (languages, traditions, or cultures), a certain conceptual structure could still be fundamental within a particular context in the sense that it is presupposed by other concepts, experience, and beliefs, such as Strawson's basic concepts.

Second, as we have discussed earlier, a conceptual scheme is closely associated with a language, but is not identical to a sentential language. However, Hacking still somehow treats a conceptual scheme as a sentential language. As Hacking notes, 'Quine's conceptual schemes are sets of sentences held to be true. Mine would be *sets of sentences* that are candidates for truth or falsity' (1982, p. 64; my italics). However, as we have argued earlier, the style of reasoning itself cannot be true or false; only the language embedded within a style of reasoning can be true or false. In fact, according to Hacking, a style of reasoning is self-authenticated and cannot be false from the perspective of its speakers. Hacking apparently confuses a style of reasoning with the language embodying the style. We need to separate a conceptual scheme (in our case, a set of M-presuppositions) from the language in which it is embedded (in our case, a P-language). As Strawson argues, a presupposition of a sentence is not a *part* of the sentence, and is not even logically *entailed* by the sentence. By the same token, an M-presupposition (such as, 'phlogiston exists') of a sentence (such as, 'The element *a* is not richer in phlogiston than the element *b*') is not *a part* of the sentence. Thus, the M-presuppositions of a scientific language, which might be important components of the corresponding scientific theory, are not parts of the linguistic set-up of the corresponding P-language.

Third, 'a conceptual scheme' is a broader notion than 'a set of concepts' if 'concept' is construed either in a normal sense, namely, an abstract entity, the meaning, the interpretation, or the disposition associated with a term ('causation', 'a tree', 'a game'), or in a functional sense, namely, whatever composes the propositional content of our assertions and beliefs. To treat conceptual schemes as a set of concepts is the approach taken by the Kantian model of conceptual schemes. It is too narrow. Instead, 'a conceptual scheme' should mean 'a scheme of what is conceptual', which includes basic concepts, categorical frameworks (lexical structures or taxonomies), modes of reasoning and justification, ways of thinking, descriptions, and explanations, and some fundamental factual commitments (universal principles, existential presumptions). To me, those are M-presuppositions of a P-language. In this sense, the presuppositional model is more comprehensive than the Kantian model.

Last but most significantly, recall the powerful insight shared by both Hacking and Rescher on scheme-transition: Scheme change and differentiation do not consist in redistribution of truth-values as the Quinean model tells us, but are semantically correlated with the redistribution of truth-value status. However, Hacking and Rescher fail to specify a badly needed truth-value condition to explain sufficiently the occurrence of truth-valuelessness and change of the truth-value status. To construe a conceptual scheme as a set of M-presuppositions promises a basic semantic framework to work out such a truth-value condition. If a conceptual scheme is a set of M-presuppositions, it is not hard to understand why the core sentences of a P-language PL_1 could be truth-valueless when viewed from the perspective of another P-language PL_2 that suspends the M-presuppositions of PL_1. As will be presented in chapter 9, a single M-presupposition of the core sentences of a P-language is *necessary* for the truth or falsity of its sentences; and the conjunction of all the M-presuppositions is *sufficient* for the truth or falsity of its sentences. This establishes that the M-presuppositions of a language constitute the truth-value conditions of its core sentences. The truth-value status of the core sentences of a P-language is determined by its M-presuppositions, which, as absolute presuppositions, are self-evident and unquestionable for its practitioners. If the speakers of PL_1 are unable to recognize and comprehend the M-presuppositions of an alien PL_2, then the core sentences of PL_2, when considered within the context of PL_1, will lack truth-values. Thus a truth-value gap occurs between PL_1 and PL_2. This is why the occurrence of a truth-value gap between two P-languages is a strong semantic indicator that the two languages embody two competing conceptual schemes.

Responses to Davidson

One might wonder how our presuppositional model of conceptual schemes fares against Davidson's criticism of the very idea of a conceptual scheme. Clearly, the model effectively sidetracks Davidson's verificationist arguments by removing its two basic assumptions (TV) and (TF). According to the presuppositional model, a conceptual scheme is not identical with a sentential language, but is a set of M-presuppositions of a P-language. Thus, sentential-language translatability can no longer be used as a criterion of the identity of conceptual schemes. In addition, an alien conceptual scheme is not a set of sentences taken to be 'largely true' to 'fit' reality. Consequently, content–scheme dualism could not be boiled down to the claim that a conceptual scheme different from the interpreter's is largely true as Davidson construed. In fact, the core sentences of an alien P-language have no truth-values—not to mention that they are by no means 'largely true'—when they are considered from the viewpoint of a competing P-language. Thus, Davidson cannot derive translatability from truth (even if his truth-conditional theory of translation can still hold up) since there are no shared truths between P-languages to begin with. Therefore, Davidson's argument from translatability against radical conceptual relativism becomes powerless in front of the presuppositional model.

As to Davidson's argument from interpretability against modest conceptual relativism, it cannot be sustained after we remove its theoretical foundation, i.e., the truth-conditional interpretation of the charity principle, which again depends upon the shared holding-true between two conceptual schemes.

'Wait a minute', a baffled critic might argue, 'if your interpretation of scheme differentiation makes sense, i.e., a truth-value gap occurs between two competing P-languages, then the result will be the same as the conclusion derived from the Quinean model: Truth-preserving translation between the two languages is impossible. If so, Davidson's argument against untranslatability still has the teeth to tear off your very idea of conceptual schemes, right?' Our hypothetical critic is right that the new criterion of scheme differentiation, i.e., occurrence of a truth-value gap between two P-languages, does logically lead to the failure of truth-preserving intertranslatability between them. But remember that the Quinean thesis of untranslatability between alternative conceptual schemes itself is not incoherent (although it is confusing and unproductive). It becomes incoherent only *in conjunction with* Davidson's interpretation of Quinean scheme–content dualism, according to which a conceptual scheme different from the interpreter's is largely true. But Davidson's truth-conditional theory of translation does not allow a divorce between truth and translation. After we break the conjunction by removing the last conjunct (Davidson's interpretation of Quinean scheme–content dualism), the thesis of untranslatability is actually harmless (and also useless). Besides, our presuppositional model does imply that two conceptual schemes are untranslatable; but it does not follow that the untranslatables are distinct conceptual schemes. It is not controversial to claim that intertranslatability is a sufficient condition for the identity of two conceptual schemes (if two languages are intertranslatable, then they embody the same conceptual scheme); so is its logical contrapositive: Untranslatability is a necessary condition for alternative conceptual schemes (two languages embedded with two different conceptual schemes are untranslatable). What is at stake in the Quine–Davidson debate is whether untranslatability is sufficient for scheme difference; or whether translatability is necessary for languagehood. The presuppositional model implies neither. This is because the failure of mutual translation between two languages does not logically lead to the occurrence of a truth-value gap between them.

Of course, our presuppositional model still faces the challenge from Davidson's criticism of Kantian scheme–content dualism. Unfortunately, our model is not able to avoid the attack by maneuvering into a different kind of conceptual schemes as we have managed to deal with Davidson's attack based on verifiability. We have to face Davidson's attack head-on, as we have done in chapter 4.

Note

1   Many writers in the Renaissance thought that geese were generated from rotting logs in the Bay of Naples and that ducks were generated from barnacles.

Chapter 7

Kuhn's Taxonomic Incommensurability: A Reconstruction

Taking the presuppositional interpretation, I puzzled over Kuhn's and Feyerabend's mature works on incommensurability. To my surprise, I found out that both pioneers of the thesis of incommensurability had already alluded to many aspects of the new interpretation in a profound way. Especially, the later Kuhn's works on taxonomic incommensurability seem to be on the edge of breaking through the barrier of the traditional way of thinking about incommensurability in Tarski's truth-functional style. In this chapter I will survey Kuhn's positions on incommensurability and reconstruct Kuhn's taxonomic interpretation contained in his later works. Feyerabend's related works will be discussed when I clarify the contents of metaphysical presuppositions in chapter 10.

1. The Early Development of Kuhn's Positions on Incommensurability

Unlike Feyerabend, Kuhn's positions on incommensurability had undergone some dramatic changes since the publication of his *Structure of Scientific Revolutions* in 1962. We can roughly divide the development of Kuhn's thoughts on incommensurability into three stages associated with the three different bearers of incommensurability: paradigm, disciplinary matrix, and lexicon. The first two stages will be presented below. The third stage will be discussed separately in the next section.

Paradigms: Normative and Semantic Incommensurability (the 1960s)

It is well known that Kuhn divides the development of natural sciences into periods of 'normal sciences' punctuated at intervals by episodes of 'scientific revolutions'. A pivotal notion, paradigm, is introduced to conceptualize this revolutionary approach toward the development of scientific knowledge. Normal science is a period of a stable development dominated by a single paradigm. A scientific revolution happens when 'an older paradigm is replaced in whole or in part by an incompatible new one' (Kuhn, 1970a, p. 92). It is at the point of a revolutionary transition between two paradigms that we encounter the phenomenon of incommensurability: 'The proponents of competing paradigms are always at least slightly at cross-purpose ... [and] fail to make complete contact with each other's viewpoints'

(Kuhn, 1970a, p. 148). Consequently, 'the transition between competing paradigms cannot be made a step at a time, forced by logic and neutral experience' (Kuhn, 1970a, p. 150).

The term 'paradigm' in *Structure* was used loosely in two different senses.

> On the one hand, it stands for the entire constellation of beliefs, values, techniques, and so on shared by the members of a given community. On the other, it denotes one sort of element in that constellation, the concrete puzzle-solutions which, employed as models or examples, can replace explicit rules as a basis for the solution of the remaining puzzles of normal science. (Kuhn, 1970a, p. 175)

As such, Kuhn's notion of paradigm is very inclusive, embracing almost all shared metaphysical, epistemic, methodological, and value commitments of a scientific community, including the components of the normative dimension, such as problems, standards, and perceptions, as well as the components of the semantic dimension, such as meaning variance. Accordingly, Kuhn's notion of incommensurability at this stage has at least four different meanings, as follows.

Normative incommensurability Two paradigms could be incommensurable due to different normative expectations built into each paradigm. The early Kuhn identifies the following three aspects of normative incommensurability.

(i) Topic-incommensurability: One definite feature of Kuhn's paradigm consists in its self-determination of what count as legitimate problems to solve. Rival paradigms do not identify the questions or problems that any adequate paradigm must solve in the same way. However, the fact that rival paradigms focus on different sets of problems, define the most basic problems differently, and assign the problems different weights of contribution to their own paradigm would not make the rivals incommensurable. To make two paradigms incommensurable, these differences have to be built into the very standards of explanatory adequacy. In this sense, topic-incommensurability is not an independent form of incommensurability. It needs to be associated with another form of incommensurability, criterial-incommensurability.

(ii) Criterial-incommensurability: According to Kuhn's theory of paradigms, it is the methodological standards of adequacy, including standards of adequate solution and adequate explanation as well as standards of comparative evaluation and rational justification that determine which problems should be solved. Those standards of adequacy that each paradigm implicitly sets for itself are sufficiently disparate from one to the next, each favoring its own achievements and research program and unfavorable with respect to the work of its rivals. For example, a criterial conception of rationality—institutionalized norms that define what is, and is not, rationally acceptable—is internal to a paradigm. Because of this, a mode of rational justification acceptable in one paradigm may not be acceptable in the other. Thus, there is no rational comparison between two competing paradigms.

(iii) Perceptual-incommensurability: Each paradigm decides its own privileged range of observational data or perceptions such that different paradigms may not acknowledge the same observational data or receive the same modes of perception. Two rival paradigms always seek to explain different kinds of observational data and lead to different perceptions in response to different agendas of problems and in accordance with different standards of adequacy. Metaphorically, 'the proponents of different paradigms practice their trade in different worlds' (Kuhn, 1970a, p. 150). This is the most controversial aspect of Kuhn's early explication of incommensurability.

Semantic incommensurability   Two paradigms are incommensurable due to meaning variance. Both sides will inevitably talk past each other when trying to resolve their conflicts since they do not speak in the same terms. This is actually the received translation-failure interpretation we have discussed in chapter 2, which I will not belabor here.

In sum, incommensurability related to the notion of paradigms is characterized as certain kinds of discontinuities between two paradigms divided by a scientific revolution. They are semantic discontinuities, topic discontinuities, criterial discontinuities, and perceptual discontinuities. The nature of these discontinuities at this stage is vague and needs further clarification. Are there any common characteristics among these different kinds of discontinuities?

Disciplinary Matrices: Models and Exemplars (the 1970s)

In the late 1960s, Kuhn clarified his notion of paradigm by replacing it with the notion of disciplinary matrix. The constituents of a disciplinary matrix include most or all components of a paradigm, which Kuhn made more specific under the new name:

> Among them would be: Shared symbolic generalizations, like 'f = ma', or 'elements combine in constant proportion by weight'; Shared models, whether metaphysical, like atomism, or heuristic, like the hydrodynamic model of the electric circuit; Shared values, like the emphasis on accuracy of prediction, discussed above; and other elements of the sort. Among the latter I would particularly emphasize concrete problem solutions, the sorts of standard examples of solved problems which scientists encounter first in student laboratories, in the problems at the end of chapters in science texts, and on examinations... . I shall henceforth describe them as exemplars. (Kuhn, 1970b, pp. 271-2)

Almost all the forms of incommensurability occurring between two paradigms can be recaptured under this new notion. However, the switch of the bearers of incommensurability from paradigms to disciplinary matrices indicates Kuhn's progress in exploring the essence of incommensurability. There are two new forms of incommensurability associated with the two essential components of a disciplinary matrix: models/metaphysical commitments and exemplars.

Models: metaphysical incommensurability   Kuhn explains his notion of models as follows:

> Models … are what provide the group with preferred analogies or, when deeply held, with an ontology. At one extreme they are heuristic: the electric circuit may fruitfully be regarded as a steady-state hydrodynamic system, or a gas behaves like a collection of microscopic billiard balls in random motion. At the other, they are the objects of metaphysical commitment: the heat of a body is the kinetic energy of its constituent particles, or, more obviously metaphysical, all perceptible phenomena are due to the motion and interaction of qualitatively neutral atoms in the void. (Kuhn, 1977b, p. 463)

Some other examples of metaphysical commitments would be the pre-Newtonian assumption that all forces act by contact, the post-Newtonian commitment to an infinite Euclidean space, and the pre-quantum mechanics commitment to the fundamental character of continuous and deterministic physical processes. The metaphysical commitments of a disciplinary matrix provide the theorists with an explicit or implicit ontology by providing answers to some fundamental questions, such as what kinds of entities exist in the domain to be explored, and how may they be expected to behave and interact with one another? Two different metaphysical commitments of two competing disciplinary matrices populate the world with different entities, properties, and different interactions. The occurrence of this kind of metaphysical discontinuity signifies communication breakdown and leads to incommensurability at a deep level, which I will call metaphysical incommensurability (incommensurability due to conflicting metaphysical commitments).

Up to now, we have explored several different forms of incommensurability in the early Kuhn's writings. What is the relationship among these different forms of incommensurability? The answer consists in the function of a metaphysical commitment: to provide the participants of a disciplinary matrix with explicit or implicit methodological guidance, which tells them what is to count as a legitimate problem in the context of this ontology, what is to count as an adequate solution to a problem, and what is to count as a satisfactory explanation. Take the Aristotelian theory of motion as an example. According to a metaphysical assumption of the Aristotelian theory, motion requires a constant acting force. Such an Aristotelian concept of motion created the problem of how to account for projectiles. The assumption not only created a problem to solve, but also determined that a legitimate explanation of projectiles had to make use of the entities, forces, etc., recognized in the *Physics*. Guided by this assumption, Buridan solved the problem with his version of the impetus theory. In contrast, once this Aristotelian metaphysical assumption was abandoned, the problems surrounding projectiles that required the postulate of an impetus either disappeared or did not make sense any more. An alternative disciplinary matrix with different metaphysical assumptions has a different list of problems and different criteria of adequacy. Their proponents use key expressions with different purposes. Hence, if the key expressions used by two rival theories happen to be the same, (e.g., 'motion' for Aristotle as opposed to 'motion' for Newton), the meanings of those expressions will differ.

Exemplars: incommensurability due to change in similarity relationships As Kuhn himself emphasizes, the most essential and novel constituent of the disciplinary matrix is exemplars. By 'exemplars', Kuhn means 'shared examples' that are concrete problem solutions accepted by a scientific community as paradigmatic. The central role of exemplars in Kuhn's disciplinary matrix is to provide members of a scientific community with a set of compatible learned expectations—although they may differ from individual to individual—about the similarities (similarity as family resemblance) among the objects and situations that populate the world they perceive. Presented with exemplars repeatedly drawn from various kinds, any member of the community can gain a learned perception of similarity relationship among different tokens, which determines the categorical structure of the language used by the community.

Kuhn's discussion on the role of exemplars in scientific categorization is in line with a new development of the studies in human categorization in cognitive science and psychology around the 1970s.[1] The question of concern is, how are objects and situations to be brought together selectively to form a category? According to this new approach, a category need not contain defining properties as necessary and sufficient conditions for membership, as the classical approach believes. If the 'categorical nature' of categories is to be explained, it appears most likely to reside in family resemblance—in the sense of Wittgenstein's[2]—among its members. Categories may be processed, formed, and evaluated in terms of their individual exemplars or prototypes.[3] The membership of a category is determined by whether a token is sufficiently similar (family resemblance) to one or more of the category's known exemplars. Accordingly, a category does not have a fixed structure with a clear-cut boundary in which all members are equivalent. Instead, a category possesses a graded structure in which the membership of each token varies in how good an example it is of the category. Of all the members, those highly similar to the prototype are typical, whereas those less similar to the prototype are less typical and those dissimilar are atypical. Thus, there are two decisive factors for each category: its *prototypes* and the *similarity relationships* (family resemblance) of instances to these prototypes. If either the prototypes or similarity relationships are different or changed, then we have different graded structures for the same category (as far as the name of a category is concerned) or even different categories (as far as the content of a category is concerned). Foremost, the studies show that our category systems and the graded structures within categories are unstable, varying widely across cultural, linguistic, societal, and historical contexts (this will be discussed in detail in chapter 10).

Kuhn draws similar conclusions from his studies in the history of science. Following Wittgenstein's notion of family resemblance, Kuhn insists that acquisition of similarity relation does not depend upon any explicit articulated rules or criteria. The knowledge of similarity can be tacitly embodied in the exemplars without intervening abstraction of criteria or generalizations. Such a family resemblance relationship determined by exemplars holds a central position in Kuhn's theory of categorization. Kuhn regards natural categories as family resemblance classes: A natural family is a class whose members resemble each other more closely than they resemble the members of the other natural families.

Similarity relationships, as a tacit classification, group similar objects in a category and separate dissimilar objects to different categories.

Kuhn finds further from his studies that different scientific languages with different disciplinary matrices at some stage of development have different category systems due to holding different networks of similarity relations among objects and situations. For example, the sun is more typical than the earth within the category of 'planet' for the proponents of Ptolemaic astronomy, since the former bears a closer resemblance relation to the ideal exemplar of the category represented by the category concept developed within the Ptolemaic tradition. The sun and the earth resemble each other and belong to the same category because both bear sufficient family resemblance to the concept of 'planet' in Ptolemaic tradition. In contrast with Ptolemaic astronomy, the earth is typical and the sun is atypical for the category 'planet' from the point of view of Copernican astronomy. This is because one is similar and the other is dissimilar to the prototype of the category of 'planet' determined by the category concept of 'planet' within Copernican tradition. The similarity relationship between the sun and the earth was changed also during the transition from Ptolemaic to Copernican astronomy. The sun and the earth no longer belong to the same category.

The above instability of graded structure of categories due to the shift of the network of similarity relations is demonstrated in full during episodes of scientific revolution. Due to the change of prototypes and similarity relationships during a revolution, a natural family ceased to be natural. Its members were redistributed among pre-existing and new-born sets. The old category system was replaced by new one. For instance, 'what had been paradigmatic exemplars of motion for Aristotle—acorn to oak or sickness to health—were not motions at all for Newton' (Kuhn, 1987, p. 19). This sort of redistribution of individuals among natural families or kinds is the central feature of the episodes labeled scientific revolutions. In this sense, a scientific revolution is, for Kuhn, characterized as a shift of patterns of similarity relationships.

> The practice of normal science depends on the ability, acquired from exemplars, to group objects and situations into similarity sets which are primitive in the sense that the grouping is done without an answer to the question, 'similar with respect to what?' One central aspect of any revolution is, then, that some of the similarity relations change. Objects that were grouped in the same set before are grouped in different ones afterward and vice versa. Think of the sun, moon, Mars, and earth before and after Copernicus; of free fall, pendulum, and planetary motion before and after Galileo; or of salts, alloys, and a sulpuhr-iron filing mix before and after Dalton. (Kuhn, 1970a, p. 200)

As Kuhn points out, one direct result of the shift of similarity relationships during a scientific revolution is a communication breakdown between two language communities involved.

> Not surprisingly, therefore, when such redistributions occur [referring to the redistribution of objects into different categories due to the change of the similarity

relations], two men whose discourse had previously proceeded with apparently full understanding may suddenly find themselves responding to the same stimulus with incompatible descriptions and generalizations. (Kuhn, 1970a, p. 201)

Especially if two category systems of two scientific languages before and after the shift of similarity relationships, which happens usually during the episodes of scientific revolutions, are incompatible—one cannot be incorporated or fully formulated in the other—then the communication breakdown between the two communities is inevitable; and a full communication between them cannot be restored. Therefore, the two languages are incommensurable. This explication of incommensurability due to change in similarity relationships is Kuhn's new unitary and highly specific surrogate for his old mixture of different forms of incommensurability. The significance of this version of incommensurability will become clear when we proceed to unearth the later Kuhn's explication of taxonomic incommensurability.

2. Taxonomic Incommensurability

From Exemplar to Lexicon

After 1983 (marked by the publication of Kuhn's essay, 'Commensurability, Comparability, Communicability'), especially after his unpublished Shearman lectures delivered in 1987 at London, Kuhn had made a series of significant progresses (in his own words, 'a rapid series of significant breakthroughs') on the explication of incommensurability. Those progresses were initiated by introducing a new bearer of incommensurability: the lexicon (the lexical structure, the taxonomy, or taxonomic structure) of a scientific language. Kuhn defines a lexicon of a scientific language as the conceptual/vocabulary structure shared by all members of the language community, or as a mental 'module in which each member of a speech community stores the kind-terms and kind-concepts used by community members to describe and analyze the natural and social worlds' (Kuhn, 1993a, p. 325). A lexicon provides the community with both shared taxonomic categories/kind-terms and shared similarity relationships among those categories/terms. The structure of a lexicon of a scientific language consists of two parts: taxonomic categories or kind-terms and similarity relationships among these categories. Using K to refer to the kind-terms and R to refer to the relationships, a lexical structure, LS, of a language L can be symbolized as: $LS_L = [K, R]$.[4]

The notion of lexicon is a further development of Kuhn's understanding of the role of scientific categorization in scientific communication, an insight that Kuhn had been pursuing since he started to focus on the notion of exemplars in the 1970s. The most significant elements of Kuhn's works on incommensurability, I believe, were originated and developed from his insight of exemplars. What the later Kuhn did, during about a quarter century of his later life, was to approach this central thesis—incommensurability due to change in similarity—from different angles, and to explore it at different levels. It is this central thesis that constitutes the spindle of Kuhn's engine of incommensurability. It is not hard to see that the later Kuhn committed himself more explicitly to this categorical approach to incommensurability. As we have mentioned in chapter 2, kind-terms in a scientific taxonomy are clothed with some expectations about their referents. These expectations are nothing but the expectations about the similarity relationships between objects and situations. 'That is the pattern of similarities that constitutes these phenomena a natural family, that places them in the same taxonomic category' (Kuhn, 1987, p. 20). The primary function of expectations about the similarity relationships is to transmit and maintain a taxonomy by passing and preserving the taxonomic categories and the structural relationships between them.

Compared with the two previous bearers, namely, the paradigm and the disciplinary matrix, which are a combination of linguistic and non-linguistic components, the lexicon is primarily a linguistic notion concerning the conceptual category system of a scientific language. By focusing on this categorical aspect of scientific language (exemplars, similarity relations, lexicons), Kuhn started to reorient his studies of incommensurability from the discussion of multiple-dimensional incommensurability, including both the normative and the semantic dimensions, to the focus on semantic incommensurability only. Incommensurability thus becomes a thesis exclusively about scientific languages. During the same process, Kuhn's formulation of semantic incommensurability was undergoing transition also. In his early formulation, semantic incommensurability between two scientific theories was attributed to change in meanings (sense and reference) of the shared terms (singular and general terms) employed by theories. The later Kuhn realized that change in the semantic values of terms is only a by-product of a deeper lexical change of the languages, that is, change in the taxonomic structures of scientific languages. As far as the taxonomic structure is concerned, moreover, what is at stake is not any terms (single and general terms), but kind-terms (especially high-level theoretical kind-terms) in the related taxonomies. In fact, the later Kuhn became more and more dissatisfied with the traditional meaning approach to incommensurability with which he was widely identified as a major advocate:

> Far from supplying a solution, the phrase 'meaning variance' may supply only a new home for the problem presented by the concept of incommensurability. ... [I]t will then appear that 'meaning' is not the rubric under which incommensurability is best discussed. (Kuhn, 1983b, p. 671)

To substitute lexicons for meanings, Kuhn wanted to 'provide an account of incommensurability that does not explicitly use even the idea of meaning' (Hacking, 1993, p. 294). Consequently, the later Kuhn restricted his attentions to one essential aspect of semantic incommensurability: taxonomic incommensurability.

As we have pointed out, for Kuhn two scientific languages are incommensurable when a necessary *common measure* is lacking between them so that the cross-language communication between their advocates breaks down.

Kuhn's different formulations of incommensurability evolved in the process of identifying such a significant common measure of full cross-language communication. It was indeed long journey home. Beginning with both the normative aspects of scientific languages (the same set of problems, compatible methodological standards of adequacy, and shared modes of perceptions) and the semantic aspects (shared meanings), Kuhn gradually pinned down the common measure as the categorical aspect of language (shared exemplars and communal perceptions of similarity relationships for scientific categorization), and eventually identified it as lexicons of scientific languages. Kuhn concluded that 'shared taxonomic categories, at least in an area under discussion, are prerequisite to unproblematic communication' (Kuhn, 1991, p. 4).

> The lexicons of the various members of a speech community may vary in the expectations they induce, but they must have the same structure. If they do not, then mutual incomprehension and an ultimate breakdown of communication will result. ... People who share a core, like those who share a lexical structure, can understand each other, communicate about their differences, and so on. If, on the other hand, cores or lexical structures differ, then what appears to be disagreement about fact (which kind does a particular item belong to?) proves to be incomprehension (the two are using the same name for different kinds). The would-be communicants have encountered incommensurability, and communication breaks down in an especially frustrating way. (Kuhn, 1993a, pp. 325-6)

Left unresolved however is an explanation of why a shared taxonomic structure is necessary for cross-language communication between two rival scientific language communities in the case of incommensurability. Presumably, to answer the question Kuhn needs to work out a theory of cross-language communication to explain the essential role of lexicons in scientific communication. Without such a theory Kuhn's taxonomic explication of incommensurability is radically incomplete. However, Kuhn appeared to be of two minds when he tried to address the issue at hand. He was torn and thus swung between the received interpretation of incommensurability as untranslatability and his more promising new interpretation in line with Hacking's.

Translation and Cross-Language Communication

Kuhn seemed to be reluctant to part with the received interpretation of incommensurability as untranslatability. He apparently thought of cross-language communication in terms of translation. In fact, as early as at the end of 1960s and during the 1970s, Kuhn, in subsequent development of his views, realized that 'untranslatable' (Quine's term) is a better word than 'incommensurable' for what he had in mind when he spoke of a communication breakdown between two rival scientific languages.[5]

> In applying the term 'incommensurability' to theories, I had intended only to insist that there was no common language within which both could be fully expressed and which could therefore be used in a point-by-point comparison between them. (Kuhn, 1976, p. 191)

'If two theories are incommensurable, they must be stated in mutually untranslatable languages' (Kuhn, 1983b, pp. 699-70). Thus, instead of saying that Aristotle's physics and Newton's physics are incommensurable, one should say that some Aristotelian statements are not translatable into Newtonian statements. Since then, in clarifying incommensurability, the issue of translation-failure between rival scientific languages became a dominant theme of Kuhn's.

After 1983, by virtue of introducing a new semantic tool of taxonomic structures, Kuhn continued to rely heavily on the notion of untranslatability in his explication of incommensurability.[6] 'Incommensurability thus becomes a sort of untranslatability, localized to one or another area in which two lexical taxonomies differ' (Kuhn, 1991, p. 5). More significantly, Kuhn seemed to rely on the notion of translatability to bridge shared taxonomic structures with cross-language communication.

> Notice now that a lexical taxonomy of some sort must be in place before description of the world can begin. Shared taxonomic categories, at least in an area under discussion, are prerequisite to unproblematic communication, including the communication required for the evaluation of truth claims. If different speech communities have taxonomies that differ in some local area, then members of one of them can (and occasionally will) make statements that, though fully meaningful within that speech community, cannot in principle be articulated by members of the other. (Kuhn, 1991, p. 4)

We can concisely formulate this reading of Kuhn (as one face of Kuhn) as follows:

(a)   Effective communication between the speakers of two competing scientific languages is possible only if translation between the languages can be carried out.[7]

(b)   Translation between two scientific languages is possible only if there is a systematic reference-mapping of the corresponding kind-terms in the two languages.

(c)   A systematic reference-mapping of the corresponding kind-terms between two scientific languages is possible only if the two languages share the same taxonomic structure.[8]

(d)   Therefore, a shared taxonomic structure is necessary for successful communication between two scientific language communities.[9]

Clearly, the above reading of Kuhn's taxonomic interpretation is one version of what we have identified as the translation-failure interpretation of incommensurability in chapter 2. According to it, incommensurability amounts to untranslatability due to radical variance of meanings of the terms in two competing scientific languages. We have argued that translation is neither sufficient nor necessary for cross-language communication. Therefore, reference to untranslatability neither identifies nor resolves the problem of incommensurability, but rather leads to confusion and misunderstanding.

The Role of Truth-Value Status in Cross-Language Communication

There is another face of the later Kuhn which has been neglected. Although one may find many reconstructions or reinterpretations of Kuhn's concept of incommensurability in the literature,[10] none have paid much attention to this aspect of Kuhn's interpretation of taxonomic incommensurability, which is arguably more coherent, tenable, and powerful. As I reread Kuhn's later works (especially the works after 1987) from the perspective of the presuppositional interpretation, I found that Kuhn had developed *implicitly* a sort of truth-value conditional theory of communication, which provides a badly needed cognitive connection between taxonomic structures of languages and cross-language communication. So construed, Kuhn's concept of incommensurability is seen not to depend upon the notion of untranslatability after all, but rather rely on a set of semantic conceptions such as taxonomic structure, truth-value status and truth-value gap, possible world, and cross-language communication.[11]

Following I. Hacking, Kuhn recognized the importance of the distinction between the notion of truth-value and the notion of truth-value status in his latest interpretation of incommensurability.

> Since that time, I have been gradually realizing (the reformulation is still in process) that some of my central points are far better made without speaking of statements as themselves being true or as being false. Instead, the evaluation of a putatively scientific statement should be conceived as comprising two seldom-separated parts. First, determine the status of the statement: is it a candidate for true/false? To that question, as you'll shortly see, the answer is lexicon-dependent. And second, supposing a positive answer to the first, is the statement rationally assertable? To that question, given a lexicon, the answer is properly found by something like the normal rules of evidence. (Kuhn, 1991, p. 9)

These two stages of theory evaluation roughly correspond to the two phases of scientific development as defined by the early Kuhn, namely, revolutionary science and normal science. The later Kuhn realized that the landmark of scientific revolutions (paradigm shifts) is not the redistribution of truth-values, but reassignment of truth-value-status.

Clearly, the later Kuhn adopted trivalent semantics in thinking of incommensurability. Recall that in trivalent semantics, a substantial number of core sentences of one scientific language, when considered within the context of a competing language, could lack classical truth-values. For Kuhn, the core sentences of a scientific language are those presupposing the lexicon of the language: 'Element *a* contains more phlogiston than element *b*' in the phlogiston theory, 'A body without external forces on it tends to seek its natural place' in Aristotelian physics, and 'Planets revolve about the earth' in the Ptolemaic astronomy, and so on. As Kuhn observes, this is what happens during scientific revolutions.

> Each lexicon makes possible a corresponding form of life within which the truth or falsity of propositions may be both claimed and rationally justified. ... With the Aristotelian lexicon in place, it does make sense to speak of the truth or falsity of Aristotelian assertions in which terms like 'force' or 'void' play an essential role, but the truth values arrived at need have no bearing on the truth or falsity of apparently similar assertions made within the Newtonian lexicon. (Kuhn, 1993a, pp. 330-31)

Newtonians find Aristotelian sentences hard to understand, not because they think Aristotle wrote *falsely*, but because they cannot attach *truth or falsity* to a great many of the Aristotelian core sentences since the Aristotelian lexicon presupposed by the sentences fails when considered within the Newtonian language. Consequently, a truth-value gap occurs between the Newtonian and the Aristotelian languages. Such occurrences of truth-value gaps abound in the history of science. 'Though the originals were candidates for true/false, the historian's later restatements—made by a bilingual speaking the language of one culture to the members of another—are not' (Kuhn, 1991, p. 9).

Along the lines of Hacking's and Rescher's reorientation of thinking of conceptual schemes in terms of redistribution of truth-value status instead of redistribution of truth-values, Kuhn's substitute for the Quinean notion of conceptual schemes is his notion of lexicons.

> What I have been calling a lexical taxonomy might, that is, better be called a conceptual scheme, where the 'very notion' of a conceptual scheme is not that of a set of beliefs but of a particular operating mode of a mental module prerequisite to having beliefs, a mode that at once supplies and bounds the set of beliefs it is possible to conceive. (Kuhn, 1991, p. 5)

Like Hacking's styles of reasoning and Rescher's factual commitments, the lexicon of a scientific language determines the truth-value status of its sentences.

3. Truth-Value Status and Taxonomic Structures

After admitting the existence of truth-value gaps between two competing scientific languages, Kuhn faces the same question that confronts Hacking and Rescher: How to explain the occurrences of truth-valuelessness or truth-value gaps based on his notion of lexicons? Unlike Hacking, Kuhn's lexical approach avoids circularity by defining lexicons independently without appeal to the notion of truth-value status. The later Kuhn seemed to come up with, as I interpret him liberally, *a sort of* truth theory that can be used to explain how a lexicon determines the truth-value status of sentences in terms of sort of truth-value conditions.

The usual theories of truth, such as the correspondence theory, are semantic theories about truth conditions and can only be used to determine the truth-value of a statement.[12] But at issue here is not whether a statement is true, but rather whether a given string of words is assertable (hence qualifying as a statement) or whether a sentence has a truth-value. What is needed is not a theory about *truth conditions*, but a theory about *truth-value conditions* (whether a sentence has a truth-value). A usual theory of truth does not help us with this. So as to distinguish such a theory of truth from the usual theories of truth, it is useful to dub the former a 'theory of truth-value'. Kuhn did not work out a complete theory of truth-value, but rather provided some clues here and there. Based on these clues, the task remains to reconstruct a Kuhnian theory of truth-value.

Taxonomic Structures and Possible Worlds[13]

Following the conventional possible-world semantics, Kuhn regards a possible world as a way our actual world might have been (Kuhn, 1988, p. 13). The problem is which concept of world is in play here. As P. Hoyningen-Huene argues persuasively, Kuhn uses the concept of world in more than one sense (Hoyningen-Huene, 1993, ch. 2). It could refer to a world that is 'already perceptually and conceptually subdivided in a certain way' (Kuhn, 1970a, p. 129), which may be called a 'world-for-us'. A world-for-us has a certain conceptual structure imposed by a certain taxonomic structure. It is a world to which we actually have conceptual access. In another sense, the concept of world could refer to the world-in-itself, namely, what is left over after we subtract all perceptual and conceptual structures imposed by human contributions from a world-for-us. Different from Kant's thing-in-itself, Kuhn stipulates the world-in-itself to be spatio-temporal, not undifferentiated, and in some sense causally efficacious. Beyond this, nothing can be said about this world.

In what sense does Kuhn use the concept of a possible world? Do possible worlds include worlds-for-us only or both worlds-for-us and the world-in-itself? Answering this question requires consideration of a controversial issue on the ontological status of possible worlds. What is the proper range of possible worlds over which quantification occurs? D. Lewis quantifies over the entire range of worlds that have been or might be *conceived*. S. Kripke, at the other extreme, quantifies over only the worlds that can be *stipulated*. Kuhn makes a general distinction between conceivable (possible) worlds from stipulatable (possible) worlds. On the one hand, not all the worlds stipulatable within a given lexicon are conceivable. A world containing square circles can be stipulated but not conceived. On the other hand, not all possible worlds conceivable by the speaker of a language are stipulatable in it. For instance, the possible worlds described by the Newtonian language are not stipulatable, although conceivable, in the language of relativity theory (Kuhn, 1988, p. 14f). Kuhn stands along with Kripke and contends that only a world that can be conceptually accessible in the sense that it can be stipulatable in some language can be a possible world (Kuhn 1988, p. 14). Since the world-in-itself is totally inaccessible conceptually to any language community, it is not a possible world in Kuhn's sense. Accordingly, the actual world in which one scientific community lives and works is a possible world that they actually perceive to be.

Notice that in Kuhn's concept of possible worlds, the alleged distinction between there being a possible world and it being conceptually accessible is blurred. For a world to be a possible world, it has to be conceptually accessible somehow by some language. Some lexical structures are prerequisite to the existence of, not just accessible to, any possible world. 'Like Kantian categories, the lexicon supplies preconditions of possible experience' (Kuhn, 1991, p. 12). To echo G. Berkeley's famous expression 'to be is to be perceived', Kuhn would say that 'to be a possible world is to be conceptually accessible'.

Nevertheless, this does not mean that a possible world is conceptually accessible to any language. To be a possible world is to be a world accessible to *certain* languages, but not to *any* language. This is because, according to Kuhn, there is a conceptual connection between the taxonomic structure of a *certain* language and its conceptual accessibility relation to *certain* possible worlds. First of all, Kuhn contends that acquisition of a *certain* taxonomic structure is prerequisite to gaining conceptual access to a *certain* possible world. Certain learned similarity–dissimilarity relationships, as the central feature of a certain taxonomic structure, is a language-conditioned way of perceiving a certain world. Until we have acquired them, we cannot perceive that world at all. Similarly, some set of kind-terms supplies necessary categories to describe a certain possible world. For example, in order to gain conceptual access to the Newtonian world, the taxonomic structure of Newtonian mechanics, especially the interrelated kind-terms like 'force', 'mass', and 'weight', must be possessed first.

Kuhn further specifies that only the possible worlds stipulatable in, or describable by, the lexicon of a language are conceptually *accessible* to the language community. This is because

> [o]nly the possible worlds stipulatable in that language can be relevant to them. Extending quantification to include worlds accessible only by resort to other languages seems at best functionless, and in some applications it may be a source of error and confusion ... [T]he power and utility of possible-world arguments appears to require their restriction to the worlds accessible with a given lexicon, the world that can be stipulated by participants in a given language-community or culture. (Kuhn, 1988, p. 14)

The question arises: 'Why must every aspect of a conceptually accessible possible world be *stipulatable* in a language? Some of the possible worlds we are interested in might have an unaccessed or inaccessible feature for which no vocabulary has been, or even could be, developed. But those possible worlds are *conceivable*. Presumably the actual world is like this.' Obviously, this objection assumes that any conceivable world would be conceptually accessible.[14] But if it is possible to conceive of there being possibilities that cannot be conceived of, then to equate conceivability with conceptual accessibility may make it impossible to deal with such cases. More to the point, the matter at issue here is a *language's*, not an individual *interpreter's*, conceptual accessibility relation to a possible world. Whether a possible world is stipulatable in a language can be determined by its taxonomic structure. But conceivability usually refers to the mental state of an individual interpreter. Any possible world would be conceivable for an interpreter if he or she wills to learn and adopt other languages that provide conceptual access to that world. Therefore, the concept of conceivability cannot help us to clarify a language's conceptual accessibility relation to a possible world. It is irrelevant for the purpose at hand.

Theoretically, a language's taxonomic structure enables the community to gain conceptual access to many, even infinite, possible worlds that are stipulatable in it. Of course, of these conceptually accessible possible worlds, only a small fraction are evidently possible for the community, which can be confirmed with experiments and observations accepted by the community. Discovering evidently possible worlds is what each scientific community undertakes to do in the course of normal science. As time passes, more and more conceptually possible worlds are excluded by requirements of internal consistency or of conformity with empirical data. Eventually, each lexicon may identify a highly limited set of possible worlds—the possible worlds that are both stipulatable and verifiable within the lexicon—and eventually a single world that the language community conceives as the actual world.

The same possible world may be conceptually accessible using different, but compatible, lexicons. Languages with incompatible taxonomic structures have access to different possible worlds.

> To possess a lexicon, a structured vocabulary, is to have access to the varied set of worlds which that lexicon can be used to describe. Different lexicons—those of different cultures or different historical periods, for example—give access to different sets of possible worlds, largely but never entirely overlapping. (Kuhn, 1988, p 11)

To sum up, according to Kuhn, conceptual accessibility to possible worlds is taxonomic-structure dependent. Only the possible worlds stipulatable in a language can be conceptually accessible to the language community. Only a world conceptually accessible to a language is a possible world for it. Since conceptual accessibility to a world is language dependent, a possible world is language dependent.[15] Therefore, a world may be possible for one language, but not possible for another. By providing the language community with a network of possibilities, the taxonomic structure of a language determines what is genuinely possible for the language.

Truth-Values and Possible Worlds

Following Putnam and many others, Kuhn contends that although the correspondence theory of truth (i.e., the idea that the substantial nature of truth consists in correspondence with the mind-independent world) has to be given up, the intuition behind it (i.e., the truth of a sentence is determined by its correspondence to a state of affairs external to the sentence) seems too obvious to be put to rest. Such an innocuous intuition can still remain at the heart of a theory of truth as long as 'a world-for-us' is substituted for 'the mind-independent world' at one side of the correspondence relationship (Kuhn, 1988, p. 24; 1991, pp. 6, 8).

> If, as standard forms of realism suppose, a statement's being true or false depends simply on whether or not it corresponds to the real world—independent of time, language, and culture—then the world itself must be somehow lexicon-dependent. (Kuhn, 1988, p. 24)

More precisely, following the Wittgenstein concept of fact-ontology that the world is the totality of facts,[16] we could treat a Kuhnian possible world as a set of internally related possible facts. Since a Kuhnian possible world is taxonomic-structure dependent, the possible facts are taxonomic-structure dependent. It is possible that some state of affairs counts as a possible fact in one language, but not in another with a sufficiently different taxonomic structure. Accordingly, a fact can be defined as the actualization of a possible fact or a possible fact verified in the actual world perceived by a language community. If a fact is the actualization of a possible fact and a possible fact is language dependent, then a fact seems to be inevitably language dependent. A fact so defined is relative to a language and subsists in a world specified by the language. There are no mind-independent facts out there waiting to be discovered.

According to the fact-based interpretation of the correspondence theory, a statement is true if and only if it corresponds to a fact. Because whether a state of affairs counts as a fact is dependent upon the taxonomic structure of a language; the same state of affairs may count as a fact in one language but not in another. Therefore, 'evaluation of a statement's truth values is, in fact, an activity that can be conducted only with a lexicon already in place, and its outcome depends upon that lexicon' (Kuhn, 1988, p. 24). Consequently, evaluation of the truth-value of a statement or a truth claim of a sentence is a correlate of a taxonomic structure (Kuhn, 1991, p. 4).[17]

Truth-Value Status and Possible Worlds

It is time to answer the question as to why some sentences that have truth-values in one scientific language lose their truth-values in another. Kuhn does not address this question explicitly. Nevertheless, based on his semantics of possible worlds and his adoption of a modified correspondence notion of truth as I have presented above, I think Kuhn would accept the following solution.

As pointed out earlier, the truth claim of a sentence P in a language L with a taxonomic structure TS consists in correspondence to a fact in the actual world perceived by TS. More precisely, take the designate of sentences as 'states of affairs'. The truth-value of P consists in whether the state of affairs designated by P corresponds to a fact admitted by L. A fact admitted by L is at least a possible fact from the viewpoint of L. Then, in order for P to have a truth-value in L, the state of affairs designated by P has to correspond to a possible fact in a possible world specified by TS of L. For Kuhn, a possible fact from the viewpoint of L has to be conceptually accessible by TS. Thus the truth-value status of P, when considered within L, is determined by whether the state of affairs designated by P corresponds to a possible fact from the point of view of L. If it does, then P is a candidate for truth-or-falsity; if it does not, then P is not a candidate for truth-or-falsity. Furthermore, if the state of affairs designated by P is not only a possible fact but also a fact specified by L, then P is true. Otherwise, it is false.

This establishes that the truth-value status of sentences is internalized within the taxonomic structure of a scientific language and becomes taxonomic-structure dependent. This is why Kuhn claims:

> Each lexicon makes possible a corresponding form of life within which the truth or falsity of propositions may be both claimed and rationally justified. (Kuhn, 1993a, p. 330)

> Where the lexicons of the parties to discourse differ, a given string of words will sometimes make different statements for each. A statement may be a candidate for truth/falsity with one lexicon without having that status in the others. And even when it does, the two statements will not be the same: though identically phrased, strong evidence for one need not be evidence for the other. (Kuhn, 1991, p. 9)

For instance, to assert the sentence,

(5) All diseases are due to the loss of the balance between the yin part and the yang part of the human body.

is to presuppose that there exist two natural forces 'the yin' and 'the yang'. This state of affairs designated by (5) corresponds to a possible fact conceptually accessible by Chinese medical lexicon. (5) thus has a truth-value, no matter whether it is actually true or false, from the viewpoint of Chinese medical theory. However, the apparently same sentence is not conceptually accessible in terms of Western medical lexicon; thereby it does not correspond to any possible fact from the viewpoint of Western medical theory. Therefore, (5) is not simply false, but rather has no truth-value.

4. The Role of Truth-or-Falsity in Cross-Language Communication

It has been argued so far that the truth-value status of sentences is, for Kuhn, taxonomic-structure dependent based on his theory of truth-value. However, Kuhn needs to explore further the role of truth-value status in cross-language communication in order to explain why, how, and in what sense a shared taxonomic structure between two scientific languages is necessary for successful communication between them.

To begin with, according to Davidson's truth-conditional theory of understanding, to grasp the factual meaning (not just meaning in general) of a sentence is to know its truth conditions. To know the truth conditions of a sentence presupposes that the sentence has a truth-value. That means that having a truth-value is a prerequisite for a sentence to be factually meaningful. If a sentence lacks a truth-value when considered within a language, then it will become factually meaningless to its speakers. This explains why Kuhn has observed that 'a historian reading an out-of-date scientific text characteristically encounters passages that make no sense' (Kuhn, 1988, p 9). This is because these out-of-date sentences, which must have been either true or false in the original text, may be impossible to be stated as candidates for truth-or-falsity from today's perspective.

Kuhn provides a different kind of argument for the necessity of truth-or-falsity in cross-language communication (Kuhn, 1991, pp. 8-10). Presumably, in order to achieve an informative use of a language, the language community has to adhere to some minimal rules of logic. Among them, the law of non-contradiction is crucial. The law claims that $(S \& \sim S)$ is logically false for any sentence S in a language L, or in symbols, $\vdash_L \sim(S \& \sim S)$. The essential function of the law is to forbid accepting both a sentence and its contrary. The law requires a choice between acceptance and rejection of a sentence in discourse. For in normal discourse (something like the stage of Kuhn's normal science), it is strongly desirable to make a choice between acceptance and rejection of a sentence in the face of evidences shared by both sides. For example, in Newtonian physics, we cannot assert both 'Event e_1 precedes event e_2' and 'Event e_2 precedes event e_1' at the same time. We cannot say that, in normal discourse, this paper is both white and not white. That is self-contradiction. Hence, it is reasonable to say that obedience to the law of non-contradiction to avoid self-contradiction is a minimal requirement for any successful linguistic communication. In this sense, the logical law of non-contradiction is one crucial minimal rule of any language game.

However, the proper function of the law of non-contradiction can be fulfilled only under some restrictions. When a sentence S has a truth-value within one language L_1, $\vdash_{L_1} \sim(S \& \sim S)$ is valid. But if S is neither true nor false within another language L_2, $(S \& \sim S)$ is not false, but rather neither true nor false. The formula $\vdash_{L_2} \sim(S \& \sim S)$ thus is invalid since $\sim(S \& \sim S)$ is untrue (neither true nor false) within L_2. This shows that the law applies only to the sentences with (actual) truth-values. It requires that the sentences involved in discourse must have (actual) truth-values or be factually meaningful for one side to communicate successfully with the other. Actually, the law of non-contradiction can and should be understood as a hypothetical law. The law entails the positive truth-value status of the sentences involved. That is, $\vdash_L \sim(S \& \sim S)$ is valid—namely, $(S \& \sim S)$ is logically false in L—only if S is either true or false. In this sense, to accept sentences in discourse as candidates for truth-or-falsity, which still allows for disagreement about their truth-values, constitutes the minimal rule of any successful linguistic communication. When one declares a sentence in discourse as a candidate for truth-or-falsity, one declares one's commitment to the law of non-contradiction, and at the same time declares oneself as an active participant in linguistic communication.

On the contrary, the law prohibits the occurrence of truth-valueless sentences in normal discourse. To deny the sentences in discourse as candidates for truth-or-falsity is to violate the law of non-contradiction, and thereby to put communication at risk. If one breaks the rule by denying that the sentences in discourse have truth-values, then one declares oneself outside the language community. If a group of members of a language community denies that the core sentences of the language have truth-values but still try to continue to claim a place in the community, then the communication between them and the rest of the community breaks down. Neither side engages in successful communication even if they seem to talk to one another.

More significantly, it follows that if the core sentences in one scientific language, which are true or false in the language, have no truth-value when considered within another competing language, then there is a truth-value gap between them. The occurrence of such a truth-value gap indicates that the communication between the two language communities breaks down in a particularly frustrating way. A fully factually meaningful sentence within one language community sounds so strange in the other that it is not factually meaningful and thereby cannot be effectively understood in the latter. Therefore, the occurrence of a truth-value gap between two languages can be used as a strong linguistic correlate of a communication breakdown between them.

Based on the above consideration, Kuhn contends that all language games are no less than true-or-false games. For Kuhn, who endorses P. Horwich's minimal theory of truth (Horwich, 1990), the truth predicates 'is true' and 'is false' exist primarily for the sake of such a logical need: to ensure that we stick to a language game. In Kuhn's own words:

> On this view [a version of the redundancy theory of truth], as I wish to employ it, the essential function of the concept of truth is to require choice between acceptance and rejection of a statement or a theory in the face of evidence shared by all. ... In this reformulation, to declare a statement a candidate for true/false is to accept as a counter in a language game whose rules forbid asserting both a statement and its contrary. ... In one form or another, the rules of the true/false game are thus universals for all human communities. (Kuhn, 1991, p. 9)

The idea that a language game is a true-or-false game can be viewed as one plausible interpretation of what Wittgenstein was driving at with his metaphor of conceptual schemes as 'language games'. It is truth-value conditions, instead of truth conditions, that are 'the rules of a language game' in this interpretation. This assertion presents a striking contrast to conventional interpretations that focus on truth conditions, such as Davidson's truth conditional theory of understanding.

5. Unmatchable Taxonomic Structures

It becomes clear now why shared or matchable taxonomic structures between two scientific languages are necessary for successful communication between them. But how can we know whether two taxonomic structures are matchable or not? Kuhn's following passages provide a hint for such a distinction:

> What members of a language community share is homology of lexical structure. Their criteria need not be the same, for those that can learn from each other as needed. But their taxonomic structures must match, for where structure is different, the world is different, language is private, and communication ceases until one party acquires the language of the other. (Kuhn, 1983b, p. 683)

> Incommensurability thus becomes a sort of untranslatability,[18] localized to one or another area in which two lexical taxonomies differ. The differences which produce it are not any old differences, but ones that violate either the no-overlap condition, the kind-label condition, or else a restriction on hierarchical relations that I cannot spell out here. Violations of those sorts do not bar intercommunity understanding. Members of one community can acquire the taxonomy employed by members of another, as the historian does in learning to understand old texts. But the process which permits understanding produces bilinguals, not translators, and bilingualism has a cost, which will be particularly important to what follows. The bilingual must always remember within which community discourse is occurring. The use of one taxonomy to make statements to someone who uses the other places communication at risk. (Kuhn, 1991, p. 5)

Full appreciation of these passages necessitates recollection of the two crucial features of Kuhn's kind-terms: the projectibility principle and the no-overlap principle. According to the former, each kind-term is clothed with expectations about its extension or referents. The expectations about a kind-term's referents are projectible in the sense that they enable members of a language community to postulate/project the use of the term to other unexamined situations, including counterfactual situations.[19] The idea that kind-terms are projectible sounds like a tautology. To be a kind-term, a term must carry with it some generality on the application of the term to its tokens. However, we tend to ignore an important characteristic of the notion of projectibility because of its platitude: *the limitation on the possible use of a kind-term.* Although one can learn and understand the kind-terms in an old or an alien language, it does not mean that one can use them *projectibly* in one's own language. One cannot speak an old or alien language while using the projectible kind-terms of the present language. Actually the same situation is faced by a bilingual who has to remind him or herself at all times of which language community he or she is in to avoid improper use of a kind-term of one language in the other language community. The no-overlap principle says that no two kind-terms at the same level of a (stable) taxonomic tree may overlap in their extensions (Kuhn, 1991, p. 4; 1993a, pp. 318-23).

The no-overlap principle is, in fact, the natural result of the projectibility principle. The expectations about a kind-term's (e.g., 'planet's') referents (e.g., planets) are usually learned in use.[20] Presented with exemplars (e.g., the sun) drawn from various examined situations, the members of a language community (e.g., the Ptolemaic community) acquire a learned expectation about the similarity relationships between the objects or situations that populate the world perceived by them. In terms of these expectations about the similarity relationship among tokens, the members of the community can tell which presentations belong to which kind and which do not (e.g., Mars belongs to the kind of planets, but the earth does not in the Ptolemaic community). Since people can acquire the same kind-term in different ways, the expectations about the referents of the same kind-term may differ from individual to individual in a language community. However, within the same language community, these different expectations are compatible in the sense that they would eventually identify the same extension for the term by effectively learning each other's expectations. On the other hand, some expectations about the referents of a kind-term may be so different in two competing language communities that they are incompatible with one another. In such a case, the members from one community (e.g., the Aristotelian community) will occasionally apply a kind-term (e.g., 'motion') to a token (e.g., the growth of an oak) to which the other (e.g., the Newtonian community) categorically denies that it applies. Usually, the non-identical extensions of the kind-term with incompatible expectations will overlap partially (e.g., for the movement of a physical object). If the speakers of the two communities use the term 'motion' separately in their own domain, no problem arises. But if they try to engage in an on-going dialogue, the difficulty arises in the region where both apply. Calling the same token 'motion' in the overlap region will always induce two conflicting expectations. Since these expectations are projectible, they cannot be only restricted within the overlap region and will be naturally extended to the respective non-overlap regions (e.g., to the growth of an oak). Therefore, ultimately the overlap is unstable and eventually only one kind-term remains within one language community (Kuhn, 1993a, p. 318).

The reason why two kind-terms at the same level of a (stable) taxonomic tree cannot overlap can be seen more clearly with some high-level theoretical kind-terms clothed with two incompatible expectations. Because these kind-terms figure importantly in fundamental laws about nature, they bring with them nomic expectations, i.e., exceptionless generalizations. In science, where they mainly function, these generalizations are usually laws of nature, such as Boyle's law for gases or Newton's laws of motion (Kuhn, 1993a, pp. 316-17). Then, if the extensions of such a kind-term with different concepts (e.g., 'planets' in Ptolemaic astronomy and in Copernican astronomy) overlap somehow and a token (e.g., 'the earth') lies in the overlap region, it would be subject to two exceptionless incompatible natural laws (e.g., Ptolemaic and Copernican laws of motion of celestial bodies). Similarly, the kind-term 'mass' in either Newton's or Einstein's language brings with it a nomic expectation in the form of a law of nature. Since the laws of nature built into the concept of mass in the Newtonian language and the Einsteinian language are incompatible, the respective expectations associated with the term are incompatible. These incompatible expectations will result in difficulties in the region where Newtonian 'mass' (mass$_n$) and Einsteinian 'mass' (mass$_e$) both apply. Calling some stuff in the overlap region 'mass$_n$' induces the nomic expectation associated with either the law of gravity or the second law of motion, while calling the same stuff 'mass$_e$' induces the incompatible nomic expectation associated with the new natural law in Einsteinian theory (general relativity theory). Hence such an overlap cannot remain stable over time (Kuhn 1988, pp. 14-23). For this reason,

> periods in which a speech community does deploy overlapping kind-terms end in one of two ways: either one entirely displaces the other, or the community divides into two, a process not unlike speciation and one that I will later suggest is the reason for the ever-increasing specialization of the sciences. (Kuhn, 1993a, p. 319)

Following from the above reading of Kuhn's insight, a primary type of unmatchable relationship between two taxonomic structures can be specified as follows:

> Two taxonomic structures are unmatchable if the extensions of some shared theoretical kind-terms in the two taxonomies overlap (but are not co-extensive) in some local area to such an extent that incorporating one into the other will directly violate the no-overlap principle.

Classical unmatchable taxonomic structures are Ptolemaic vs Copernican astronomy (overlapping extensions of 'planets'), Aristotelian vs Newtonian mechanics (overlapping extensions of 'motion'), and so on.

Notice that the no-overlap prohibition only applies to the theoretical kind-terms with nomic expectations. The principle does not apply to the extensions of singular terms (names and definite descriptions) and must be weakened for low-level empirical kind-terms with normic expectations (the generalizations that admit exceptions). For low-level empirical kind-terms, only terms that belong to the same contrasting set are prohibited from overlapping in extensions. Therefore, the no-overlap prohibition is restricted within some local area of two overlapping taxonomies and leaves most parts of them open for possible overlapping. That means that a major overlap between two unmatchable taxonomies is still possible.

In addition to the above primary type of unmatchable relationships between two taxonomies, there exists at least another type:

> Two taxonomic structures could be mismatched to such an extent that they are either totally disjointed or lack any major overlap.

Sometimes two competing languages may categorize a domain so differently that there is virtually no major overlapping between their taxonomies. B. Whorf has shown how different language communities might categorize the world around themselves in very different ways. Chinese medical theory and Western medical theory belong to such cases. It is hard to locate any major overlap between them since they employ two totally disjointed category systems at the theoretical level. Also, it is perfectly conceivable that two alien cultures may have two disjointed taxonomies although it is very unlikely that this will happen frequently.

6. Truth-Value Gaps and Incommensurability

It is time to put all the threads together for an overall picture of Kuhn's taxonomic interpretation of incommensurability.

(a) Human categorization is determined by different contextual factors, such as cultural, historical, and linguistic factors, and varies widely across different contexts. The taxonomic structures of different scientific languages about the same subject matter can be totally different. The taxonomic structures of two successive rival scientific languages before and after a scientific revolution may change dramatically so that two structures become unmatchable.

(b) Different taxonomic structures gain conceptual access to different sets of possible worlds consisting of different possible facts. The truth-value status of sentence P of one scientific language L_1 is determined by whether P, when considered within another competing scientific language L_2, describes a possible fact in a possible world conceptually accessible by L_2. Therefore, truth-value conditions of sentences are taxonomic-structure dependent.

(c) When the taxonomic structures of two competing scientific languages are unmatchable to one another, the two sets of possible worlds specified by them will be disjointed. Many possible facts within one set of possible worlds specified by one language would not count as possible facts when considered within the other language. Since the truth-value status of sentences is possible-fact dependent, there would be a truth-value gap occurring between the two languages. This is exactly what happens in the episodes of scientific revolutions. During these periods, scientific development turns out to depend on transitions to another disjointed set of possible worlds due to a switch to another unmatchable taxonomic structure. Is it, in these circumstances, appropriate to say that the members of the two communities live in different worlds? (Kuhn, 1988, pp. 13-15, 22-4)

(d) Accepting sentences in discourse as candidates for truth-or-falsity is an essential ingredient of any unproblematic linguistic communication. If there is a truth-value gap between two languages, this minimal logical rule of any language game is violated. Thus, the occurrence of a truth-value gap between two languages indicates that the communication between them is problematic and inevitably partial. 'The would-be communicants have encountered incommensurability' (Kuhn, 1993a, p. 326).

The above is the argument for a particular reconstruction of Kuhn's new interpretation of incommensurability, which is a combination of a logical-semantic theory of taxonomy, a semantic theory of truth-value, and a truth-value conditional theory of communication. According to this truth-value interpretation of incommensurability, two scientific languages are incommensurable when core sentences of one language, which have truth-values when considered within its own context, lack truth-values when considered within the context of the other due to their unmatchable taxonomic structures.

Notes

1   To know more about this approach, please refer to E. Rosch, 1973, 1975, 1978; E. Rosch and C. Mervis, 1975; L. Barsalou, 1987; L. Barsalou and D. Sewell, 1984.

2   It is well known that Wittgenstein argues that the tokens of a type need not have common elements, from which a general rule or a set of criteria can be derived, in order for the type to be understood and used in the normal functioning of a language. He suggests that, rather, a family resemblance relation might be what linked the various tokens of a type into a similarity set (similarity in the sense of family resemblance). A so-called family resemblance relationship is a primitive relationship among items in which each item has at least one element in common with one or more other items, but no, or few, if any, elements are common to all items. For instance, a set of items consisting of the following different combinations of letters is a family resemblance set: ABC, BCD, CDE, DEF.

3   'Prototype' is a term introduced by E. Rosch. By a prototype of a category, Rosch means the clearest cases of category membership defined operationally by people's judgments of goodness of membership in the category. In Kuhn's hands, 'exemplar' means originally 'shared examples' that are concrete problem solutions accepted by a scientific community as paradigmatic. In the case of categorization, 'exemplar' simply means the most typical token of a category according to a language community.

4   For a detailed explanation of Kuhn's notion of lexicon or taxonomy, please refer to chapter 2 and I. Hacking, 1993.

5   See Kuhn, 1970a, 1970b, 1976, and 1979.

6   Kuhn, 1983b, pp. 669-70; 1988, p. 11; 1991, p. 5.

7   Kuhn, 1977a, p. 338; 1983b, p. 683; 1988; 1991, p. 5.

8   Kuhn, 1983b, p. 683; 1988, p. 22; 1991, p. 5, 1993a, p. 324.

9   Kuhn, 1988, p. 16; 1991, pp. 4-5; 1993a, pp. 325-6.

10   Such as W. Balzer, 1989, M. Biagioli, 1990, H. Brown, 1983, G. Doppelt, 1978, D. Fu, 1995, P. Hoyningen-Huene, 1990, H. Hung, 1987, M. Malone, 1993, B. Ramberg, 1989, and H. Sankey, 1991, to mention only a few.

11   Perhaps my following readings of Kuhn cannot be found explicitly in Kuhn's writings. It is a reconstruction that uses various hints from Kuhn's writings that, I think, reveal his mature understanding of incommensurability. What I try to do is to construct from these hints a reasonably clear and coherent theory of incommensurability. Inevitably, the reader will find that my interpretation of Kuhn is heavily influenced by my presuppositional perspective. Actually, maybe the contrary is true: My presuppositional interpretation is inspired by Kuhn's insights revealed in his later writings.

12   Here we are only concerned with the so-called epistemic dimension of truth (about truth conditions), not about the semantic dimension of truth (about the metaphysical nature of truth). See M. Devitt, 1984 for the distinction.

13   Kuhn, 1970b, pp. 268, 270-71, 274; 1983b, p. 683; 1987, pp. 20-21; 1988, pp. 11, 13-14, 22-4; 1991, pp. 5, 10, 12; 1993a, pp. 319, 330-31.

14   Conceivability is at most (and should be) a necessary condition of conceptual

accessibility. A world that cannot be conceived of, such as a world containing square circles, is not a conceptually accessible possible world.

15   Careful readers might notice some slippage between 'there being a possible world' and 'it being conceptually accessible'. In fact, when I say 'a possible world is language dependent' or 'a taxonomic structure specifies/determines a possible world', I actually mean that 'whether a possible world is conceptually accessible is language dependent' or 'a taxonomic structure specifies/determines whether a possible world is conceptually accessible to the language'. After this clarification, I will, for simplicity, continue to use the former way of speaking.

16   For a good defense of Wittgenstein's fact-ontology as opposite to Aristotelian thing-ontology, see H. Gaifman, 1975 and 1976.

17   To claim that the evaluation of the truth-value of a statement or a truth claim is lexicon dependent does not mean that truth itself is relative to language. The core of Kuhn's taxonomic relativist view can be fully preserved by the claim that incompatible taxonomic structures may yield different truth claims. Readers may profit from a similar interpretation, in S. Hacker, 1996, of conceptual relativism based on the distinction between truth itself and truth claims.

18   Although Kuhn continued to use the term 'untranslatability' in his explication of incommensurability after 1987, he used the term in a different sense from the traditional notion of truth-preserving translation. It is the notion of truth-value-preserving translation that the later Kuhn had in mind when he connected untranslatability with incommensurability.

19   Strictly speaking, to say that a kind-term is projectible is to say that the expectation about a kind-term's referents, rather than the kind-term itself, is projectible.

20   Kuhn, 1987, pp. 20-21; 1988, pp. 14-23; 1993a, pp. 317-18, 325-6.

Chapter 8

A Defense of the Notion of Semantic Presupposition

It has become clear that Collingwood's and Strawson's notion of semantic presupposition, along with its two twin notions—truth-value status and truth-valuelessness (truth-value gaps)—are the very foundation of my presuppositional interpretation of incommensurability. However, the notion of semantic presupposition has been under constant attack. The attacks come primarily from two directions.[1] On one front, some critics attack the notion indirectly by undermining the central notion of any theory of semantic presuppositions: the notion of truth-valuelessness. It has been a contentious debate whether we should admit the existence of truth-valuelessness in our natural languages, and whether we should introduce it into semantics. For many philosophers, not only do our natural languages not admit truth-valuelessness, but also the notion itself is highly suspect. For some, the notion is 'such a creeping infection that when we allow it to emerge in semantics, it will smite it'. On the other front, other critics attack the notion of semantic presupposition head-on, either arguing that the notion itself is not theoretically coherent or contending that the notion, although it is theoretically coherent, is in fact empty since it cannot be exemplified. It is thereby not a philosophically interesting notion. For these reasons, the notion of semantic presupposition is 'a contemporary myth in semantics' and 'needs to be brought to light' (Böer and Lycan, 1976).

In particular, Böer (1976) and Lycan (1976, 1984, 1987, 1994) present lengthy and sophisticated arguments against the concept of semantic presupposition from both directions and their arguments have not been sufficiently rebutted. Some of their arguments are now being repeated by others and having a great deal of influence.

To clear these barriers in the way of the presuppositional interpretation, I need to clarify and defend the notions of truth-valuelessness and semantic presupposition against a variety of objections, especially Böer and Lycan's. To do so, I will first formally present a coherent and integrated notion of semantic presupposition in terms of a formal treatment of a three-valued language. I will then turn to two central arguments against the notion of semantic presupposition presented by Böer and Lycan. At last, the notion of truth-valuelessness is defended against two critical objections.

1. The Duel between Russell and Strawson over Vacuous Sentences

Consider the following three sets of sentences.

(i) Existential sentences and presuppositions

    (16) The present king of France is bald.
    (16n) The present king of France is not bald.
    (16a) The present king of France exists.

    (17) Some unicorns in the African jungle are hairless.
    (17n) Some unicorns in the African jungle are not hairless.
    (17a) At least one unicorn exists in the African jungle.

(ii) Sortal sentences and presuppositions

    (18) My soul is red.
    (18n) My soul is not red.
    (18a) My soul is/can be colored.

    (19) Some planets travel around the earth.
    (19n) Some planets do not travel around the earth.
    (19a) The earth is not a planet.

(iii) State-of-affair sentences and presuppositions

    (20) Astronomical event e_1 precedes event e_2.
    (20n) Astronomical event e_1 does not precede event e_2.
    (20a) There is an absolute time ordering.

There are two common features to each of the above sets of sentences: (a) It is clear that the first two sentences in each group somehow strongly 'suggest' or 'imply' the third sentence. This kind of 'felt implication' relationship between these sentences needs explanation. (b) When asked whether the first two sentences are true or false when the third sentence is not true, responders usually hesitate to give an affirmative or negative answer. They cannot simply choose one of the two classical truth-values (true or false) on the spot. Either answer seems to set a trap for them.

It is well known that B. Russell (1905, 1957) and P. Strawson (1950, 1952) gave different interpretations of the two features observed above. The debate has continued over the truth-value status of vacuous existential sentences like (16) since then. Russell and Strawson agreed that appeal to ordinary intuitions is not sufficient to determine whether sentence (16) is false or neither-true-nor-false when sentence (16a) is not true. We need to appeal to some theoretical considerations:

(a) For Russell, the felt implication between (16) and (16a) is nothing but the classical logical entailment relation in a subtle way. For Strawson, in contrast, the felt implication is one of semantic presuppositions in the sense that whenever both (16) and (16n) are true or false, (16a) has to be true.

(b) For Russell, when (16a) fails to be true, (16) is simply false. There is no need to appeal to truth-value gaps. The classical bivalent semantics is thus preserved. For Strawson, in contrast, when (16a) fails to be true, both (16) and (16n) would be neither true nor false. Obviously, we have to introduce some kind of trivalent semantics to accommodate the occurrence of truth-value gaps in our languages.

Before we move further, let us make a few important distinctions. First, a sentence could be truth-valueless either due to the failure of one of its presuppositions or due to purely syntactic or semantic matters. Non-declarative sentences (such as questions, imperatives, and performatives), some declarative sentences containing unspecified hidden parameters (for instance, sentences with vague predicates such as 'It is a heap'; sentences dealing with moral judgments such as 'Burning a cat for fun is morally wrong'), and ill-formed meaningless sentences ('sat Kanrog subbppp on') are commonly accepted as truth-valueless. But the real controversy arises when we ask whether a well-formed, meaningful, declarative sentence free of unspecified hidden parameters could be truth-valueless. Since this kind of sentences look like fact-stating sentences, we can call them fact-stating sentences in short. In the following discussion, by 'sentences' I will usually mean fact-stating sentences unless clarified otherwise. With the above qualification in mind, the issue to be addressed here is: Are there (fact-stating) sentences that are truth-valueless due to the failure of semantic presuppositions underlying them?

Second, the term 'presuppositions' has been used to describe pragmatic as well as semantic presuppositions. Very roughly put, presuppositions are pragmatic 'iff the implications in question arise only in virtue of contextual considerations, the roles of the relevant sentences in standard speech acts, Gricean conversational matters, simple matters of background knowledge on the part of particular speakers, etc.' (Lycan, 1984, pp. 79-80). For instance, the following notion of presupposition is pragmatic: A sentence *A* presupposes (or invites the inference of) *B* if and only if, given certain background beliefs that we have, we would have some warrant for assuming that if one utters *A*, then one will act as if one is willing to be regarded as having committed oneself to the truth of *B*. On the other hand, presuppositions are semantic 'iff the implications in question are a function of semantic status, semantic properties, propositional content, or logical form', but not a function of context (Lycan, 1984, p. 79). In this chapter, we restrict our discussion to semantic presuppositions only, especially existential presuppositions. Semantic presuppositions include logical presuppositions (defined by logical implication within an uninterpreted language such as existential presuppositions) and analytical presuppositions (defined by analytical implication within an interpreted language such as sortal and state-of-affair presuppositions) (Martin, 1979, p. 268 f).

Considering that it is widely accepted that semantics deals with the relationship between language and the world, I prefer the following definition of semantic presupposition:

> A presupposition is semantic if and only if the implication in question is a contingent factual presumption about the way the world is around the speaker and hearer.

Since the contingent factual presumptions about the way the world is can manifest themselves in different forms, we can classify semantic presuppositions into different categories, such as existential presuppositions (about the entities existing in the world), sortal presuppositions (about categorical frameworks of the world), and state-of-affair presuppositions (about some specific state of affairs).

As the title of the chapter indicates, I will restrict our discussion to semantic presuppositions only. My purpose in this chapter is to argue that the basic notion of semantic presupposition is sound. For this limited purpose, I will only focus on one special kind of semantic presupposition, namely, the so-called existential presupposition. I do not intend to give a full analysis to other kinds of semantic presuppositions, such as analytical presupposition, which will be touched on in chapter 11. Nevertheless, the basic conclusion drawn from the analysis of existential presuppositions is applicable to other kinds of semantic presuppositions.

2. A Formal Three-Valued Language

Let us first set up a formal treatment of a trivalent language.[2] This treatment will serve as a basic framework for the formal presentation of a trivalent version of semantic presupposition in the next section.

Language

> Def. An uninterpreted language L is any pair <Syn, Val> such that Syn is a syntax and Val (a set of admissible valuations for L) is a set of functions mapping the sentences of Syn into truth-values.

Here, Syn is a structure containing (a) sets of expressions or descriptive terms that have no fixed meanings; (b) a connected set of logical terms that have fixed senses and are paired one-to-one in accordance with formation rules; (c) a series of formation rules that connect descriptive terms with logical terms to form well-formed formulae. 'Val' represents the set of all logically possible worlds/interpretations consistent with the intended reading of the logical terms of Syn. Val does not assign any specific meaning to descriptive terms in Syn.

Truth Operator, Falsity, and Truth-value Status

We need to state a truth predicate explicitly in order to formulate the notion of semantic presupposition in language L. Unfortunately, the truth of sentences within a given language is undefined in that language according to the proof of Gödel and Tarski. We can define truth for L in M, but not in L, truth for M in M', but not in M, and so on. Therefore, we have to extend L to M in which a truth predicate can be explicitly stated. We can achieve this by adding a sentential operator to L, defined as follows:

$T_L(A) = _{df.}$ It is true in L (from the viewpoint of L) that A.

Here A represents any sentence.

In classical two-valued semantics, falsity is defined as the absence of truth by taking truth and falsity as contradictory concepts. Falsity is simply equal to non-truth. That is,

$F_L'(A) = _{df.}$ It is not the case that it is true that A, or in symbols, not-$T_L(A)$.

In contrast, in our three-valued semantics, non-truth is further divided into falsity and neither-truth-nor-falsity. Falsity is defined as the truth of the negation of a sentence A. That is,

$F_L(A) = _{df.}$ It is true that the negation of A, or in symbols, $T_L(\text{not-}A)$.

I adopt the definition of falsity in the three-valued semantics for obvious reasons. $F_L(A)$ is read as 'It is false in L that A'. Table 8.1 is the truth-value table definition of the truth operator and the falsity operator in our three-valued semantics ('n' represents neither-true-nor-false):

Table 8.1 A truth-value table of truth and falsity operators

A	$T(A)$	$F(A)$	$F'(A)$
t	t	f	f
n	f	f	t
f	f	t	t

In our system, 'A is true or false from L's viewpoint' can be expressed as $T_L(A)$ v $F_L(A)$. 'A is neither true nor false from L's point of view' can be expressed as not-$(T_L(A)$ v $F_L(A))$ or (not-$T_L(A)$ & not-$F_L(A)$).

Other Sentential Operators

In our three-valued semantics, there are two notions of negation depending on what the designated truth-values are. Truth is always designated and falsity is never designated in any system of three-valued semantics. Whether neither-truth-nor-falsity is designated depends on whether one wishes to preserve truth or to preserve non-falsity in a valid inference. If one's intention is to preserve truth, only truth is designated; if to preserve non-falsity, then both truth and neither-truth-nor-falsity are designated. If truth is the only designated truth value in L, we have a notion of unconditional negation:

Def.   The unconditional negation of a sentence, briefly, $\sim A$, is true if and only if the sentence denied is false.

Correspondingly, if non-falsity is the designated truth-value in L, we have a notion of conditional negation:

Def.   The conditional negation of a sentence A, briefly, $\neg A$, is true if and only if the sentence denied is not true.

The corresponding truth-value table of these two concepts of negation is given in table 8.2.

Table 8.2 A truth-value table definition of conditional and unconditional negations

A	$\sim A$	$\neg A$
t	f	f
n	n	t
f	t	t

In addition, let us extend the distinction between contradictories and contraries in traditional two-valued logic to our three-valued semantics.

Def.   Two sentences are contradictories of one another if and only if they cannot both be true and they cannot both be false, although they may both be neither true nor false.

Def.   Two sentences are contraries of one another if and only if they cannot both be true, but they can both be non-true.

Both unconditional and conditional negations are negations in the sense of contradictories.

Finally, conjunction, disjunction, and material implication can be defined in the following strong matrix (Kleene's strong matrix):

Table 8.3 Conjunction, disjunction, and material implication

	&			v			→		
	t	f	n	t	f	n	t	f	n
t	t	f	n	t	t	t	t	f	n
f	f	f	f	t	f	n	t	t	t
n	n	f	n	t	n	n	t	n	n

Entailment and Formal Implication

Although logical entailment is essentially a notion of classical two-valued semantics, we can easily define it in our three-valued semantics:

Def. A entails B in a language L = <Syn, Val>, briefly, $A \vdash_L B$, iff for any V in Val, (a) if $V(A) = T$, then $V(B) = T$; and (b) if $V(B) = F$, then $V(A) = F$.

A crucial feature of entailment is that it preserves the principle of contraposition, that is:

A entails B if and only if $\sim B$ entails $\sim A$, or in symbols, $A \vdash B$ iff $\sim B \vdash \sim A$.

In addition, $\vdash_L A$ means that A is unconditionally valid in L, or A is true in all valuations of L. For example, $\vdash_L T_L(A) \lor F_L(A)$ means that A is true or false in L unconditionally. Furthermore, we can use material implication and a truth operator to formulate the entailment relation as defined above as follows:

$A \vdash_L B$ iff $\vdash_L T_L(A) \rightarrow T_L(B)$ and $\vdash_L F_L(B) \rightarrow F_L(A)$.

Corresponding to the entailment relation, which is essentially a notion of two-valued logic, we can introduce the notion of formal implication in our three-valued semantics to represent the logical inference relationship:

Def. A formally implies B in a language L = <Syn, Val>, briefly, $A \vDash_L B$, iff for any V in Val, if $V(A) = T$, then $V(B) = T$.

Unlike entailment, formal implication does not preserve the principle of contraposition. $\sim B \vDash \sim A$ does not necessarily follow from $A \vDash B$. Actually, the principle of contraposition is the principle of two-valued logic, which is dropped in any three-valued semantics. Furthermore, $\vDash_L A$ means that A is unconditionally valid in L, or more precisely, A is never false in all valuations of L (A is always either true or neither-true-nor-false). $\vDash_L T_L(A) \lor F_L(A)$ means that A is unconditionally true or false in L. Similarly, we can formulate formal implication in terms of material implication and a truth operator:

$$A \models_L B \quad \text{iff} \quad \models_L T_L(A) \rightarrow T_L(B).$$

A Few Useful Logical Rules[3]

R1. $\models (F(A) \rightarrow T(\sim A)) \,\&\, (T(\sim A) \rightarrow F(A))$
R2. $\models (F(\sim A) \rightarrow T(A)) \,\&\, (T(A) \rightarrow F(\sim A))$
R3. If $A \models C$ and $B \models C$, then $A \vee B \models C$
R4. $A \models B$ iff $T(A) \models T(B)$
R5. $T(A) \vee T(B) \models T(A \vee B)$ and $T(A \vee B) \models T(A) \vee T(B)$

Especially, from R3, R4, and R5, we can infer that

if $A \models B$ and not-$A \models B$, then $\models (T(A) \vee T(\text{not-}A)) \rightarrow T(B).$

This inference will be very useful in our formulation of semantic presuppositions later. The same inference holds for entailment also:

if $A \vdash B$ and not-$A \vdash B$,   then $\vdash (T(A) \vee T(\text{not-}A)) \rightarrow T(B).$

3. A Definition of Semantic Presupposition

Is The Notion of Semantic Presupposition Trivial?

Some critics contend that the notion of semantic presupposition itself is theoretically trivial. According to an intuitive interpretation of the notion, a sentence B is a semantic presupposition of another sentence A if and only if whenever A is true or false B must be true. That means that both A and its negation entail B. However, this interpretation makes presupposition B tautologous: If A entails B and the denial of A entails B, then their disjunction, i.e., $(A \vee \neg A)$, entails B. But $(A \vee \neg A)$ is a tautology. Since a tautology cannot entail a non-tautologous sentence, B must be a tautology as well. Then a presupposition of any sentence can never be untrue (Böer and Lycan, 1976, p. 6). However, a presupposition should be *contingent*, not a tautology. Otherwise, any logical truth B is semantically presupposed by any sentence A; since a logical truth is entailed by any sentence. In other words, any sentence has an unrelated tautologous sentence as its presupposition. In this way, the notion of semantic presupposition is trivialized.

   To rebut the above charge and other related criticisms, I need to construct a non-trivialized definition of semantic presupposition based on the trivalent semantics introduced in the last section. For obvious reasons, our following discussion will focus on logical presuppositions only.

Adequacy of the Notion of Semantic Presupposition

Let us set up the following necessary conditions for any satisfactory notion of semantic presupposition.

Conforming to Strawson's rules The debate between Strawson and Russell on the notions of semantic presupposition and truth-valuelessness emerged from their different intuitive readings of sentences with non-denoting subject terms like (16). Both Russell and Strawson agree that (16) somehow implies (16a) in the sense that if (16) is true, then the truth of (16a) will necessarily follow. But they diverge when (16a) is false. Russell conceives the case in traditional two-valued semantics. Hence, the principle of contraposition holds between (16) and (16a) since (16) entails (16a). That means that (16) is necessarily false when (16a) is false. On the contrary, Strawson treats the case in a three-valued semantics in which the principle of contraposition does not hold. Sentence (16) should be neither true nor false when (16a) is not true. Furthermore, for Strawson, both (16) and the negation of (16), i.e., (16q), bear a special relation to (16a). If either (16) or its negation, (16q), is true, then (16a) is true as well.

For comparison, we can formulate Russell's and Strawson's intuitions, which I call Russell's or Strawson's rules, in table 8.4.

Table 8.4 Russell's and Strawson's rules of semantic presupposition

Strawson's Rules	Russell's Rules	Comparison
Rule I: $\models T(16) \rightarrow T(16a)$	$\vdash T(16) \rightarrow T(16a)$	agree
Rule II: $\models T(16n) \rightarrow T(16a)$	$\vdash T(16n) \rightarrow T(16a)$ or $\vdash T(16n) \rightarrow F(16a)$	(partially) disagree
Rule III: $\models F(16a) \rightarrow \sim(T(16) \vee F(16))$	$\vdash F(16a) \rightarrow F(16)$	(completely) disagree

Any satisfactory formal account of semantic presupposition has to validate Strawson's rules.

Making a sound distinction between two kinds of non-truths[4] Although the truth-value of (16), when (16a) is not true, is controversial, it is widely accepted that both sides should agree on the following claims. Given a sentence,[5]

(21)   The current president of China is bald.

Then,

(a)   Although both (16) and (21) are non-true, (16) is non-true in a way which differs from the non-truth of (21). The non-truth of the former is due to failure of the denotation of the subject while the latter due to the falsity of the predicate of the subject.

Consequently,

(b)   It is possible for the negation of (16) to be non-true if (16) is non-true (when (16) is neither true nor false). But the negation of (21) is true if (21) is non-true ((21) is actually false).

In general, there is an intuitively recognizable distinction to be drawn between two kinds of non-true sentences. Such a distinction needs an explanation. A satisfactory account of semantic presuppositions should regiment the data in (a) and (b).

Giving a non-trivialized notion of semantic presupposition   The semantic presuppositions of a sentence should be contingent, not tautologous. In other words, a semantic presupposition of a sentence may fail to be true under some interpretations or models. This requirement is intended to exclude the possibility that any logical truth (sentences that are true under all possible interpretations) is semantically presupposed by any sentence.

A Variety of Formulations of Semantic Presupposition

The following are some typical definitions of semantic presupposition:

(P1)   *A* semantically presupposes *B* if and only if both *A* and not-*A* imply *B*.
(P2)   *A* semantically presupposes *B* if and only if both *A* and its logical contrary imply *B*.
(P3)   *A* semantically presupposes *B* if and only if *A* entails *B* and the negation of *A* entails *B*.
(P4)   *A* semantically presupposes *B* if and only if both *A* and the negation of *A* materially imply *B*.
(P5)   *A* semantically presupposes *B* if and only if *A* necessitates *B* and $\models F(B) \rightarrow (\sim T(A) \ \& \sim F(A))$.
(P6)   *A* semantically presupposes *B* if and only if *A* necessitates *B* and the negation of *A* necessitates *B*.
(P7)   *A* semantically presupposes *B* if and only if *B* is a necessary condition of the truth or falsity of *A*. Or, whenever *A* is true or false, *B* is true.

Each of these definitions has some flaws and cannot meet our conditions of adequacy of the notion of semantic presupposition. For clarity, I will divide these definitions, according to their structures, into three groups and examine each group below.

Definitions P1, P2, P3, and P4

P1, P2, P3, and P4 define semantic presupposition in a broadly similar way. They share a common structure represented by P1. Actually, P1 can be used as a schema to represent most of formal accounts of semantic presupposition. For this purpose, let us reformulate P1 as Schema P:

> Schema P: A sentence A semantically presupposes B in a language L, briefly, $A \Rightarrow B$, if and only if both A and its *negation*, not-A, *imply B* in L.

Schema P not only represents some existing formal formulations of semantic presupposition, like P2, P3, and P4, but also covers some potential candidates for it. By discussing Schema P, we will touch on a rather broad range of possible interpretations of the notion of semantic presupposition.

Different theories of semantic presupposition differ radically as to how to unpack Schema P, especially in their interpretations of the negation 'not' and implication relation in it. This is because Schema P has two undetermined parameters. One is the sense of the negation 'not-A', the other the sense of 'imply'. Let us examine each parameter in turn.

Negation: conditional versus unconditional negation, contradictory versus contrary/subcontrary Suppose that A presupposes B. If 'not-' in Schema P is understood as a conditional negation, then from Schema P, we have:

> (a) A implies B and (b) $\neg A$ implies B.

If 'implies' means 'formally implies' (It will not affect our argument if 'implies' is read as 'entails' or 'materially implies'), then (c) follows from (a) and (b):

> (c) $\models (T(A) \lor T(\neg A)) \rightarrow T(B)$.

By contraposition, we have:

> (c') $\models \neg T(B) \rightarrow \neg(T(A) \lor T(\neg A))$.

Since $T(\neg A) \neq F(A)$, when B is not true, A is not necessarily neither true nor false. This proves that the definition of semantic presupposition in terms of conditional negation does not conform to Strawson's rule III. Therefore, this account is too weak.

However, perhaps the defender of conditional negation would protest that the notion of semantic presupposition in terms of conditional negation is intended to avoid truth-valuelessness. Then the fact that it does not conform to Strawson's rule III should be regarded as its merit instead of a flaw. Let us accept this defense for the sake of argument. Nevertheless, this defense still cannot eliminate another serious problem faced by defining semantic presupposition in terms of conditional negation. If B is untrue, then the antecedent $\neg T(B)$ is true. In order to make (c′) unconditionally valid, the consequent $\neg(T(A) \vee T(\neg A))$ has to be true. But the consequent is logically false. That establishes that A's presupposition, B, can never fail to be true. Thus, this notion of semantic presupposition based on conditional negation is trivialized.

As I have pointed out before, either unconditional or conditional negation of a sentence is the contradictory of the sentence (in the sense that a sentence and its contradictory cannot both be false or both be true). When faced with the threat of trivializing semantic presupposition, one way around it is to employ the notion of contrary, instead of contradictory, in defining semantic presupposition. This is the rationale of P2. The basic idea behind P2 is quite simple. Let us say that every sentence not only has a negation in the sense of logical contradictory (no matter whether as a conditional or as an unconditional negation), but also has a logical contrary. Then we can define semantic presupposition in terms of logical contrary instead of logical contradictory. If a statement and its logical contrary both imply a common statement, then they presuppose that statement. Let us formalize P2 in our formal system (taking 'imply' as formal implication for a reason that will become clear later). That is,

A presupposes B iff $\models T(A) \rightarrow T(B)$ and $\models T(*A) \rightarrow T(B)$.

Here '$*A$' represents the logical contrary of A. From this definition, we have:

(d) $\models (T(A) \vee T(*A)) \rightarrow T(B)$ and   (e) $\models \sim T(B) \rightarrow \sim (T(A) \vee T(*A))$.

When B is untrue, the antecedent of (e), $\sim T(B)$, is true. Since A and $*A$ can both be false, the consequent of (e), i.e., $\sim(T(A) \vee T(*A))$, may be true when B is false. Therefore, it is possible for B to be untrue while (e) is unconditionally valid. No truth-value gap necessarily occurs. P2 appears to be a decent solution that avoids trivializing presupposition and preserves bivalent logic as well.

According to the notion of logical contrary in traditional two-valued logic, the definition of logical contrary is clear. It has to be defined in such a way that when two sentences are contraries of one another they can both be non-true although they cannot both be true. Now, the real problem is how to formulate the logical contrary, $*A$, of a typical presupposing sentence A. G. Englebretsen (1973) suggests treating the logical contrary of a sentence A as the sentence with a negation occurring within A, which Russell calls the secondary occurrence of negation. In contrast, the logical contradictory of A, which Russell calls the primary occurrence of negation, is the sentence with a negation outside *A*. For example, the contrary of a universal subject-predicate sentence,

    (22)   All S is P,

is

    (∗22) All S is not P, or No S is P.

But its logical contradictory would be

    (~22) It is not the case that all S is P, or in symbols, not-(all S is P) = Some S is not P.

For a singular subject-predicate sentence,

    (23)   S is P,

its logical contrary is

    (∗23) S is not P,

while its contradictory is

    (~23) It is not the case that S is P, or in symbols, not-(S is P).

It will become clear that such a distinction between logical contradictory and contrary is nothing but the distinction between external negation and one kind of internal negation (the internal negation of a sentence as the contrary or as the subcontrary of that sentence).

Since I will address the problem of analyzing the notion of semantic presupposition with respect to the distinction between external and internal negations in detail later, I will leave my criticism of P2 until then. Nevertheless, the following two points should suffice to show the inadequacy of P2. On the one hand, not all presupposing sentences (such as particular subject-predicate sentences) have their corresponding contraries; on the other hand, the real trouble with P2 is that it does not conform to Strawson's rule III. It makes no sense to call a 'felt implication' a semantic presupposition if it does not support the notion of truth-valuelessness; it would be more appropriate to regard such an implication as another version of entailment.

From the above analyses we know that the negation of a presupposing sentence *A*, i.e. not-*A*, cannot be either *A*'s conditional negation or its logical contrary. An appropriate candidate of the negation in question seems to be the unconditional negation of *A*: ~*A* within three-valued logic. Suppose that *A* presupposes *B*. Taking the negation of *A* as an unconditional negation, according to Schema P (taking 'implies' as formal implication), we have:

(f) $\models (T(A) \vee T(\sim\!A)) \to T(B)$

(g) $\models \sim\!T(B) \to \sim\!(T(A) \vee T(\sim\!A)).$

A has to be neither true nor false when *B* is not true.

Nevertheless, I have to point out that $\sim\!A$ is not the only candidate for the negation that could be used to define semantic presupposition. We know that some sentences not only have their contradictories but also have their subcontraries. We can define the notion of logical subcontrary as follows:

> Def.   Two sentences are subcontraries of one another if and only if they cannot both be false but can both be non-false.

For example, the internal negation of a particular subject-predicate sentence is the subcontrary of that sentence:

(24)   Some S is P.

(#24)   Some S is not P.

Suppose that *A* presupposes *B*. Let us take the negation of *A* in Schema P as the subcontrary of *A* and use the symbol '#*A*' to represent it. Then, from Schema P (taking 'implies' as formal implication), we have:

(h) $\models (T(A) \vee T(\#A)) \to T(B)$

(i) $\models \sim\!T(B) \to \sim\!(T(A) \vee T(\#A)).$

According to formula (i), *A* has to be neither true nor false when *B* is not true. The possibility of *A* to be true or to be false is ruled out when *B* is not true. Otherwise, *B* cannot be untrue, and is thereby trivialized.

The requirement of the negation of a presupposing sentence in Schema P as the subcontrary of the sentence is weaker than the requirement of the negation as the contradictory. But the problem with this is that for some presupposing sentences, there is no corresponding subcontrary (for example, a universal subject-predicate sentence, *All S is P,* does not have a subcontrary.). From now on, I will take the negation in Schema P, i.e., 'not-*A*', as either the subcontrary of *A*, i.e., #*A* or as the contradictory of *A*, i.e., $\sim\!A$, if the subcontrary is not available. Since the notion of subcontrary is more comprehensive than the notion of contradictory, we may read a contradictory as one case of subcontrary. In the following discussion, for clarity and simplicity of formal treatment,[6] I will only use $\sim\!A$ in the related formulae unless indicated otherwise. But please remember that reading 'not-*A*' in Schema P as the subcontrary of *A* is more precise.

Table 8.5 A comparison of negations 'not-*A*'

A	subcontrary/contrary		contradictory	
	#*A*	*A*	~*A*	¬*A*
t	**t** / f	f	f	f
n	n	n	n	t
f	t	**f** / t	t	t

Implication: material, formal implication, or entailment? P3 takes 'implies' in Schema P as entailment. The major problem with P3 is that if we define the notion of semantic presupposition in terms of entailment, then the notion would become trivialized since the presuppositions of a sentence based on such a definition can never be false. This can be shown easily by means of the formal system we have introduced earlier: If $A \vdash B$ and $\sim A \vdash B$, then by the definition of entailment, we have

(a) $\vdash T(A) \rightarrow T(B)$ (b) $\vdash F(B) \rightarrow F(A)$
(c) $\vdash T(\sim A) \rightarrow T(B)$ (d) $\vdash F(B) \rightarrow F(\sim A)$

By R2, the combination of (b) and (d) leads to

(e) $\vdash F(B) \rightarrow (T(A) \,\&\, F(A))$.

That means that if B is a semantic presupposition of A, then B cannot be false. Otherwise, the formula (e) cannot be valid since its consequent, $T(A) \,\&\, F(A)$, can never be true (it is logically false). Furthermore, from (a) and (c) we have

(f) $\vdash \sim T(B) \rightarrow (\sim T(A) \,\&\, \sim F(A))$.

(f) is what we expect. But it is not consistent with (e).

If a theory interprets the implication in Schema P as material implication, we may call such a presupposition material presupposition by the analogy of material implication. P4 defines such a material presupposition. The problem with P4 is easy to see. If $\vdash (A \lor \sim A) \rightarrow B$, then, by contraposition, $\vdash \sim B \rightarrow (A \,\&\, \sim A)$. When B is false, the antecedent $\sim B$ is true, then the consequent $A \,\&\, \sim A$ has to be true since the formula is unconditionally valid. But the consequent $A \,\&\, \sim A$ cannot be true. That means that the presupposition B of A cannot fail to be true, and is thereby trivialized.

The real reason why a definition of semantic presupposition by virtue of entailment or material implication makes the notion trivialized is not very hard to see. This is because both relations, although they can be defined in three-valued logic, still preserve the principle of contraposition. The principle of contraposition is essentially the principle valid in classical two-valued logic. Any three-valued semantics rejects the principle of contraposition. It is plain now that the upshot from the failure of defining semantic presupposition in terms of classical entailment or material implication is that a notion of semantic presupposition requires a strict implication that does not preserve the principle of contraposition. A non-classical implication is called for. The formal implication defined in our three-valued semantics is what we need.

Definition P5

Böer and Lycan correctly diagnose that defining semantic presupposition in terms of entailment relation would trivialize presupposition because entailment supports contraposition. They claim that the proper way to analyze the notion of semantic presupposition is to employ a model-theoretic notion of strict implication that does not support contraposition. That is the notion of necessitation. It is clear that, in Böer and Lycan's hands, the notion of necessitation functions as the notion of formal implication as defined earlier. 'A sentence $S1$ necessitates a sentence $S2$, roughly, just in case there is no model relative to which $S1$ is true and $S2$ is untrue' (Böer and Lycan, 1976). So, a sentence A necessitates another sentence B if and only if $\models T(A) \rightarrow T(B)$. Then P5 can be formulated as:

(a) $\models T(A) \rightarrow T(B)$ (b) $\models F(B) \rightarrow (\sim T(A) \ \& \ \sim F(A))$.

The real problem with P5 is that it cannot justify Strawson's rule II, which can be restated here as $\models T(\text{not-}A) \rightarrow T(B)$. From (b), by contraposition, we have

(c) $\models (T(A) \ v \ T(\sim A)) \rightarrow \sim F(B)$.

(c) holds if and only if both

(d) $\models T(A) \rightarrow \sim F(B)$ and   (e) $\models T(\sim A) \rightarrow \sim F(B)$

hold. In our three-valued logic, $\sim F(B)$ does not imply $T(B)$. Hence, Strawson's rule II cannot be derived from P5. In fact, when A is false but B is untrue, Strawson's rule II is falsified while the same valuation validates (a) and (b). This establishes that, supposing that A presupposes B, according to P5, the truth of the negation of A does not necessarily imply the truth of B. This directly violates Strawson's rule II.

Definition P6 and P7

If we read necessitation relation as that defined by Böer and Lycan, P7 can be easily derived from P6, and vice versa. From P6, we have

$\models T(A) \rightarrow T(B)$ and $\models T(\sim A) \rightarrow T(B)$.

By R3, $\models (T(A) \text{ v } T(\sim A)) \rightarrow T(B)$. By R2, P6 eventually becomes

$$\models (T(A) \text{ v } F(A)) \rightarrow T(B).$$

Thus, it is nothing but P7.

P7 seems to have an appealing character free from the theory of negation. It appears that without explicitly mentioning negation, we would avoid much confusion caused by different readings of negation. However, this appealing feature is only superficial. It is more convenient to use the concept of falsity directly as far as Strawson's rule III is concerned. But in formulating and testing any definition of semantic presupposition, we have to deal with Strawson's rule II. In doing so, we need to explicitly use a specific reading of negation. For example, according to P7, sentence (16) presupposes (16a) if and only if both the truth of (16) and the falsity of it formally imply (16a). How can we find whether the falsity of (16) implies (16a), if the definition of formal implication only specifies the relation between the truth of an implying sentence and the truth of the implied sentence? We have to convert the falsity of (16) into the truth of its contradictory, namely, $F(16) = T(\sim 16)$. Therefore, the employment of negation in defining presupposition is inevitable. For this reason, we can regard P7 as one version of Schema P.

Besides, P7 does not specify the notion of falsity. As we have mentioned before, there are two senses of falsity: one in classical logic, the other within three-valued logic. If the falsity of A in P7 refers to the bivalent notion of falsity, i.e., $F'(A)$, then P7 would be trivialized. For then P7 is equivalent to

$$\models \sim T(B) \rightarrow \sim(T(A) \text{ v } F'(A)).$$

Since $T(A) \text{ v } F'(A)$ is a tautology ($F'(A) = \sim T(A)$), the consequent $\sim(T(A) \text{ v } F'(A))$ is logically false. Thus, B cannot fail to be true and is trivialized. Therefore, we cannot leave the notion of falsity in the definition of semantic presupposition unspecified. The only way out is to define falsity in three-valued logic. This feature is reflected in Schema P by $T(\text{not-}A)$ that represents the notion of falsity in three-valued logic.

A Definition of Semantic Presupposition

The general conclusion drawn from the above analyses of P1, P2, P3, P4, P5, P6, and P7 is plain: The best candidates for the two parameters of Schema P are: (a) reading 'not-A' as the subcontrary (including contradictory) of A; (b) reading 'imply' as formal implication. Then Schema P can be refined as follows:

> Schema P: A sentence A semantically presupposes a contingent sentence B[7] in a three-valued language L, briefly, $A \Rightarrow B$, if and only if both A and its subcontrary, $\#A$, (or its unconditional negation, $\sim A$, if the subcontrary is not available) formally imply B in L.

That is, $A \Rightarrow B$ iff $\models (T(A) \vee T(\#A \,/\, {\sim}A)) \rightarrow T(B)$.

From now on, I will use this modified Schema P as our formal definition of semantic presupposition.

Let us test Schema P against our three requirements of the notion of semantic presupposition. First, it is possible for B to fail to be true in our definition. Actually, when B is not true, no matter whether we take the negation of A as the subcontrary or as the contradictory of A, A has to be neither true nor false. That is,

$$\models {\sim}T(B) \rightarrow {\sim}(T(A) \vee T({\sim}A \,/\, \#A)).$$

Therefore, we have a non-trivialized notion of semantic presupposition. This not only takes care of the non-trivialization requirement, but also meets Strawson's rule III. Second, sentence (16) ('The present king of France is bald') is neither true nor false; for its presupposition (16a) ('The present king of France exists') fails while sentence (21) ('The current president of China is bald') is false since a presupposition of (21) ('The present president of China exists') is true. In this way, we make a reasonable distinction between two kinds of non-true sentences by assigning them different truth-value status.

Third, in order to meet the rule II, we need to show that both a presupposing sentence and its negation (in the sense of unconditional negation or subcontrary) bear a special relation to a third sentence, i.e., their presupposition. If A is a particular subject-predicate sentence, say, (24), then it is obvious that both A and its subcontrary $\#A$, say, (#24), formally imply their presupposition

(24a) There exists at least one S.

Now the problem is whether both a singular subject-predicate sentence (23) and its contradictory (~23) (or its subcontrary (#23)) imply their presupposition (23a), respectively, that is,

(23a) S exists.

It is plain that when (23) is true, (23a) has to be true. So (23) $\models$ (23a). However, there is a doubt whether the negation of (23) formally implies (23a) (Böer and Lycan, 1976). I will argue below that the negation of (23) does formally imply (23a).

Actually, our three-valued language L (which is a representation of our natural language) permits us to conclude that both (23) and (~23) formally imply (23a). What we need to do is to specify further its semantics Val.[8] Val is often defined by virtue of a set of formed structures called models. A model is a subset of possible worlds or interpretations. Such a model consists of two parts: one is the domain D of the discourse, the other the function f that maps the predicates in Syn into the elements in D. Valuations are then defined to represent the models by assigning specific truth-values to each sentence under a specific model.

Following J. Martin (1975, p. 257), let us specify a model M for Syn of L as any pair <D, f>. Syn is the syntax of L with a singular subject-predicate sentence, 'S is P'. Syn also contains a logical term 'exists', the existential predicate. Here, 'D' is a non-empty domain. 'f' is a function on all predicates and some subjects such that (a) for any predicate P, f(P) is a subset of D; (b) for any denoting subject S, f(S) is in D; (c) f(exists) = D. The set Val representing the model <D, f> maps sentences of Syn into truth-values in the following way: For any singular subject-predicate sentence, 'S is P', it is true if f(S) is in f(P) ('S' refers to something within the extension of 'P'); 'S is P' is false if f(S) is in D but not in f(P) ('S' refers to something that is not within the extension of 'P'); 'S is P' is neither true nor false otherwise ('S' does not refer or f(S) is not in D). It follows from the above valuation that

S is P $\models$ S exists or (23) $\models$ (23a) and   ~(S is P) $\models$ S exists or (~23) $\models$ (23a)

since they are theorems of L under the model M.

In conclusion, our definition of semantic presupposition meets all the three requirements of any satisfactory notion of semantic presupposition. This shows that the notion of semantic presupposition is theoretically coherent and integrated.

4. Argument from the Distinction between Internal and External Negation

The remaining problem is whether a theoretically coherent notion of semantic presupposition as I have defined it can be exemplified as a practically feasible notion. Böer and Lycan believe that it cannot although they accept it as a theoretically coherent notion. In this and the next sections, I will consider Böer and Lycan's two central arguments against the notion of semantic presupposition: the argument from the distinction between internal and external negation, and the argument from counterexamples. In terms of those two arguments, Böer and Lycan's were intended to show that the notion of semantic presupposition, although it could be made to be theoretically coherent, is actually empty.

A Dilemma: Internal versus External Negation

Böer and Lycan's first critical argument against semantic presupposition has the following two basic components.

The distinction between internal and external negation with respect to scope It is believed that negation in our natural language is ambiguous not only due to two different readings of negation with respect to their senses (i.e., unconditional negation and conditional negation), but also due to different scopes of negation. For example, Russell's paraphrase of a grammatically simple sentence (16) is a logically complex sentence (16'),

(16′) There exists one and the only one person who is the present king of
France, and this person is bald. Put in symbols,

∃x (Bald(x) & ∀y (King(y) ↔ x = y)).

According to Russell, the negation of (16) is ambiguous with respect to the scope
of the negation. The negation can attach to the widest possible scope (the primary
occurrence of negation). That is,

(ex-16′) It is not the case that there exists one and the only one person who is
the present king of France, and this person is bald. In symbols,

~ ∃x (Bald(x) & ∀y (King(y) ↔ x = y)).

On the other hand, the negation can attach to the narrow scope (the secondary
occurrence of negation). That is,

(in-16′) There exists one and the only one person who is the present king of
France, and this person is not bald. In symbols,

∃x (~Bald(x) & ∀y (King(y) ↔ x = y)).

Böer and Lycan adopt Russell's two readings of negation with respect to
scope, and call ex-(16) its external negation and in-(16) its internal negation.

> The distinction between external and internal negation is a scope distinction, a negation
> being external when it has wide scope, internal when it occurs within the scope of the
> 'presupposition'-generating locution. (Lycan, 1984, p. 91)

Presumably, based on our analysis of negation in Schema P, external negation corresponds to the logical contradictory. For this reason, I will use ~*A* to represent the
external negation of *A* later on. The notion of internal negation corresponds to
either the notion of contrary or the notion of subcontrary depending on the structure
of the sentence in question.

Two essential requirements of presupposition   Suppose *A* presupposes *B*.
According to the adequacy requirement of the notion of semantic presupposition,
there are two essential requirements for the negation of *A*. First, the negation of *A*
has to formally imply *B*. That is, not-*A* ⊨ *B* or ⊨ T(not-*A*) → T(*B*). Second, the
negation of *A* has to be the logical contradictory of *A*. If either one of these two
conditions is not met, then *A* cannot be said to presuppose *B*.

The first requirement is obvious. The second appears to be convincing if we
realize that in the following formula,

⊨ ~T(*B*) → ~(T(*A*) v T(not-*A*))

'not-*A*' has to be the contradictory of *A*; otherwise, when *B* is not true, *A* would not necessarily be neither true nor false. Actually, if *A* and not-*A* can both be false at the same time, then *A* would be false when *B* is not true.

According to Böer and Lycan, the distinction between external and internal negation itself gives rise to an inescapable dilemma for the champion of semantic presupposition (Böer and Lycan, 1976, p. 77). Suppose that *A* presupposes *B* and *A* is a logically complex sentence. The alleged dilemma goes as follows:

(a)   There are two essential requirements for the negation of *A*: it has to be the contradictory of *A*, and it must formally imply *B*.

(b)   The negation of *A* can be read *only* in two ways, either as the external negation of *A*, i.e., ~*A*, or as the internal negation of *A*, i.e., in-*A*.

(c)   If 'not-*A*' is read as the external negation of *A*, then it does not formally imply *B*.

(d)   If 'not-A' is read as the internal negation of *A*, then it is not the logical contradictory of *A*.

Therefore,

(e)   In either case, the two requirements of semantic presupposition cannot be fulfilled at the same time. Either way, semantic presupposition is ruled out.

The general conclusion drawn from the above dilemma is that the notion of semantic presupposition, although it is theoretically coherent, is in fact empty since it cannot be exemplified (Böer and Lycan, 1976, p. 10). We cannot even give any concrete sentence that presupposes another sentence in Strawson's sense. Take sentence (16) again as an example. The external negation of (16), i.e., (~16), does not formally imply (16a) since the following sentence is consistent,

(~16&~16a) It is not the case that the present king of France is bald and there is not any present king of France (Böer and Lycan, 1976, p. 59).

On the other hand, the internal negation of (16) is not the contradictory of (16). Therefore, (16) does not presuppose (16a), but only entails it.

In my judgment, this argument presents one of the most serious challenges to the tenability and integrity of the notion of semantic presupposition. If it worked, then the notion of semantic presupposition would become useless. However, the argument does not work in the way Böer and Lycan expected. I argue below that the alleged dilemma is a fallacy. It does not rule out an interesting notion of semantic presupposition that can be properly exemplified in many interesting cases.

Can A Semantic Presupposition Be Exemplified?

As I have argued earlier, treating the negation as contradictory is not the only appropriate reading for the negation in an appropriate definition of semantic presupposition (Schema P). Actually, taking the negation of a presupposing sentence *A* as the contradictory of *A* is too strong in many cases. A more appropriate reading of the negation of *A* is the subcontrary of *A*. If we take a negation of *A* as its subcontrary in general with its contradictory as one version of the subcontrary as I have suggested, then semantic presupposition can be properly exemplified. For instance, for a particular existential sentence (17), its subcontrary is,

(#17)   Some unicorns in the African jungle are not hairless.

Then, according to Schema P, both (17) and (#17) imply (17a), respectively. Furthermore, when (17a) is false, we have

$$\models \sim T(17a) \rightarrow \sim (T(17) \vee T(\#17)).$$

There is no case in which T(17) and T(#17) can both be false since one is the subcontrary of the other. Although (17) and (#17) can both be true, this possibility is ruled out. Otherwise, the formula cannot be unconditionally valid when (17a) is not true. Thus, the only possible truth-value for (17) is neither true nor false when (17a) is not true. It is clear that the relationship between (17) and (17a) meets our three requirements of semantic presupposition. Therefore, (17) semantically presupposes (17a). This shows that the premise (a) of the alleged dilemma is not justified.

The power of the alleged dilemma depends upon a basic assumption that there is a distinction between external and internal negation for every presupposing sentence. However, the distinction is not universally applicable to many presupposing sentences. For a singular subject-predicate sentence (23), the alleged distinction between external and internal negation is blurred. As every student of logic knows, we cannot read (23) as a particular sentence (24), otherwise (23) would lose its universal aspect. The more proper way is to read (23) as the conjunction of a corresponding universal sentence (22) and a particular sentence (24). That is,

(23)   S is P $=_{def.}$ (All S is P) and (Some S is P).

Then the external and the internal negation of (23) are respectively (~23) and in-(23),

(~23)   ~(S is P) = ~(All S is P) or ~(Some S is P) = (Some S is not P) or (No S is P)

(in-23)   in-(S is P) = in-(All S is P) or in-(Some S is P) = (All S is not P) or (Some S is not P) = (No S is P) or (Some S is not P)

This shows that there is no real difference between the external and the internal negation for a singular subject-predicate sentence. If so, then the external negation of (23) (or (16)) formally implies (23a) (or 16a) just as any internal negation of a presupposing sentence formally implies the same presupposition as that sentence itself does (I have proved this earlier). In addition, (~23) (or (~16)) is the contradictory of (23) (or (16)). According to Schema P, (23) (or (16)) presupposes (23a) (or (16a)).

The opponents of the notion of semantic presupposition might point out that if we adopt Russellian treatment to paraphrase (23), which looks like a grammatically simple sentence, into a logically complex sentence, then the distinction between the internal and the external negation of (23) should be very clear. It is the distinction between internal and external negation with respect to scope as follows.

(23′) $\exists x \, (P(x) \, \& \, \forall y \, (S(y) \leftrightarrow x = y))$
(in-23′) $\exists x \, (\sim P(x) \, \& \, \forall y \, (S(y) \leftrightarrow x = y))$
(~23′) $\sim \exists x \, (P(x) \, \& \, \forall y \, (S(y) \leftrightarrow x = y))$

When its presupposition (23a), or in symbols,

(23a′) $\exists x \forall y \, (S(y) \leftrightarrow x = y),$

is false, (23′) and (in-23′) are both false. Then (23′) entails, instead of presupposes, (23a′). Thus the dilemma stands.

The same strategy can be used against the two kinds of internal negations I have made, namely, internal negation as contrary or as subcontrary. Take sentence (17) (the same for sentence (24)) as an example. We should read (17), which appears to be a grammatically simple sentence, as a logically complex sentence. In other words, we should transfer the short surface form of (17) into the following logical form of conjunction:

(17′)   There exist some unicorns in the African jungle and they are hairless, or in symbols, $\exists x \, (Unicorn(x) \, \& \, Hairless(x))$.

Then the so-called subcontrary of (17′) would be,

(#17′) There exist some unicorns in the African jungle and they are not hairless, or in symbols, $\exists x \, (Unicorn(x) \, \& \sim Hairless(x))$.

Besides, we have

(17a′) $\exists x \, Unicorn(x)$.

We can see from the two formulas (17′) and (#17′) that if (17a′) is false, (17′) and (#17′) can both be false at the same time. That means that (#17′) is not the subcontrary of (17′). There is hence no real case for subcontrary. The requirement of negation as subcontrary fails. If so, even when we read the negation in Schema P as the subcontrary, (17′) is still not necessarily neither true nor false when (17a′) is not true since (17′) can be false in the following formula,

$$\models \sim T(17a') \rightarrow \sim (T(17') \lor T(\#17')).$$

Hence (17′) fails to presuppose (17a′).

Generally put, the essence of the above treatment of a presupposing sentence is as follows. Suppose that a sentence *A* presupposes *B*. We can always treat *A* as exponible into a conjunction with its presupposition *B* as one conjunct and a sentence *C* about the property of *B* as the other conjunct as shown in sentence (25),

(25)   *B* & *C*.

Furthermore, we can treat the internal negation of (25) as attaching a negation not to the presupposed conjunct *B*, but to the other conjunct *C* [9] as shown in (in-25),

(in-25)   *B* & ~ *C*.

It is obvious that both (25) and (in-25) imply *B*. Then if the presupposition *B* of (25) and (in-25) is false, (25) and (in-25) must both be false. In this way, (25) is false when its presupposition *B* does not hold (J. Martin, 1979, pp. 251-2, 268).

I have two responses to this objection. First, there is a grave defect in the syntactic structure of exponible sentences. The method of exponibilia treats a simple grammatical form as masking a complex logical form. For example, for a simple identity sentence (26),

(26)   The king of France is the king of France, or in symbols, k = k

if we adopt the Russellian reading, we should paraphrase (26) as (27),

(27)   The one and only one person has the property ascribed to the king of France and that person is self-identical, or in symbols,
         $\exists x \, (x = x \, \& \, \forall y \, (King(y) \leftrightarrow x = y \,))$.

It is objected, by D. Kaplan (1975) and others, that we should not invoke hidden complexity unless there is a good reason to do so. That is, we should not read (26) as (27) until we have investigated the options of identifying (27) with (26) and found (26) to be unworkable (S. Lehmann, 1994, p. 309). In addition, all things being equal, if we have to paraphrase a simple grammatical sentence, its logical form should correspond as closely as possible to the surface form of the sentence. It is after all the surface form of a sentence in our natural language that is being used and explained. However, Russellian treatment frequently requires such extensive rewriting of the surface form of a simple sentence that its syntax becomes too complicated to be understood. For example, it is not convincing to construe the simple sentence (26) as the very complex sentence (27). A theory that treats (26) as a simple identity sentence would be better (J. Martin, 1979, p. 253).

A more crucial problem for the issue in hand is that treating the simple grammatical form of a sentence as the disguised complex logical form is a typical method employed by classical logic only. In fact, after a simple presupposing sentence is translated into a conjunction with its presupposition as one conjunct, it naturally follows that the sentence must be false when its presupposition is false. Therefore, accepting the Russellian reading of a presupposing sentence would amount to adopting Russell's treatment of a non-denoting sentence. In this sense, whether we should accept the method of exponibilia is a crucial issue at stake. Adopting it without any convincing argument is to beg the question from Strawson's point of view. For this reason, non-classical theories of presuppositions should not employ this method, not just because of its grave defect in syntax level, but because adopting it amounts to dropping Strawson's notion of semantic presupposition from the outset.

Finally, the premise (c) of the argument is false. Suppose A formally implies B. Whether the external negation of A formally implies B depends on specific structures of sentences A and B. We cannot claim in general that any external negation of A does not formally imply B. Here is a counterexample. As I have argued earlier, the external negation of a singular subject-predicate sentence, namely, ~(S is P), formally implies the sentence 'S exists' as the original sentence 'S is P' does.

5. Argument from Counterexamples

Another major critical argument raised by Böer and Lycan against the notion of semantic presupposition claims that it is easy to provide many perfect counterexamples to an enormous number of alleged semantic presuppositions. They contend that semantic presupposition as species of formal implication must hold universally without conceivable counterexamples. So if they can give some counterexamples in which the so-called semantic presuppositions are cancellable, then the notion of semantic presupposition itself cannot be held consistently. In fact, there would be no genuine instance of semantic presuppositions if such prefect counterexamples can be found. The strategy of Böer and Lycan is then to make up counterexamples in which the alleged semantic presuppositions can be canceled.

This argument runs as follows. Suppose that A presupposes B. That means, according to the general definition of semantic presupposition (the initial Schema P) that both A and its negation formally imply B. That is,

(a) $\models T(A) \rightarrow T(B)$ and   (b) $\models T(\text{not-}A) \rightarrow T(B)$.

Let us focus on the formula (b) only. It is obvious that B cannot be false if not-A is true since not-A formally implies B. That means that the possibility that

(c)   The negation of A is true but B is false, or in symbols, T(not-A) & F(B)

is ruled out by the very definition of semantic presupposition since (c) is *self-contradictory* or logically false if A really presupposes B. Therefore, if we can show some cases of alleged presuppositions in which (c) can be held *without contradiction*, then the alleged semantic presupposition B is canceled.

Böer and Lycan give a few counterexamples in which (c) can be held without contradiction. Consider the following set of sentences:

(28a)   It is false that the present king of France is bald because[10] there is not any present king of France.

(28b)   It is false that it was John who caught the thief because no one caught the thief.

(28c)   It is false that my soul is red because my soul is not colored.

(28d)   It is false that John managed to solve the problem because this problem is so easy to solve.

According to Böer and Lycan, it is important to notice that these sentences are fully intelligible and are clearly *not contradictory*. In this way, the various 'presuppositions' carried by these negations of original presupposing sentences can be easily canceled. No semantic presuppositions are involved in these cases. Again, they reach the same conclusion as they drew from the argument from the distinction between internal and external negation: The notion of semantic presupposition is empty since it cannot be exemplified.

I grant that semantic presuppositions should be held universally. Hence, (c) would be a contradiction and a counterexample to semantic presuppositions if A did presuppose B. However, I am wondering whether the sentences given are really valid counterexamples, or whether they are genuine instances of (c). I will argue that these cases given in (28a, 28b, 28c, and 28d) are not genuine instances of (c). Therefore, they are not valid counterexamples to semantic presuppositions.

The crucial issue here is how to interpret the negation 'not-A' in our definition of semantic presupposition. As we have mentioned earlier, there are a variety of readings of the negation 'not-A' in our natural language. For instance, 'not-A' may be read as the external negation or the contradictory of A that in turn includes an unconditional negation or a conditional negation; 'not-A' may be read as the internal negation of A that may again be read either as the contrary of A or as the subcontrary of A. It seems to be clear that, for Böer and Lycan, 'not-A' here is read as the external negation of A or the contradictory of A. If so, this argument shares the same assumption with the argument from the distinction between external and internal negation: not-A has to be the contradictory of A. However, if not-A is the contradictory or the external negation of A, then not-A would not imply B.

Therefore F(B) & T(not-A) would not involve self-contradiction.

However, as I have argued, 'not-A' in the definition of semantic presupposition should be read as the subcontrary of A or the contradictory of A if A's subcontrary is not available. Let us examine what the genuine instances of (c) are in our reading of 'not-A', starting with (28b). The cleft sentence, 'It was John who caught the thief', can be read as 'Someone who caught the thief was John'. The subcontrary of the sentence is, 'Someone who caught the thief was not John'. Then the instance of (c) with respect to the sentence, 'It was John who caught the thief', would be,

> (28b′) Someone who caught the thief was not John because no one caught the thief.

(28b′) involves self-contradiction since the first conjunct formally implies the falsity of the second conjunct. In contrast, (28b) is not the genuine instance of (c) because the first conjunct of (28b) is the external negation of the original sentence, 'It was John who caught the thief', instead of the internal negation of that sentence as it should be. The first conjunct, 'It is false that it was John who caught the thief', does not imply the falsity of the second conjunct. This is the reason why (28b) does not involve self-contradiction.

Turn now to the singular sentence (28a). If we understand 'not-A' as the internal negation of A (more precisely, the subcontrary of A), then the genuine instance of (c) with respect to the sentence, 'The present king of France is bald', is,

> (28a′) The present king of France is not bald because there is not any present king of France.

Since the first conjunct of (28a′) formally implies the falsity of the second conjunct, (28a′) is a self-contradictory sentence. On the other hand, as I have argued before, there is no real distinction between external and internal negation with respect to singular sentences like (28a). The external reading of (28a), that is, 'It is false that the present king of France is bald', still formally implies the falsity of the second conjunct. Hence, (28a) is a self-contradictory sentence. Either way, (28a) does not constitute a valid counterexample of semantic presuppositions. A similar analysis can be applied to other alleged counterexamples given by the critics.

The conclusion drawn from the above analyses is plain: The set of sentences (28) does not provide valid counterexamples to semantic presuppositions. No refutation of semantic presupposition is established.

6. Is the Notion of Truth-valuelessness Untenable?

The notions of truth-valuelessness and semantic presupposition are twin concepts. The former is engendered by the latter. Explaining the occurrence of truth-valueless (fact-stating) sentences is an essential utility of semantic presuppositions. On the other hand, truth-valuelessness is an inevitable notion in any semantic theory of semantic presupposition within trivalent semantics. However, for many philosophers, the notion of truth-valuelessness is highly suspect since it is not in line with common wisdom in semantics: it contradicts either the accepted notion of falsity or the received principle of bivalence.

Truth-valuelessness and Falsity

Let us first consider a rather simple-minded argument against the notion of truth-valuelessness: By falsity we mean non-truth (lack of truth). To say that a sentence is neither true nor false is to say that it is true and untrue, which is a self-contradiction. Therefore, the notion of truth-valuelessness is not a coherent notion. Although the argument is overly simplistic as it is, many really do take it seriously. In defense of his minimal theory of truth, P. Horwich (1990, p. 80) gives us a sophisticated version along the same line of reasoning. According to Horwich, the simplest deflationary strategy is to define falsity as the absence of truth as stated in (MT1):

(MT1)   A proposition that P is false if and only if P is not true; or in symbols, $F(P) \leftrightarrow \neg T(P)$.

Given any logic that licenses the principle of contraposition, from (MT1) we have

(MT2) P is not false $\leftrightarrow P$ is not not true; namely, $\neg F(P) \leftrightarrow \neg \neg T(P)$.

Therefore, if P is neither true nor false, then

(MT3) P is not true and not false $\rightarrow P$ is not true & P is not not true; that is, $(\neg T(P) \& \neg F(P)) \rightarrow (\neg T(P) \& \neg \neg T(P))$.

Thus we cannot claim of a proposition that has no truth-value; for to do so would imply a contradiction.

It is obvious that the above argument begs the question since it already assumes the notion of falsity in classical bivalent semantics (the two-valued falsity). If we define the notion of falsity in three-valued semantics (the three-valued falsity),

(MT4)   A sentence S is false if and only if the unconditional negation of S is true; or in symbols, $F(S) \leftrightarrow T(\sim S)$,

then, when S is neither true nor false, we have:

(MT5) $(\sim T(S) \& \sim F(S)) \rightarrow (\sim T(S) \& \sim T(\sim S))$.

There is no contradiction involved at all.

Of course, Horwich realizes clearly that the contradiction in (MT3) does not derive solely from his minimal theory of truth, but depends also on the notion of falsity and negation in bivalent semantics. The reasons why he endorses such a notion of falsity are, first, that it conforms to ordinary usage; and second, that there can be no theoretical reasons, from his theory of minimal truth, to prefer a three-way distinction of truth-values. Although I disagree with his justification of the endorsement of the notion of falsity in bivalent semantics, I share with Horwich the same insight that we need an independent justification for the endorsement of a specific notion of falsity. We cannot in general suppose that we have given a proper account of a concept of falsity by simply describing the usage of the word 'falsity', or by describing those circumstances in which we do, and those in which we do not, make use of the word. We must give an account of the point of the concept 'falsity' as we use it. We need to explain what functions we use the word for and what philosophical consequences we can draw from this usage. Otherwise, the debate between Russell and Strawson, as Russell contends, is nothing but a verbal dispute. I will argue that it is far from clear why falsity should be defined as non-truth. In fact, there are a number of convincing or at least equally weighty conceptual reasons for dividing non-truth into two subcategories—falsity and neither-truth-nor-falsity—as there are for bivalence.

As I have mentioned earlier, it is very natural to distinguish between two kinds of non-truths of a singular subject-predicate sentence: The sentence is untrue because either the subject has a denotation but the predicate does not apply to it or the subject lacks a denotation. Even the critics of semantic presupposition admit that there is an intuitive difference between two sorts of non-truths. For example, Böer and Lycan admit that respondents hesitate in issuing classical truth-values (truth or falsity) to sentence (16) ('The present king of France is bald'). The common sense of falsity in bivalent semantics does not work under such a situation. To simply say that sentence (16) is false is inappropriate and misleading (since it violates Grice's Maxim of Strength), but to say that sentence (21) ('the current president of China is bald') is false will be appropriate. Although we cannot draw, from the hesitation of respondents in issuing truth-value judgment, a conclusion that the respondents have a natural notion of truth-valuelessness, we can safely say that the respondents have an intuition of the difference between two kinds of non-truths. Such an intuitively recognizable difference in non-truths needs explanation.

I have shown that the opponents of the notion of semantic presupposition are unable to give a satisfactory explanation of two kinds of non-truths without falling into self-contradiction. Trivalent semantics, by classifying non-truths into falsity and neither-truth-nor-falsity, not only makes a sound distinction between two sorts of non-truths, but also is able to locate the real source of such a distinction based on the notion of semantic presupposition. This shows that the debate between Russell and Strawson over the truth-value status of non-denoting sentences is not merely a verbal dispute, but is theoretically productive. To distinguish different non-truths gives us one sufficient theoretical reason to introduce the notion of truth-valuelessness.

One objection to introducing the notion of three-valued falsity is that this notion does not conform to our ordinary usage of falsity in common situations. We have no 'pure intuitions' of truth-valuelessness. It is not an ordinary, commonsensical notion, but a theoretical artifact of linguistic and philosophical semantics. It contradicts our standard ordinary usage.

It is true that the notion of three-valued falsity is not a commonsensical notion. But from this, it does not follow that the notion contradicts our standard ordinary usage or our intuitions of the notion. The reason why we have not gotten used to this notion is that we usually do not have occasions to develop relevant intuitions and to practice the usage of three-valued falsity, not because it contradicts our common sense. We seldom, for example, even encounter sentences like (16), (17), or (18) in our everyday conversation. However, we do have a strong intuition of two kinds of non-truths whenever we encounter a pair of sentences like (17) and (21). We feel something is wrong to simply answer, 'it is false', in response to (16) although we feel secure to give the same answer to (21). This intuition can be used as a starting point to build up a new standard usage of the notion of three-valued falsity. Imagine a linguistic society in which people encounter sentences whose presuppositions do not hold all the time. In order to ensure successful communication, they have to make a distinction between two kinds of non-truths (perhaps the distinction was made initially by some linguistics). After long-time development of the usage, the notion of truth-valuelessness would become a perfectly commonsensical notion in this society eventually.

Furthermore, the assumption that there is a standard ordinary usage of falsity in general (falsity as lack of truth) is very doubtful. All cases we have for the alleged standard usage of falsity are easier cases in which the corresponding presuppositions are held, in which there is no need to employ the other subcategory of non-truth, i.e., the notion of neither-truth-nor-falsity. But for other cases in which the corresponding presuppositions fail, we do not have a chance to form a standard usage yet. In other words, there is no standard way of dealing with sentences with false presuppositions. Therefore, it is unfair to regard the standard way of dealing with bivalent sentences as the standard usage of falsity in general.

Since there is no standard usage of falsity in general, we have to depend on the theoretical usage of it. Nevertheless, there is no established standard theoretical usage of falsity in general either, although there have been long traditions both for and against treating falsity as lack of truth since Aristotle. I conclude that the notion of three-valued falsity is not in violation of both ordinary and theoretical usage of 'falsity'.

Truth-valuelessness and the Principle of Bivalence

Perhaps the most serious challenge to the notion of truth-valuelessness comes from the tradition of bivalence in semantics. According to the principle of bivalence, each well-formed declarative sentence is either true or false. If so, it is impossible for a sentence to be neither true nor false. In fact, an attempt to save the principle of bivalence from the threat of truth-valuelessness is what really motivates many critics of the notion of truth-valuelessness and semantic presupposition.

I am not intending to give a comprehensive examination and critique of the principle of bivalence here; it is not necessary to do so for our current limited purpose. In the following, I will focus on only one logical argument for the principle of bivalence that is intended to reject the notion of truth-valuelessness.

Many critics of the notion of truth-valuelessness argue that the acceptance of truth-valuelessness amounts to rejecting the principle of bivalence. The rejection of the principle would force us to give up one basic classical logical law—the law of the excluded middle. This is because the principle of bivalence is equivalent or necessarily follows from the law of the excluded middle. But we have no convincing reasons to give up the law of the excluded middle. So we had better stick to the principle of bivalence. For example, one of the arguments for the principle of bivalence presented by P. Horwich (1990, pp. 80-82) runs as follows. The logical law of the excluded middle claims that, if P is a declarative sentence, the proposition that $(P \vee \text{not-}P)$ is necessarily true. That is,

(M) $\vdash P \vee \neg P$ [11]

According to Tarski's Convention T, for each sentence S of an object language L_O, a metalinguistic sentence (T-sentence) in the corresponding metalanguage L_M can be given in the following form:

(T) S is true in L iff P.

Here 'S' stands for a specific sentence of L_O and 'P' stands for a corresponding sentence of L_M. The T-sentence gives us the truth conditions of sentence S in L_O. If L_M includes L_O and 'true' is treated as a sentential operator, then the T-sentence becomes:

(T) It is true that P iff P,   or $T(P)$ iff P.

If 'iff' in (T) is understood as material equivalence, then we have schema T'

(T') $\vdash T(P) \leftrightarrow P$ and $\vdash T(\neg P) \leftrightarrow \neg P$.

By the rule of constructive dilemma, from (M) and (T'), we have

$\vdash T(P) \vee T(\neg P)$.

Since $T(\neg P) = T(\neg T(P)) = F(T(P))$, and by (T'), $F(T(P)) = F(P)$, $T(\neg P) = F(P)$. Then,

(B) $\vdash T(P) \vee F(P)$

can be derived from (M) and (T). (B) reads 'It is true that P or it is false that P', which is presumably the principle of bivalence. This establishes that it is inconsistent to deny the principle of bivalence but retain the law of the excluded middle since the latter entails the former.

The above argument is misleading due to a misinterpretation of T-sentence. As a matter of fact, whether we can derive (B) from (M) in terms of Convention T all depends on how we read T-sentence. Normally, there are at least two different interpretations of T-sentence due to two different readings of 'if and only if'. In one interpretation, 'if and only if' means *material equivalence* as shown in schema T'. In the other, 'if and only if' means *deductive equivalence* and can be read as implication relation (either classical entailment or formal implication). Then the corresponding schema T'' would be

(T'') $T(P) \models P$ (or $T(P) \vdash P$) and $P \models T(P)$ (or $P \vdash T(P)$).

One problem with schema T' is that it is only valid in classical two-valued logic. When we move from two-valued logic to three-valued logic, (T') fails. In fact, when P is neither true nor false, (T') is not valid since the left hand of the equivalence, sentence $T(P)$, is false but the right hand of the equivalence, sentence P, is neither true nor false. Furthermore, schema T' says that P and $T(P)$ have exactly the same truth-value in any situation. This in turn, by the truth-conditional theory of meaning, will be the case if and only if P and $T(P)$ have the same meaning or express the same proposition. But many philosophers have denied that there is an identity of meanings between P and $T(P)$. For instance, 'Tom is tall' and 'It is true that Tom is tall' have different meanings on the grounds that the latter is in some sense metalinguistic while the former is not. This shows that Schema T' is not an appropriate formulation of T-sentence. Besides, there is another more appropriate reading of T-sentence, namely, Schema T''. Schema T'' only says that when the premise P or $T(P)$ is true so is the conclusion $T(P)$ or P; and it says nothing about the truth-value of the conclusion if the premise is not true (false or neither-true-nor-false). For this reason, Tarski's convention T is satisfied even in three-valued logic.

According to schema T'', $P \vdash T(P)$ and $\neg P \vdash T(\neg P)$. From schema T'' and $\vdash$ $(P \lor \neg P)$, we have $\vdash T(P) \lor T(\neg P)$. But within three-valued logic, $T(\neg P) =$ $F(T(P)) \neq F(P)$. Therefore, $\models T(P) \lor F(P)$ cannot be derived from (M). This establishes that the semantic principle of bivalence must be distinguished from the logical law of the excluded middle. It is a tenable position to deny the former while retaining the latter within three-valued semantics.

Notes

1   For some criticisms of the notion, please see M. Bergmann, 1981, S. Böer and W. Lycan, 1976, W. Lycan, 1984, 1987, G. Englebretsen, 1973, R. Kempson, 1975, W. Sellars, 1954, D. Wilson, 1975, J. Orenduff, 1979, G. Gazdar, 1979, J. Atlas, 1989, and many others.

2   The present chapter was directly inspired by J. Martin's 1979, which has helped me a great deal in straightening out my own thought on the topic. Especially in my formal treatment of the notion of semantic presupposition, I borrow many analytic tools from it.

3   From now on, for simplicity, whenever I use T(A), it is implicitly assumed that A is a sentence in a language L. So I will omit the explicit mention of L in $T_L(A)$. The similar treatment applies to ' $\models_L$ ' and ' $\vdash_L$ '.

4   See M. Bergmann, 1981.

5   Let us put the issue of vague predicates aside, and suppose that '. . . is bald' is not a vague predicate. Actually, we can easily avoid a vague predicate by using other predicates, such as, '. . . is female'.

6   The notions of subcontrary and contrary, unlike that of contradictory, are not well-defined truth-functional operators, though we could make them so by splitting '#' (or '*' into '#$_t$' and '#$_f$' (or '*$_t$' and '*$_f$').

7   The requirement of B as a contingent sentence is intended to exclude an extremely trivialized notion of semantic presupposition according to which any logically true sentence B is presupposed by any sentence A.

8   S. Lehmann points this out to me.

9   Treating the internal negation of (25) as attaching a negation to the presupposed conjunct B, namely, ($\sim B$ & C), would make the internal negation of (25) not imply B.

10   'Because' is not a well-defined truth-functional operator (in fact, it is not even a truth-functional operator). Presumably Böer and Lycan use 'because' here as the conjunction 'and'. I assume the reasoning behind this usage is something like this: 'C because D' implies 'C and D'. Hence, if 'C and D' are contradictory (a logically false sentence), so must 'C because D' be. For this reason, I will treat 'because' as 'and' in analyzing their argument.

11   The reason for using conditional negation '¬' instead of unconditional negation '∼' in the formulation of the law of the excluded middle is obvious. In fact, if 'not-' is read as unconditional negation within three-valued logic, the proposition (P ∨ not-P) would not be necessarily true; since when P is neither true nor false the proposition is neither true nor false.

Chapter 9

The Structure of
A Presuppositional Language

Based on the formal treatment of semantic presupposition in the previous chapter, we are ready to formalize two crucial notions that we have introduced informally so far: the notions of truth-value gap and presuppositional language.

1. A Theory of Truth-Value

As we have observed repeatedly in chapters 5, 6, and 7, a communication breakdown between two P-language communities is often signified by the occurrence of a truth-value gap between the two languages. How can we explain semantically the occurrence of such a truth-value gap? Presumably, we need some sound semantic theory on *truth-value conditions*. Although I. Hacking's styles of reasoning and N. Rescher's factual commitments are constructed to provide the explanation, they do not work out a satisfactory semantic theory to back up their ideas. Kuhn seems to try to construct a truth-value condition based on possible facts to explain truth-value gaps. But his theory is apparently piggybacking on a dubitable correspondence theory of truth and the notion of possible world and possible fact. Besides, semantically, the linguistic formation of the notions of fact and possible fact is based on the notion of truth and truth-value status: A possible fact is the state of affairs described by a sentence with a truth-value and a fact is what is described by a true sentence. A potential circularity lurks in the background when Kuhn tries to define truth-value status based on the notion of possible fact.

The notion of a P-language based on the notion of semantic presupposition introduced in chapter 6 provides us with a basic conceptual framework for the explanation of a truth-value gap between two P-languages. If some absolute M-presuppositions underlying one P-language PL_1 (such as the traditional Chinese medical theory) cannot be grasped or are not accepted by the speakers from the other P-language PL_2 (such as contemporary Western medical theory), then many core sentences of PL_1 are truth-valueless when considered within the context of PL_2. Hence, a truth-value gap between PL_1 and PL_2 occurs. Now what we need is to work out a truth-value condition based on the notion of semantic presupposition.

The Usual Theories of Truth and Truth Conditions

The usual theories of truth are supposed to answer the question, 'What is truth?' However, this question is ambiguous and misleading. The question is ambiguous since it can be taken as a question about either the nature of truth ('What is the nature of truth?') or the conditions of truth ('Under what conditions is a sentence true?'). M. Devitt calls the former a constitutive issue of truth, which concerns the semantic dimension (more precisely, the metaphysical dimension) of truth, and the latter is an evidential issue concerning the epistemic dimension of truth (Devitt, 1984).

As far as the metaphysical dimension of truth is concerned, there are two opposite views concerning the nature of truth: inflationary versus deflationary accounts of truth. The two accounts disagree on whether Tarski's Convention T suffices to account for the nature of truth. According to the inflationary account of truth, Tarski's Convention T is incomplete at explaining the entire conceptual and theoretical roles of truth, for it does not reveal the essential nature of truth. So Convention T has to be inflated with other semantic, epistemic, or pragmatic substantial properties, such as the metaphysical property of corresponding with reality, the epistemic property of being verified in ideal situations, the pragmatic property of facilitating successful activity, or even the logical property of belonging to a harmonious system of beliefs. As P. Horwich (1990) points out, there is a common doctrine shared by all the inflationists: The predicate 'is true' seems to attribute a substantial property such as the property of corresponding with reality to assertions. Truth has some hidden structure awaiting our discovery. Such a hidden structure is supposed to be the essential nature of truth. In other words, truth is an ingredient of reality whose underlying essence will, it is hoped, be revealed someday somehow. This essence of truth is by no means revealed by Convention T.

According to the deflationist point of view, such a doctrine about the substantial nature of truth is unjustified, and is actually false. It is caused by a mistaken analogy of the predicate 'is true' to some physical predicate, like 'is magnetic'. But, unlike most other natural properties, 'being true' is not used to attribute any sort of property to assertions. There exists no underlying essence of truth to be discovered. This is the reason why Strawson contends that the question, 'What is truth', is misleading—it sounds like it only concerns the nature of truth and while doing so it smuggles in an essentialistic notion of truth. In contrast, according to the deflationist account of truth, Tarski's Convention T is complete. The entire conceptual and theoretical roles of truth are fully captured by the initial triviality of Convention T. Truth is no more than what is presented in Convention T. There is no hidden naturalistic essence underlying the notion of truth. Of course, this does not mean that, after rejecting the idea that the truth predicate designates a naturalistic property, the predicate is no longer philosophically significant. It only means that the function of truth is no longer metaphysical as the inflationists expect. Truth predicates exist mainly for the sake of a certain logical need since we frequently have occasions to endorse one another's statements, to reaffirm our own statements, and so on. In this sense, truth is still a property of some sort (a sort of logical property). As Kuhn puts it, 'the essential function of the concept of truth is to require choice between acceptance and rejection of a statement or a theory in the face of evidence shared by all' (1991, p. 9).

As far as the epistemic dimension of truth is concerned, a theory of truth first assumes, based on the principle of bivalence, that a fact-stating sentence has a (classical) truth-value, and then attempts to specify which truth-value it has in terms of some specific truth conditions. For example, according to one version of the correspondence theory of truth, a sentence is true if and only if it corresponds to a fact. According to the coherence theory of truth, a belief is true if and only if it belongs to a coherent system of beliefs.

It is easy to see that the two dimensions of the notion of truth are conceptually connected. The truth conditions specified by a theory of truth are in accordance with the nature of truth identified by the theory. For instance, based on the nature of truth identified by the correspondence theory of truth—namely, that truth is a substantial property of corresponding with reality—the truth conditions of a sentence consist in whether the sentence corresponds to reality. If truth is the essential epistemic property of being verified in ideal situations, then a sentence is true if and only if it can be verified in an ideal situation. However, the above way of thinking seems to encourage a misconception[1] that the truth conditions specified by a theory of truth are determined by the very nature of truth identified by the theory. To me, the relation between the nature of truth and the truth conditions is the other way around. It is the truth conditions specified by a theory that indicate the nature of truth to which the theory commits. For example, since the truth conditions of a sentence, according to the correspondence theory, consist in whether the sentence corresponds to reality, the essential property of truth consists in corresponding with reality. In this sense, we can say that the truth conditions are more basic than the nature of truth. Therefore, I shall take the notion of truth as primarily an epistemic (in Devitt's sense) notion concerning the truth conditions.

For this reason, I prefer to read the original question of truth, 'What is truth?' as the question, 'By virtue of what is a sentence true *if it has a truth value?*' The conditional clause of the question leaves room for the notion of truth-value status to figure in. According to this reading, the usual theories of truth—such as the correspondence theory, the coherence theory, or the pragmatic theory, etc.—are theories about the truth conditions. As to whether Convention T reveals all the essential natures of truth (if any), I prefer to leave this issue open considering our limited purpose here.

Truth-Value Conditions

Since a usual theory of truth concerns the truth conditions of a sentence, it can only be used to determine the truth-value of the sentence by assuming that it has a truth-value. But what we need to know in the case of incommensurability under discussion is whether or not a sentence of concern has a truth-value. A usual theory of truth does not help us with this. For this reason, I would like to introduce a theory of truth-value to specify the truth-value conditions.

According to Tarski's semantic theory of truth, a theory of truth for language L is a set of axioms that entail, for any sentence in L, a statement of conditions under which that sentence is true. If we have a definition of the truth predicate 'is true in L' satisfying Tarski's Convention T,

(Con-T) s is true in L iff p

we have a theory of truth for L. When 's' is replaced by a canonical description of a sentence S in an object language L_O and 'p' by a sentence P of a metalanguage L_M, the corresponding T-sentence,

(T) S is true in L iff P,

gives us the truth conditions of sentence S in L_O.

Accordingly, we can define a semantic theory of truth-value as a theory that specifies, for any presupposing sentence in language L, the conditions under which it has a truth-value. I call such conditions *the truth-value conditions*. A satisfactory theory of truth-value should meet the following requirements: (a) Giving truth-value conditions to every presupposing sentence (i.e., the sentences that carry semantic presuppositions) within language L; (b) Not depending on any specific theory of truth so we can effectively separate our theory of truth-value from a theory of truth; (c) Validating our partial understanding of the notion of truth-valuelessness; (d) Giving an account of truth-value conditions that makes no use of unexplained semantic concepts (such as meaning, reference, synonymy, translation, interpretation, non-linguistic fact, state of affairs, etc.) and propositional attitudes (intention, belief, etc.) of the speaker. Consequently, the proposed truth-value conditions should be equally available to both the speaker and the interpreter involved in linguistic communication.

Strawson's notion of semantic presupposition provides us with a basic theoretical framework for such a theory of truth-value. As I have argued in chapter 8, for any presupposing sentence, when considered within language L, the sentence would be truth-valueless if its presupposition failed to be true in L. Take the contrapositive of the same thesis: A presupposing sentence is true or false only when its presupposition is true in L. That means that the truth of a presupposition is *necessary* for truth-or-falsity of the sentence carrying that presupposition. For example, sentence (16) ('The present king of France is bald') presupposes sentence (16a) ('The present king of France exists'). (16) is true or false only when (16a) is true; otherwise (16) is neither true nor false.

A presupposing sentence may have many different presuppositions. For example, sentence (16) has at least three different presuppositions, i.e., (16a), (16b) ('There is a country named France'), and (16c) ('A person can have hair'). If we can identify all the presuppositions of a presupposing sentence, then the truth of the conjunction of these presuppositions will be *sufficient* for truth-or-falsity of the sentence. For example, the conjunction of (16a), (16b), and (16c) (maybe plus other presuppositions), if true, is sufficient for (16) to be true or false. Here is another example. 'The box is red' presupposes 'The box exists' and 'The box is colored'. If these two presuppositions are true, then we know that the sentence, 'The box is red', is either true or false.

I will call the conjunction of all the presuppositions of a presupposing sentence its sufficient presupposition, which can be defined as follows:

> A sentence B is a sufficient presupposition for a sentence A when considered within language L iff $\models_L T_L(B) \rightarrow (T_L(A) \vee F_L(A))$.

It should be clear that if B is a sufficient presupposition of A, then the truth of B is not only sufficient but also necessary for the truth-or-falsity of A.

Let us formalize the above reasoning. Suppose B_i (i = 1, 2, . . . , n), when considered within language L, is a complete list of all the presuppositions of sentence A. Then its sufficient presupposition is $B = (B_1 \& B_2 . . . \& B_n)$. Recall that if A presupposes B_i, we have $\models_L (T_L(A) \vee F_L(A)) \rightarrow T_L(B_i)$. Since B is a sufficient presupposition of A, we have the following formula:

$$\models_L (T_L(A) \vee F_L(A)) \leftrightarrow T_L(B)$$
$$\text{Or} \quad \models_L (T_L(A) \vee F_L(A)) \leftrightarrow T_L(B_1 \& B_2 . . . \& B_n).$$

If any of the presuppositions B_i is untrue, then A will be neither true nor false; if all the presuppositions are true, then A is true or false.

Unlike Tarski's Convention T that is used to determine the truth-value of a sentence in a language, we can use the above formula to determine the truth-value status of a presupposing sentence. I call it 'Convention P'. Putting in a format analogous to Convention T:

> (Con-P) A presupposing sentence A is true or false when considered within language L if and only if A's sufficient presupposition B is true in L. That is, $\models_L (T_L(A) \vee F_L(A)) \leftrightarrow T_L(B)$, or in brief, $A \Rightarrow_s B$.

Here the subscript 's' indicates that B is a sufficient presupposition of A. In other words, A is true or false from the perspective of L if and only if every presupposition of A is true in L.

Here is an example of the application of Convention P. Consider the fate of a Newtonian sentence (29) when viewed from the perspective of Leibniz's language of space L_Z.

> (29) An object in an otherwise empty space could have been located at any number of different spatial points.

(29) presupposes (29a),

> (29a) There exists a self-existing empty spatial continuum so that all physical and geometrical properties of it exist independently of different configurations of physical bodies within it.

(29a) is false from the perspective of L_Z. Therefore, according to Convention P, (29) is neither true nor false when considered within L_Z. Putting the inference into our formal formulation:

$$(29a) \quad \models_L (T_L(29) \text{ v } F_L(29)) \leftrightarrow T_L(29a).$$

But,

$$\models_L \sim T_L(29a).$$

Therefore,

$$\models_L \sim (T_L(29) \text{ v } F_L(29)).$$

Our Convention P entails the following rules of presupposition:

(CP1) If $A \Rightarrow_S B$, then $\sim A \Rightarrow_S B$.

(CP2) If $A \Rightarrow_S B_1, B_1 \Rightarrow_S B_2 \ldots$ and $B_i \Rightarrow_S B_{i+1}$, then $A \Rightarrow_S (B_1 \& B_2 \& \ldots \& B_i \& B_{i+1})$.

(CP3) If $A \Rightarrow B$ and $B \Rightarrow C$, then $A \Rightarrow C$.

'$A \Rightarrow_S B$' means that B a sufficient presupposition of A; '$A \Rightarrow B$' means that B is a necessary and sufficient presupposition of A.

A few points about Convention P should be kept in mind. First, similar to Davidson's notion of truth, the notion of truth-value status denotes a primitive irreducible concept, which can be explained and exemplified in a specific linguistic context, but it cannot be given a general definition. Just as Tarski's Convention T does not define the notion of truth, but defines the truth predicate 'is true in L', what we define in Convention P is not the notion of truth-value status, but rather the truth-value predicate 'is true-or-false in L'.

Second, to avoid extraneous problems about the notion of truth, I use the conditions of *holding-true* in Davidson's sense, instead of the conditions of *being-true*, as the truth conditions in determining the truth-values of the presuppositions of sentences in Convention P. To say that a sentence is held to be true in a language is to say that the language community accepts or believes the sentence to be true based on some available sufficient reasons. In this way, the truth-value of a sentence is attainable for the language community (Davidson, 1984, pp. 152, 173, 224). Truth-values are thereby language dependent and become a language community's shared beliefs. However, our adoption of the conditions of holding-true does not make the notion of truth subjective. Davidson has argued that the fact that the speaker of a language holds a sentence to be true, under observed circumstances, is *prima facie* evidence that the sentence is true under those circumstances.[2] Moreover, truth-values of sentences are relative to a specific language. Therefore, different languages can assign different truth-values to the same sentence. But the generic primitive notion of truth can still remain the same for upholders of different languages, for example, the notion defined by Tarski's Convention T. In addition, it is essential to have attainable or recognizable truth conditions in order for truth-value conditions to be available to both the speaker and the interpreter involved in linguistic communication.

Third, as we can see from Convention P, whether a sentence S has a truth-value when considered within the context of language L depends on the truth-value of S's sufficient presupposition in L. This establishes that the notion of truth-value status is a relational concept in the sense that the truth-value status of a sentence is internal to a specific language within which the sentence is considered. It is the language itself that creates the possibility of truth-or-falsity. A sentence may be a candidate for truth-or-falsity in one language but not in another. For example, sentence (30),

(30)   Element a contains more phlogiston than element b,

has a truth-value when considered within the language of phlogiston theory, but has no truth-value when considered within the language of modern chemistry. This is because (30) presupposes (30a),

(30a) Phlogiston exists.

(30a) is held to be true within phlogiston theory, but untrue within modern chemistry. For this reason, (30) is true or false when considered from the perspective of phlogiston theory, but neither true nor false from the perspective of modern chemistry.

Fourth, to say that a sentence is true (or false) in terms of some given truth conditions presupposes that the sentence has a positive truth-value status in terms of some given truth-value conditions. In this sense, a theory of truth-value (the notion of truth-value status and truth-value conditions) is more fundamental than a theory of truth (the notion of truth and truth conditions). The notion of truth presupposes the notion of truth-value status. Therefore, to say of a sentence that it is true (or false) means that it has two properties. The first is that it satisfies some given truth conditions. The second is that it satisfies some given truth-value conditions. This second property is omitted entirely in the usual theories of truth. Classical concepts of truth have succeeded only at the expense of ignoring truth-value conditions by taking the positive truth-value status for granted. By including truth-value conditions as a precondition of truth, our double-truth-condition treatment of truth (the combination of truth conditions and truth-value conditions) is, I believe, a more rigorous development of a philosophical analysis of the notion of truth in general.

2. The Occurrence of Truth-Value Gaps

A Truth-Value Gap Regarding Individual Sentences

Suppose that a sentence B when considered within L_1 is the sufficient presupposition of a sentence A. Assume further that B is held to be true in L_1:

$$\models_{L_1} T_{L_1}(B) \rightarrow (T_{L_1}(A) \vee F_{L_1}(A)), \text{ and } \models_{L_1} T_{L_1}(B).$$

Then we have $\models_{L_1} T_{L_1}(A) \vee F_{L_1}(A)$. This means that A is true or false when considered within L_1. By contrast, suppose that B is held to be untrue in another competing language L_2:

$$\models_{L_2} \sim T_{L_2}(B) \rightarrow \sim(T_{L_2}(A) \vee F_{L_2}(A)), \text{ and } \models_{L_2} \sim T_{L_2}(B).$$

Then, $\models_{L_2} \sim(T_{L_2}(A) \vee F_{L_2}(A))$. This establishes that A is neither true nor false when considered within L_2. In this case, we say that there is a truth-value gap occurring between the two languages regarding an individual sentence A.

The occurrence of a truth-value gap regarding individual sentences is not just a theoretical hypothesis, but what has actually occurred in the history of science. Take sentence (30) as an example again. (30) presupposes both (30a) and (30b),

(30b) '*a*' and '*b*' are not empty (have proper denotations).

Take L_1 as the language of phlogiston theory, L_2 as the language of modern chemistry. In either language, (30a) and (30b) are presuppositions of (30) and the conjunction of (30a) and (30b) is a sufficient presupposition of (30). We know that (30a) is held to be true in phlogiston theory, but untrue in modern chemistry. For simplicity, let us further suppose that '*a*' and '*b*' do denote real objects. According to Convention P, from

$$\models_{L_1} T_{L_1}(30a \ \& \ 30b) \rightarrow (T_{L_1}(30) \vee F_{L_1}(30)) \text{ and } \models_{L_1} T_{L_1}(30a \ \& \ 30b),$$

we have $\models_{L_1} T_{L_1}(30) \vee F_{L_1}(30)$. Therefore, (30) is true or false for the advocates of phlogiston theory. On the other hand,

$$\models_{L_2} \sim T_{L_2}(30a) \rightarrow \sim(T_{L_2}(30) \vee F_{L_2}(30)) \text{ and } \models_{L_2} \sim T_{L_2}(30a).$$

From these two formulae, we derive that $\models_{L_2} \sim(T_{L_2}(30) \vee F_{L_2}(30))$. It means that (30) is neither true nor false for the advocates of modern chemistry theory. Consequently, there is a truth-value gap occurring between the two languages regarding (30).

Of course, a sentence of a language could be truth-valueless even when it is considered within the context of the same language. For example, (30) could be neither true nor false even when considered within phlogiston theory if '*a*' does not denote ((30b) is untrue). But there is an essential difference between the case in which (30) is truth-valueless because (30b) fails to be true and the case in which (30) is truth-valueless because (30a) fails to be true. The falsity of (30b) can be granted in phlogiston theory without endangering the integrity of the theory. However, the rejection of (30a) will endanger the integrity of phlogiston theory, which is not permissible by the theory.

A Truth-Value Gap between Two P-Languages

The above analysis of the occurrence of truth-value gaps is restricted to individual sentences. However, the more significant cases are not the cases in which only a few isolated sentences are truth-valueless due to the failure of a few separated assumptions presupposed by these sentences, but rather the cases in which a substantial number of sentences of one language, which have truth-values in this language, lack truth-values when considered within the context of another competing language due to the failure of one or more shared presuppositions. In other words, it is possible that the occurrence of truth-valueless sentences in a language due to the failure of semantic presuppositions may spread throughout the whole language if those semantic presuppositions are absolute presuppositions of the language. In fact, such languages are what we call P-languages. Recall that by a P-language we mean a language whose core sentences share one or more *absolute* presuppositions.

Based on Convention P, the possibility that many core sentences of one P-language, when considered within the context of another P-language, might be truth-valueless is perfectly conceivable. Suppose that a set of sentences of one language PL_1, $< S_i >$, i = 1, 2, ... n, presupposes the same assumption B when those sentences are considered within another competing language PL_2. Suppose further that B is held to be false in PL_2:

$$\models_{L2} (T_{L2}(S_i) \vee F_{L2}(S_i)) \rightarrow T_{L2}(B) \quad \text{and} \quad \models_{L2} \sim T_{L2}(B), \quad i = 1, 2, \ldots n.$$

Then according to Convention P, all the sentences in $<Si>$, when considered within the context of PL_2, are truth-valueless. That is, $\models_{L2} \sim (T_{L2}(Si) \vee F_{L2}(Si))$. In this case, we say that a truth-value gap occurs between languages PL_1 and PL_2.

For illustration, let us reconsider the debate between the Newtonian language and the Leibnizian language of space. What causes the truth-value gap between the two P-languages of space? Based on our formal treatment of the two languages, the sentences in class S_N-S_L of the Newtonian language presuppose the existence of absolute space. One can speak of absolute changes of locations and absolute motion. Hence it is factually meaningful to talk about a comparison of the locations of a given body at different times as described in sentences (9) ('The body *b* at time t could have located in a different place') and (10) ('The spatial location of the body b at time t_1 is different from its location at time t_2'). The assumption that there exists a self-existing absolute space is a shared universal assumption underlying most sentences in class S_N-S_L. Therefore, to accept the truth-or-falsity of sentences in S_N-S_L amounts to accepting Newtonian absolute space. Suspending or rejecting this assumption will change the truth-value status of these sentences from having truth-values to having no truth-values. Therefore, the different attitudes toward the truth-value status of sentences in S_N-S_L indicate different attitudes toward the assumption of Newtonian absolute space. This is exactly what happened when switching from the Newtonian language to the Leibnizian language. Due to the denial of the assumption of absolute space, the sentences in S_N-S_L when considered within the context of the Leibnizian language were truth-valueless. Consequently, a truth-value gap occurred between two languages.

3. The Structure of a Presuppositional Language

A P-language is an interpreted language PL consisting of, but not limit to, at least the following two essential components:

(PL1) An uninterpreted language L_u = <Syn, Val> such that Syn is a syntax and Val is a set of logically possible valuations/interpretations that, by specifying a set of *logically* possible contexts, maps the sentences in Syn into *logically* possible truth-values.

(PL2) A set of postulates of truth-value status <PT> relative to L_u = <Syn, Val> such that <PT>, by specifying a set of *conceptually* possible contexts, maps the sentences in Syn into *conceptually* possible truth-values.

Syn is the syntax of PL that contains logical symbols with fixed meanings, descriptive symbols without fixed meanings, formation rules connecting logical symbols with descriptive symbols to form well-formed formulae, and sentences formed in terms of language formation rules. Take the Newtonian language of space L_N as an example. In L_N, logical symbols are quantifiers and truth-functional connectives. The descriptive symbols in L_N include time terms and space terms (constants and variables), the proper names of physical bodies, temporal predicates, equality symbols, and two special function symbols.

Possible Truth-Values

The notion of possible truth-value refers to some possible usage of a sentence to say something true or false in a certain possible context. This is because whether a sentence can be used to make a statement depends upon certain contexts. For example, sentence (16) ('The present king of France is bald'), when uttered by someone in the reign of Louis XV, was used to make a statement since (16) is actually true or false in this context. In contrast, the same sentence (16) might not be used to make a statement in another context. If (16) is uttered by someone today, then the sentence is neither true nor false since its presupposition that (16a) ('The present king of France exists') fails in this context.

Obviously, the notion of 'context' here does heavy duty. Although the notion as used here is a broad one, it is primarily a linguistic notion. For instance, the above utterances of (16) are two different speech acts occurring in two different linguistic contexts. Specifically, we can treat a language as a whole as a context (the context of a language). As we have pointed out earlier, the truth-value status of a sentence is language dependent. A sentence (say, sentence (30)) might be true or false when considered within the context of one language (say, the language of phlogiston theory), but lacks a truth-value within another (say, the language of modern chemistry). This confirms that the truth-value status of a sentence is (linguistically) context dependent. It is possible that a sentence could be neither true nor false when considered within one fixed context, but true or false when considered within another possible context. Especially, the same sentence could have a different truth-value status when considered within different languages.

The thesis of linguistic-context dependence of truth-value status implies that one sentence that is actually neither true nor false in a current context might be true or false in a different possible context. If (16) is uttered today, then it has no truth-value. Nevertheless, it could have a truth-value in some possible context, say, when uttered in the reign of Louis XV. Sentence (30), when considered within the context of modern chemistry, has no truth-value. However, it does not prevent the same sentence from having a potential truth-value when considered within some other possible contexts, say, within the context of phlogiston theory. For this reason, we need to distinguish the *actual* truth-value status of a sentence in a fixed context and its *potential* truth-value status in other possible contexts:

> To say that a sentence has a *possible* truth-value means that it could be used to make a statement in some possible context C because its (sufficient) presupposition *could be* held to be true in C.

Presumably, whether the above definition of possible truth-value status makes sense depends on how to specify a possible context in which a sentence could be used to make a statement. Whether such a context is possible is relative to a particular language. Otherwise, anything is possible. So our question becomes: What counts as a possible context in which a sentence could be used to make a statement relative to a language? To clarify the notion of possible context, we need to make use of a common distinction between logical possibility and conceptual possibility.

Logically possible contexts One effective way to define logically possible contexts in which sentences could be used to make statements is to appeal to the notion of interpretation. A possible context in which a sentence could be used to make a statement relative to a language is a state of affairs that can be described by the language. Whether a state of affairs can be described by the sentences of a particular language depends on the interpretations that the language could assign to the descriptive terms of its sentences. Therefore, the possible contexts relative to a language can be specified by the so-called possible interpretations or valuations that could be given to the descriptive symbols of the language. As long as we can identify a set of possible interpretations relative to a language, we can identify the corresponding set of possible contexts.

> An interpretation of a sentence S is *logically possible* relative to a language containing L_u = <Syn, Val> if the meanings assigned to the descriptive symbols in S are consistent with the intended readings of the logical terms in Syn and S is formed in terms of the formation rules of Syn.

This is exactly the function of Val in our P-language PL. Val does not assign any specific meanings to the descriptive symbols in Syn. Instead, it represents a range of all the logically possible interpretations consistent with the intended senses of the logical symbols and the formation rules in Syn. By assigning these logically possible meanings to the descriptive symbols in Syn and by following the formation rules, the interpreted language describes a set of logically possible contexts in which the sentences could be used to make statements. In this sense, we say that Val is a set of functions that, by specifying a set of logically possible contexts, map the sentences in Syn into logically possible truth-values. Suppose I $\ni$ Val and S $\ni$ Syn. Then, $I(S)$ = a logically possible truth-value of S. Roughly speaking, as long as a sentence in Syn is in a good syntactic and semantic order according to Val, it could have a logically possible truth-value.

Conceptually possible contexts   However, to say that a sentence has a logically possible truth-value for a language does not mean that such a possible truth-value is *conceptually available* for the speaker of the language. This is because to say that a context specified by a logically possible interpretation of a language is logically possible for a language does not mean that the context is conceptually accessible or recognizable to the speaker of the language. For illustration, let us consider an example given by M. Schlick. Schlick (1991) invites us to imagine an opponent who holds that 'within every electron there is a nucleus which is always present, but which produces absolutely no effect outside'. When the sentence,

(31)   Electrons have eternally hidden nuclei,

is considered within the context of modern atomic physics, it is logically possible for there to exist a context in which (31) could be used to say something true or false; since the sentence, not like the expression, 'hidden UFO eternally electrons', is in a good syntactic and semantic order. However, an atomic physicist cannot conceptually identify and comprehend a possible context in which the truth-value status of (31) can be verified (in terms of conceptual verifiability or testability). In other words, the truth-value conditions of (31), although they might be logically possible, are not conceptually accessible to the physicist. A similar analysis applies to sentence (4) ('The association of the yin and rain makes people sleepy'). Western physicians cannot identify and comprehend the pre-modern Chinese mode of reasoning underlying (4). Therefore, they are unable to figure out a possible context in which the sentence could be used to make a statement. In this sense, the truth-value conditions of the sentence, although they are logically possible, are not conceptually accessible to them.

On the other hand, some contexts in which a sentence is put forward to make a statement are both logically and conceptually accessible to the interpreter. Take the Leibnizian language of space as an example. When considered within the Leibnizian language of space L_Z, the sentences in S_N-S_Z, such as sentences (9) and (10) in chapter 5, have logically possible truth-values in a Newtonian space since the meanings of the terms in (9) and (10) are consistent with the intended meanings of the logical symbols, and these terms are connected according to the formation rules in Syn of L_Z. In addition, the truth-value conditions of the sentences in S_N-S_Z are conceptually accessible to a Leibnizian. A Leibnizian is able to identify and comprehend the underlying presupposition of the sentences, i.e., the existence of Newtonian absolute space.

Presumably, to know the conceptually possible contexts in which a sentence could be used to make a statement is to know conceptually its possible truth-value conditions. As long as one can identify and comprehend conceptually the truth-value conditions of a sentence, one knows the possible contexts in which a sentence could be used to make a statement. Thus, an effective way to define (conceptually) possible contexts is to appeal to one's knowledge of the truth-value conditions.

> A context in which a sentence could be used to make a statement is *conceptually possible* for the interpreter of a language if he or she is able to identify and comprehend its truth-value conditions.

According to Convention P, the truth-value status of a presupposing sentence S is determined by its presuppositions. To know those underlying presuppositions of S is to know its truth-value conditions. For a P-language, the most significant presuppositions are shared absolute presuppositions, which we call metaphysical presuppositions (M-presuppositions). Thus, to be able to identify and comprehend M-presuppositions of a P-language is essential for the interpreter to identify and comprehend the truth-value conditions of its core sentences. Thus, we can define conceptually possible contexts for a P-language as follows:

> A context in which a core sentence S of a P-language PL_1 could be used to make a statement is *conceptually possible* to the interpreter of another P-language PL_2 only if he or she is able to *identify and comprehend* (not necessarily accept) the M-presuppositions of PL_1.

It should be clear now that M-presuppositions are what we have specified as the postulates of truth-value status, i.e., <PT> of a P-language PL. <PT> singles out a specific set of conceptually possible contexts from all the logically possible contexts described by Val of L_u = <Syn, Val>. By specifying a set of conceptually possible contexts, <PT> maps the sentences in Syn into conceptually possible truth-values: <PT> (S) = a conceptually possible truth-value of S.

Actual Truth-Values

One might be able to identify and comprehend the truth-value conditions of a sentence, i.e., its underlying presuppositions, without accepting them as true. Therefore, interpreters from one language might accept that a sentence has a conceptually possible truth-value (since they can recognize its underlying presuppositions), but categorically deny that it has an actual truth-value (since the presuppositions, from the perspective of their own language, are false). The Leibnizians can grasp very well the truth-value conditions of Newtonian sentences in S_N-S_Z, namely, the existence of Newtonian absolute space, but categorically deny that it fits reality. Hence, the sentences in S_N-S_Z do not have actual truth-values to them. For the Newtonians, however, the sentences (9) and (10) have actual truth-values, no matter whether they are actually true or not, since their presupposition (namely, the existence of Newtonian absolute space) is held to be true. But for the Leibnizians, (9) and (10) have only conceptually possible truth-values, not actual truth-values since the same underlying presupposition is not held to be true for them. To demonstrate, we can surely grasp the set of presuppositions underlying the sentences in the fairy story *Snow White* without committing ourselves to the reality of the story. Therefore, we need to distinguish possible truth-values from actual truth-values. To say that a sentence has an *actual* truth-value in a fixed context C (usually within a context of a language) means that the sentence is actually used to make a statement (a statement is either true or false) in C because its presuppositions are actually held to be true in C. By definition,

> A sentence has an *actual truth-value* for the interpreter who speaks a language L if the presuppositions underlying the sentence are held to be true from the perspective of L.

4. Some Features of a Presuppositional Language

M-Presuppositions and Assertorial Content of P-Languages

Between the two components of a P-language, a set of postulates of truth-value status <PT>, which are the M-Presuppositions of the language, is the hallmark of the P-language. Two P-languages differ in just this regard, having different M-presuppositions. The conceptual core of a P-language consists in its M-presuppositions, which are contingent factual presumptions about the world perceived by the language community. This is why Rescher calls them factual commitments. To claim M-presuppositions of a P-language to be factual presumptions, even to be factual commitments, assumes that a language can have assertorial content. However, one might challenge this basic premise: How can a language have assertorial content or a point of view? A language is supposed to be mere means of making statements. Statements, and only statements, are supposed to have assertorial contents and hence truth-values, while a language should be assertorially neutral. One language might lack some means (such as some words and/or a category system) to express something that can be expressed by another language. But it does not follow that the users of the language categorically deny anything that their language cannot describe. The absence of 'tiger' in one language does not mean that the user of the language does not admit the existence of tigers. Moreover, the fact that one language has some means (words and category systems) to describe a state of affairs does not mean that the users of the language commit themselves to the existence of the state of affairs described. The presence of 'unicorn' in one language does not imply a commitment to the existence of unicorns.

The above point may be illustrated by the following pair of imaginable languages.[3] Imagine that L_A is the language of a tribe living deep in the Amazon, while L_F is the language of a group of Fiji islanders. L_A but not L_F has terms for the vegetation making up the jungle canopy, terms for monkeys, for snakes, for lizards, for ungulates and carnivores, terms for tree spirits (a superstitious belief), as well as the corresponding category systems for these jungle plants and animals. L_F but not L_A has a word for the ocean, terms for waves and tides, for ocean fishes, for sailboats, fishing nets, for sea dragons, etc. as well as the corresponding category systems for them. The Amazonians can describe a state of affairs of a monkey picking a fruit from a tree for which the Fiji islanders lack expressions to describe. In contrast, a Fiji islander can say 'Utu's fishing boat has a big sail', which the Amazonians cannot say at all. It seems absurd to claim that the Amazonians categorically deny the existence of sea dragons while the Fiji islanders deny the existence of tree spirits. It is not fully justified also to claim that the Amazonians commit themselves to the existence of tree spirits while the Fiji islanders commit themselves to sea dragons. The moral behind this imaginable case has been theorized by I. Scheffler (1967, p. 36) in terms of the distinction between a vocabulary on the one hand and a body of assertions on the other, between categories or classes on the one hand and expectations or hypotheses as to category membership on the other. In Scheffler's words, 'categorization provides the pigeonholes; hypothesis makes assignments to them'. The existence of certain pigeonholes does not compel the user to direct letters to them.

Though the above doctrine of assertorial neutrality of language might be applicable to some natural languages—languages that have nothing to do with scientific theorization—it is surely not applicable to the languages of our concern here, namely, scientific languages, or P-languages in general. Scientific languages behave like theories. When one adopts a scientific language, whether for communication, expressions, or description, one does make some assertorial commitments.

First, a scientific language has assertorial content by making assertorial commitments to the existence of certain theoretical entities. To adopt the language of phlogiston theory, one commits oneself to the existence of phlogiston. To adopt the Newtonian language of space, one commits oneself to the reality of Newtonian absolute space. To speak of the language of traditional Chinese medical theory, one makes a commitment to the reality and function of the yin and yang in the universe.

Second, some category systems adopted by some scientific languages do have assertorial content, for they make predictions and are therefore subject to the test of observation.[4] By putting the earth, as a star, in the center of the universe, the Ptolemaic language makes different predictions and calculations of the movement of planets than the Copernican language. These predictions based on the Ptolemaic taxonomy are subject to empirical tests. In this sense, the Ptolemaic taxonomy has assertorial content by leading to testable observations. More importantly, as I will argue later, the taxonomy of a P-language actually functions as a set of sortal presuppositions of some substantial sentences of the language. The truth-values of these sortal presuppositions determine the truth-value status of these sentences. Because of this, we can even say, to some extent, that some category systems do qualify as being either true or false.

Third, a language can have assertorial content in the sense that it places restrictions on the possible events that can be expressed. By this, I do not mean the platitude that some possible states of affairs can be described by one language, but cannot be described by another because the latter lacks words to do so. The second language can be used to express the possible state of affairs in question if we simply enrich it by adding necessary words to it. Recall that the Amazonians cannot express an event of Utu's fishing boat having a big sail. But that does not mean that the Amazonian's language L_A excludes the possible enrichment of the language such that the enriched language can be used to describe the state of affairs in question. What we should do is simply to add a set of terms for fishing boat, sail, ocean, etc. However, in many cases in which two languages involved are inconsistent[5] with one another, some states of affairs that are describable by one language are not describable in principle by the other language or its enrichment. As Kuhn (1993a, pp. 330-31) has argued, with the Newtonian language of mechanics in place it does not make sense to speak of Aristotelian assertions in which terms like 'force' and 'void' play an essential role. This is because there is no way, even in an enriched Newtonian vocabulary, to convey the Aristotelian assertions regularly misconstrued as asserting the proportionality of force and motion or the impossibility of a void. Using the Newtonian lexicon, these Aristotelian assertions cannot be expressed. Therefore, some events or states of affairs described by the Aristotelian language are unable to be described by the Newtonian language. In this sense, the Aristotelian language places some restrictions on the possible events that can be described (recall Kuhn's notion of possible worlds as lexicon-dependent).

In conclusion, it is no longer a novel idea that scientific languages and concepts are themselves laden with a variety of assertorial commitments about the world around us. They are no longer seen as a neutral vehicle for making substantive factual commitments but themselves are loaded with such commitments. A scientific language can and does have assertorial content.

Core Sentences

The M-presuppositions of a P-language, functioning as semantic presuppositions, underlie many sentences of the language. But not all these sentences presuppose M-presuppositions directly, although some do. For example, sentence (30) ('Element *a* contains more phlogiston than element *b*') directly presupposes (30a) ('Phlogiston exists'), which is a M-presupposition of the language of phlogiston theory. In contrast, sentence (32),

(32)   Element *a*, when burning, releases more heat than element *b*,

when considered within the language of phlogiston theory, does not directly presuppose (30a). Instead, (32) when considered within the language of phlogiston theory directly presupposes (30). According to rule CP3 of Convention P, (32) indirectly presupposes (30a) within the language of phlogiston theory. Since the sentences that presuppose directly the M-presuppositions of a P-language are more close to the theoretical core of the language, I call them the core sentences of the P-language. By definition,

A sentence *S* of a P-language PL is a core sentence of PL if *S* directly presupposes some M-presuppositions of the language.

The Semantic Status of Cross-Language Sentences

We have argued that truth-value status is language dependent. It is likely for one sentence *S* of one language L_1, which is true or false within the context of L_1, to lack a truth-value when considered within the context of the other competing language L_2. But this way of speaking of a cross-language sentence and its truth-value status seems to cause some confusion: (a) Since *S* is not a sentence of L_2, how can the interpreter from L_2, who is not supposed to understand L_1, consider its truth-value status? What is the semantic status of *S* anyway? Is *S* the original sentence in L_1, the translation of the original sentence in L_1 into a corresponding sentence *S'* in L_2, or the content or meaning of the original sentence?[6] (b) Presumably, according to common wisdom, the semantic rules of a language, whatever they are, should be able to determine whether a sentence of the language has a truth-value. How, then, can the truth-value status of *S* in one language be determined by some facts or semantic rules about another language that is entirely external?[7]

To clarify this confusion, I have to clarify the notion of sentence and the meaning of 'sentence' that I have in mind. The term 'sentence' can be used either loosely and (one might argue) uncritically or strictly and critically in both philosophical discussion and everyday discourse. Strictly speaking, sentences are syntactic objects or well-formed linguistic symbols existing in a particular language. Sentences in this strict sense can be called *uninterpreted* sentences, such as {Snow is white}[8] in which the term 'snow' does not have a fixed reference and the predicate 'is white' does not have a fixed extension. More precisely, 'sentences' here mean what we usually called sentence-types, syntactic forms that are exemplified by sentence-tokens (particular utterances, individual sounds and marks located in particular region of space and time). In contrast, loosely speaking, 'a sentence' can mean an *interpreted* sentence formed by assigning specific semantic values to the constituents of a corresponding uninterpreted sentence. For example, 'Snow is white'[9] is an interpreted sentence if the term 'snow' and the predicate 'is white' have fixed meanings as we use them in English. Furthermore, an interpreted sentence is either asserted or not asserted depending on the context in which it is considered. For example, the sentence 'The present king of France is bald' was asserted when it was uttered in the seventeenth century, but is not asserted when it is uttered today. The *content* asserted by an interpreted sentence is usually called an assertion, a statement, or a proposition.[10] Different interpreted sentences, such as 'John is loved by Jenny' and 'Jenny is falling in love with John', can be used to make the same statement that Jenny loves John; and the same interpreted sentence, such as 'I am in love' uttered by Jenny and the same sentence uttered by John can be used to make different statements in different contexts.

Uninterpreted sentences have neither truth-values nor truth-value status. They are pure linguistic entities. But the notion of truth is a semantic notion that links language to the world. As far as the bearer of truth-value status is concerned, interpreted sentences have truth-value status, being either true-or-false or neither-true-nor-false. As far as the bearer of truth-values (in trivalent semantics) is concerned, interpreted sentences are true, false, or neither-true-nor-false. Statements, as asserted (interpreted) sentences, are always either true or false. In this sense, a statement has only one truth-value status, i.e., being true-or-false. So far, I have been using 'sentence' as interpreted sentence. I will continue to do so unless we specify otherwise.

Based on the above distinction, there is no confusion when we are only dealing with a single sentence within a single language. When we say that a sentence S of a language L is true or false, we actually mean that S is *asserted* (thereby used to make a statement) within L. When we say that S is neither true nor false, we actually mean that S is *not asserted* (thereby does not have a cognitive content) within L. But the troubles seem to emerge when we are dealing with cross-language sentences: If S refers to a sentence in L_1, then how can interpreters who do not understand L_1 makes a judgment on the truth-value status of S (would S be trivially truth-valueless since it is just a noise for them)? If S refers to the translation of S into a corresponding sentence S' in interpreters' own language L_2, then how can such translation get off the ground since they do not understand L_1? Last, if S refers to the statement made by the sentence, then how can a statement lack a truth-value?

These questions arise from confusion between scientific languages or P-languages and natural languages. To see this, let us consider two different cases of cross-language understanding based on the distinction between scientific language and natural language. In one case, suppose that two scientific languages $L(T_1)$ and $L(T_2)$ (say, the Newtonian language and the Einsteinian language) are coded in the same natural language L (say, English). In this case, to say that a sentence S of $L(T_1)$ (which is also a sentence of L), which has a truth-value within the context of $L(T_1)$, has no truth-value when considered within the context of $L(T_2)$ is to say that the same sentence S, which is asserted or used to make a statement within the context of $L(T_1)$, cannot be asserted or used to make a statement within the context of $L(T_2)$. In the other case, suppose that two scientific languages PL_1 and PL_2 (say, the language of contemporary Western medical theory and that of traditional Chinese medical theory) are coded in two different natural languages L_1 and L_2, respectively (say, the former in English and the latter in Chinese). Then when we claim that a sentence S in PL_1, which has a truth-value in PL_1, is neither true nor false, we actually mean that the translation of S of L_1 into a corresponding sentence S' in L_2 (not the translation into PL_2) cannot be asserted or used to make a statement within the context of PL_2.

These formulations clear up the confusions as long as we realize the following two points: (a) The interpreter who does not understand the language of a scientific theory T, namely, $L(T)$, can know very well the natural language L in which T is coded. (b) The translation we need in the second case is between two natural languages (L_1 and L_2) used to code two scientific languages ($L(T_1)$ and $L(T_2)$), not the translation between the two scientific languages. For interpreters from one natural language L_1 to translate another natural language L_2 into L_1 might require them to understand L_2. Nevertheless, such a translation does not require understanding of the scientific theory $L(T_2)$ coded in L_2.

At last, let us turn to the problem of truth-value status of cross-language sentences. It is true that whether or not a sentence S of a scientific language $L(T)$ has a truth-value, when considered within the context of the same language $L(T)$, is determined by the semantic rules of $L(T)$. But when the same sentence S (or the translation of S into the natural language in which another scientific language $L(T')$ is coded) is considered within the context of $L(T')$, due to the different semantic rules involved in $L(T')$, S will have a different truth-value status. We are not saying that the truth-value status of a sentence S of $L(T)$ is determined by another language $L(T')$. What we are claiming is that when S is considered within the context of $L(T')$, S cannot be used to make a statement in $L(T')$. To claim that the truth-value status of sentences is language dependent is to claim that whether sentences can be *used to make statements* or to assert is language dependent.

Notes

1   M. Devitt (1984) seems to commit himself to this misconception when he emphasizes that the semantic dimension of truth should be distinguished from the epistemic dimension and that the former is more fundamental than the latter.

2   Davidson, 1984, pp. 152, 168-9, 200-201.

3   Austen Clark points out this imagined case to me.

4   This point is made in detail by H. Hung, 1981a and 1981b.

5   Inconsistency here means conflict in assertorial contents.

6   Anne Hiskes raises this concern.

7   An issue raised by Austen Clark.

8   I use { ... } to mark uninterpreted sentences.

9   I use single quotations ' ... ' to mark interpreted sentences.

10 For my limited purpose, I will not make a further distinction between statement, assertion, and proposition here. There are two possible ways to make a distinction between proposition and statement/assertion, either by means of the theory of speech acts or in terms of the theory of intentionality. According to J. Searle's theory of speech acts, a proposition is the common content expressed by a few closely related (interpreted) sentences. For example, (interpreted) sentences (a), (b), and (c),

> (a) John will love Jenny.
> (b) Jenny will be loved by John.
> (c) John will fall in love with Jenny.

express the same proposition that John will love Jenny. A proposition does not have any special force attached to it, while a statement is a speech act that is a proposition with an assertorial force attached to it. For instance, the following three speech acts,

> (d) I state that John will love Jenny.
> (e) I question whether John will love Jenny.
> (f) I promise that John will love Jenny.

express the same propositional content that John will love Jenny. Only (d) is a statement. In other words, a statement/assertion is a proposition the truth of which the speaker has committed him or herself to based on reasonable evidence.

Chapter 10

Existential Presumptions
and Universal Principles

The hallmark of a P-language is its M-presuppositions, the absolute presuppositions underlying core sentences of the language, which determine the (conceptually or actually possible) truth-value status of its sentences. As contingent factual presumptions about the world perceived by the language community, M-presuppositions of a P-language can manifest themselves in many different ways. It is almost impossible to give a complete list of all forms of M-presuppositions since they may vary with different P-languages. Nevertheless, we can still specify three exemplars of M-presuppositions based on the three corresponding forms of semantic presuppositions identified earlier:

(i) Existential presumptions—existential presuppositions;
(ii) Universal principles—state-of-affairs presuppositions;
(iii) Categorical frameworks—sortal presuppositions.

We will discuss existential presumptions and universal principles here, and categorical frameworks in chapter 11.

1. Existential Presumptions

An existential sentence, such as (16) ('The present king of France is bald'), presupposes the existence of the denotation of the subject of the sentence. In a similar way, many sentences of a scientific language presuppose the existence of the theoretical entities postulated by the corresponding theory. For example, the existence of phlogiston—as an existential presumption about the existential state of the world around the language community of phlogiston theory—is presupposed by the theory. Consider how sentence (30) ('Element a contains more phlogiston than element b') and its negation (~30) ('It is not the case that element a contains more phlogiston than element b') presuppose (30a) ('Phlogiston exists'). (30a) is the core ontological commitment of phlogiston theory. Suspension or denial of (30a) would render both (30) and (~30) truth-valueless. Similarly, the existence of the yin and the yang as well as the Five Elements, as existential presumptions of traditional Chinese medical theory, underlies core sentences of the language, such as sentence

(4) ('The association of the yin and rain makes people sleepy'). Rejection of the yin and the yang would undermine Chinese medical theory and render those sentences presupposing them truth-valueless. We will call those ontological commitments to the existence of some theoretical entities by a scientific language its existential presumptions. The existential presumptions of a scientific language function as shared existential presuppositions underlying numerous core sentences of the language.

2. Universal Principles

The term 'universal principles' is used by Feyerabend in his mature explication of the concept of incommensurability. Initially, Feyerabend calls universal principles 'fundamental rules or laws'. By 'fundamental rules or laws', Feyerabend means some very basic assumptions underlying a language. Feyerabend assumes that

> The rules (assumptions, postulates) constituting a language (a 'theory' in our terminology) form a *hierarchy* in the sense that some rules *presuppose* others without being presupposed by them. A rule R' will be regarded as being more fundamental than another rule R'', if it (is) presupposed by more rules of the theory, R'' included, each of them being at least as fundamental as the rules presupposing R''. It is clear that a change of fundamental rules will entail a major change of the theory, or of the language in which they occur. Thus a change in the spatiotemporal ideas of Newton's celestial mechanics makes it necessary to redefine almost every term, and to reformulate every law of the theory, whereas a change of the law of gravitation leaves the concepts, and all the remaining laws, unaltered. The former ideas are more fundamental than the law of gravitation. (Feyerabend, 1965a, p. 114; the second italics is mine)

Similarly, the notion of impetus depends upon the Aristotelian principle that all motion is the result of the continuous action of some force, which constitutes the fundamental universal principle of the impetus theory.

Although Feyerabend speaks of incommensurability in many domains of the sciences and illustrates his thesis of incommensurability by many celebrated case studies from the history of science, his prize example, in my reading, is one from a domain outside science: his case study on the shift from the 'archaic style' to 'the classical style' in ancient Greek art (Feyerabend, 1978, pp. 230-48, 260-77). It is in this case study that Feyerabend gives a detailed analysis of the notion of universal principles.

Feyerabend recognizes that every formal feature of a style (a theory or a language) corresponds to (hidden or explicit) assumptions inherent in the underlying cosmology. Therefore, although there is no one-to-one correspondence between a style and its underlying cosmology, it is possible to unearth the cosmology of a style (which is a precise account of the world as it is seen by the participants of the style) by means of analyzing its formal features. The cosmology of a style, for Feyerabend, consists of basic elements (entities, the concept of objects, and the concept of facts) as well as universal principles. In terms of analyzing the formal features of 'the archaic style' and 'the classical style', Feyerabend unearths their underlying cosmologies respectively and proceeds to describe the transition from the universe of the archaic Greeks (cosmology AG hereafter) to the substance-appearance universe of the classical Greeks (cosmology CG hereafter).

The transition from one cosmology to an incommensurable one involves two major changes: the change of elements and the change of the universal principles. Let us, following Feyerabend, examine these two changes in detail.

To begin with, cosmologies AG and CG are built from totally different elements before and after the transition. The transition introduces new entities and new relations between entities, thereby introducing a totally different, new concept of objects. Cosmology AG contains things, events, and their parts; it does not contain any appearances. The entities of cosmology AG are relatively independent parts of objects that enter into external relations without changing their own intrinsic properties. The concept of an object in cosmology AG therefore is none other than the concept of an aggregate of equi-important perceptible parts or entities. The entities and their relations constitute the object; when they are given, then the object is given as well. Thus, complete knowledge of an object is complete enumeration of its parts and peculiarities. On the contrary, cosmology CG that arose in the seventh to the fifth centuries BC distinguishes between reality and appearances. The elements of cosmology CG fall into two classes: essence and appearances (of objects). No enumerations of aspects or parts of an object are identical with the object. Therefore,

> [T]he concept of an object has changed from the concept of an aggregate of equi-important perceptible parts to the concept of an imperceptible essence underlying a multitude of deceptive phenomena. (Feyerabend, 1978, p. 264)

Accordingly, the transition from cosmology AG to cosmology CG involves a very different way of conceptualizing facts by introducing a new concept of fact. After the transition, it is not just that the participants of the new cosmology describe the same state of affairs differently, but rather that they describe altogether *different* states of affairs. Cosmology CG does not contain a single element of cosmology AG, neither a single object nor a single fact of AG. Moreover, there is no way of incorporating an AG-fact into the CG-picture and vice versa. This is because cosmology CG (or cosmology AG) has some structural properties that prevent the co-existence of AG-facts and CG-facts. These structural properties are set up by the underlying universal principles of cosmology CG (or cosmology AG). This brings us to the second major change in the transition from cosmology AG to cosmology CG: the change of the universal principles of each cosmology.

For Feyerabend, 'universal principles' is another name for 'fundamental rules or laws' as introduced earlier in his contextual theory of meaning. According to Feyerabend, the essential function of the universal principles of a cosmology is to set up a set of (conceptually) possible worlds in which some possible facts can be described. Violation of a universal principle of a cosmology (theory or language) results in suspending all the possible facts that presuppose the universal principle. In Feyerabend's own words,

> We have a point of view (theory, framework, cosmos, mode of representation) whose elements (concepts, 'facts', pictures) are built up in accordance with certain principles of construction. The principles involve something like a 'closure': there are things that cannot be said, or 'discovered', without violating these principles (which does not mean contradicting them). Say the things, make the discovery, and the principles are suspended. Now take those constructive principles that underlie every element of the cosmos (of the theory), every fact (every concept). Let us call such principles universal principles of the theory in question. Suspending universal principles means suspending all facts and all concepts. (1978, p. 269)

Now, if the transition from one cosmology (a theory, a language, a framework) to another involves the transition between two incompatible universal principles, then mutual violation of each other's universal principles will mutually suspend each other's presumptive facts.

> Let us call a discovery, or a statement, or an attitude *incommensurable* with the cosmos (the theory, the framework) if it suspends some of its universal principles. (Feyerabend, 1978, p. 269; italics as original)

> It is true that incommensurable frameworks and incommensurable concepts may exhibit many structural similarities—but this does not remove the fact that universal principles of the one framework are suspended by the other. It is *this* fact that establishes incommensurability despite all similarities one might be able to discover. (Feyerabend, 1978, p. 277; italics as original)

This process is what happens during the transition from cosmology AG to cosmology CG.

Feyerabend's analysis of the incommensurable relation between cosmology AG and cosmology CG due to mutual suspension of each other's universal principles applies to many other classical transitions in the history of science, such as those from classical mechanics to quantum mechanics, from the impetus theory to Newton's mechanics, etc. For instance, the quantum theory constitutes facts in accordance with the uncertainty relations,[1] one crucial universal principle of the quantum theory. This principle is suspended by the classical approach.

The transition from Newtonian mechanics to the theory of relativity can be used to illustrate further Feyerabend's point. Numerous sentences of the language of Newtonian mechanics presuppose one universal principle: The properties such as shapes, masses, and periods inhere in objects and change only by direct physical interactions. This universal principle is suspended by the theory of relativity. The relativity theory implies that these inherent properties do not exist (at least in the interpretations of Einstein and Bohr), and that shapes, masses, and periods are relations between physical objects and co-ordinate systems so that they may change, without any physical interaction, when the coordinate system is replaced by another. Furthermore,

> The theory of relativity also provides new principles for constituting mechanical facts. The new conceptual system that arises in this way does not just *deny* the existence of classical states of affairs, it does not even permit us to *formulate statements* expressing such states of affairs. It does not, and cannot, share a single statement with its predecessor—assuming all the times that we do not use the theories as classificatory schemes for the ordering of neutral facts. ... Using classical terms we assume a universal principle that is suspended by relativity which means it is suspended whenever we write down a sentence with the intention to express a relativistic state of affairs. Using classical terms and relativistic terms in the same statement we both use and suspend certain universal principles which is another way of saying that such a statement does not exist: the case of relativity vs. classical mechanics is an example of two incommensurable frameworks. (Feyerabend, 1978, pp. 275-6; italics as original)

It is worth pointing out that Feyerabend's notion of universal principles is closely related to Kuhn's notion of metaphysical commitments of a disciplinary matrix. The metaphysical commitments of a disciplinary matrix provide the theorists with an explicitly or implicitly formulatable ontology, and provide answers to some fundamental questions concerning the existing state of the world around a language community. Two different metaphysical commitments of two competing disciplinary matrices populate the world with different properties and entities with different interactions. To use an example given by Kuhn:

> After 1630, for example, and particularly after the appearance of Descartes' immensely influential scientific writings, most physical scientists assumed that the universe was composed of microscopic corpuscles and that all natural phenomena could be explained in terms of corpuscular shape, size, motion, and interaction. That nest of commitments proved to be both metaphysical and methodological. As metaphysical, it told scientists what sorts of entities the universe did and did not contain: there was only shaped matter in motion. As methodological, it told them what ultimate laws and fundamental explanations must be like.... (Kuhn, 1970a, p. 41)

Feyerabend's universal principles of a scientific language are one kind of M-presupposition of a P-language, an absolute state-of-affairs presupposition. According to Feyerabend, the essential function of the universal principles of a theory or language is to construct a set of (conceptually) possible worlds in which some possible facts can be described. Violation of a universal principle of a theory or language results in suspending all the possible facts that presuppose the universal principle. A possible fact is the state of affairs described by a sentence with a truth-value. Therefore, it is a universal principle of a language that determines whether a sentence has a truth-value, and thereby determines whether a state of affairs described by the sentence is a possible fact. Rejecting the universal principles of a language amounts to denying the states of affairs described by the sentences of the language as possible facts, thereby denying that these sentences have truth-values.

Therefore, it seems to me to be appropriate to reconstruct Feyerabend's concept of incommensurability as follows: A theory T (a language, a conceptual scheme) is incommensurable with another theory T′ (another language, conceptual scheme) if T suspends T′'s crucial universal principles that are presupposed by crucial facts. The conflict between universal principles is an essential nature of incommensurability.

To sum up, universal principles are fundamental factual assumptions underlying a P-language about the existential state of the world around the language community: The postulate of 'absolute space and time' in Newtonian physics; 'Fermat's conjecture' in classical arithmetic; the principle that any motion is due to the action of some kind of force in Aristotle's theory of motion; the 'inertial law' in the impetus theory of motion; the presumption that shapes, masses, and orbital periods are changed only by physical interactions in Newton's mechanics; uncertainty relations in the quantum theory, and so on. These assumptions function as the fundamental shared state-of-affairs presuppositions of the language. They construct a set of conceptually possible contexts in which the sentences of the language could have conceptually possible truth-values. Some universal principles are so essential to a P-language that the denial of them is unacceptable by the users of the language. Failing to identify and comprehend them would deprive the conceptually possible truth-values of many substantial sentences of the language. Using the traditional Chinese medical theory as an example, for Western physicians, many core sentences of the Chinese medical language, such as sentence (4) ('The association of the yin and rain makes people sleepy'), do not describe any conceptually possible state of affairs, and therefore do not possess any conceptually possible truth-values. This is because the underlying universal principles (the yin-yang doctrine and the pre-established harmony between human affairs and nature) presupposed by those sentences are beyond Western physicians' conceptual reach. They cannot identify and comprehend possible contexts in which these sentences have truth-values.

Furthermore, suspending or rejecting the universal principles of a P-language would make its core sentences actually truth-valueless. For instance, the Leibnizians are able to identify a conceptually possible context in which the sentences like (9) ('The body b at time t could have located in a different place') and (10) ('The spatial location of body b at time t_1 is different from its location at time t_2') could have conceptually possible truth-values. This is because the Leibnizians can identify the truth-value conditions of the sentences, that is, the existence of Newtonian absolute space. However, the Leibnizians deny the reality of those possible contexts. Therefore, for the Leibnizians, although (9) and (10) could have conceptually possible truth-values, they are actually truth-valueless.

3. Modes of Reasoning and Hidden Universal Principles

Modes of Reasoning

As we have introduced in chapter 6, I. Hacking presents a historical thesis that there are different styles of scientific reasoning existing within the Western scientific tradition. Recall that we have identified two distinctive features of Hacking's styles of scientific reasoning. First, a style of reasoning is not supposed to be a set of beliefs or assumptions about the nature of the world, but is *the way* beliefs or propositions are *proposed and defended*. Second, a style of reasoning is not a truth-preserving rule of formal logic, like deduction and induction, but the way of *determining what are taken to be legitimate candidates for truth-or-falsity*.

C. Taylor, in his comment on P. Winch's interpretation of Zande magical rites, hits upon a concept of modes of thinking or ways of understanding substantially similar to Hacking's concept of styles of scientific reasoning. In Taylor's interpretation, Winch describes that the Zande were trying to, by virtue of their magic rites, express an attitude toward contingencies in their rites, rather than trying to gain control of these contingencies (Winch, 1979). The problem with Winch's interpretation, Taylor argues, is that it assumes that the Zande were able to make the distinction between expressing and controlling contingencies. According to Taylor's interpretation, however, the Zande were not able to make what seems to be a fundamental distinction for us. Therefore, the right thing to say about the Zande is that they tried to accomplish the function of controlling contingencies also when they seemed only to express them in their rites, since, for the Zande, expressing and controlling are intertwined. Taylor further points out that the absence of the distinction between expressing and controlling can be better understood by grasping a traditional way of understanding the world or a mode of thinking that makes no distinction between understanding the world and coming into attunement with it:

> We don't understand the order of things without understanding our place in it, because we are part of this order. And we cannot understand the order and our place in it without loving it, without seeing its goodness, which is what I want to call being in attunement with it. Not being in attunement with it is a sufficient condition of not understanding it, for anyone who genuinely understands must love it; and not understanding it is incompatible with being in attunement with it, since this presupposes understanding. (Taylor, 1982, pp. 95-6)

Taylor identifies a traditional mode of thinking or way of understanding that is frequently congenial to the human mind in our past: the intertwinedness between understanding and attunement. Taylor believes that this unique mode of thinking appeared in the European cultural tradition before the advent of modern science. Actually, Taylor's observation is applicable to other cultural traditions also. We can find a similar way of understanding embodied in the past of other cultural traditions, including the pre-modern Chinese intellectual tradition, which I will turn to shortly.

By focusing on the relationship between understanding the world and being in attunement with it, Taylor actually explores a unique mode of thinking or way of understanding. It is a distinctive mode of thinking existing in the past of some cultural and intellectual traditions. This mode of reasoning has the following three aspects: (a) The attitudes toward the world that attunement involves: There are different versions of attunements specified by different attitudes toward the world. For instance, 'Taylor's characterization of attunement as involving love of the world seems best suited to Platonic Christianity. The Chinese version of attunement involves acceptance and appreciation rather than anything we could characterize as love' (D. Wong, 1989, p. 148). (b) The understanding of the order of the world and the place of human beings in it: How human beings, as part of this order, become attuned to the world by keeping certain relations to it. (c) The connection between understanding and attunement: Whether or not understanding the world is closely associated with achieving attunement with it. For example, for the Zande, attunement is intertwined with understanding. Expressing the recognition that life is subject to contingencies, as a version of attunement to the world, is intertwined with the Zande way of understanding the world that makes control possible.

There is a common formal feature shared by both Hacking's styles of reasoning and Taylor's modes of thinking. That is, they are not supposed to be *the premises of reasoning* or a set of basic beliefs about the world, but rather are assumed to be *the form of reasoning* or the way of justification. This kind of form of reasoning has two critical features: (a) It is the form of reasoning that determines what facts count as legitimate evidence for justification, in what ways they count, and the weight of evidence in support of propositions; (b) It is the form of reasoning that determines whether a sentence has a truth-value. For example, within the Renaissance tradition, the form of reasoning embodied in the tradition regards states of affairs dealing with the mutual interaction between the human body and celestial bodies and other chemical elements as alleged facts. The very form further determines that the associations between mercury, the planet Mercury, the market-place, and syphilis count as evidence for the claim that mercury salve is good for syphilis. Within this form of reasoning, a Paracelsan sentence like 'Mercury salve is good for syphilis' has a truth-value.

Both Taylor's concept of modes of thinking or ways of understanding and Hacking's concept of styles of reasoning concern a special kind of forms of reasoning from different perspectives. From now on, I will use the concept of *modes of reasoning* to refer to both Taylor's modes of thinking and Hacking's styles of reasoning without further distinction.

From Modes of Reasoning to Hidden Universal Principles

According to Hacking, the style of reasoning is a form of reasoning rather than the premises of reasoning. It is not an usual form of reasoning, but a form of reasoning that determines the truth-value status of the sentence. Roughly put, a form of reasoning is an inference process that can pass some specific semantic values of a set of sentences (such as the truth-values of the premises in an argument) to another related sentence (such as the conclusion in an argument). The mode of reasoning that Hacking has in mind is supposed to be a special kind of form of reasoning that can pass the truth-value status of a set of sentences to the truth-value status of another sentence. If so, the classical bivalent forms of reasoning, such as deduction, induction, or reduction, will not do for this purpose. Since these classical forms of reasoning are truth preserving, they can only be used to pass truths, instead of truth-value status, from a set of sentences to another related sentence. Hacking clearly realizes this and claims that his styles of reasoning are not the forms of reasoning in the traditional sense. But he does not specify at all what alternative kind of forms of reasoning he has in mind except to claim that his forms of reasoning are supposed to determine the truth-value status of sentences.

I fail to see how a form of reasoning, no matter what it is, can determine the truth-value status of sentences. More precisely, how can a mode of reasoning, as a form of reasoning, determine whether a sentence has a truth-value unless it can be linked to certain premises or assumptions?[2] According to our Convention P, the truth-value status of a sentence is determined by the truth-value of its presuppositions. A presupposition of a sentence is a factual assumption about the existential state of the world around a language community. This means that as far as the function of truth-value status determination is concerned, a mode of reasoning should be conceived as the premises of reasoning, instead of as the form of reasoning. In fact, Taylor's mode of thinking seems to be more like a set of basic premises about the world than a form of reasoning. On the other hand, however, a mode of reasoning should not be reduced to a premise of reasoning. According to Hacking, a mode of reasoning is the way beliefs or propositions are proposed and defended, not the content of beliefs. According to Taylor, a mode of reasoning is a specific way of thinking or understanding, not the content of a belief to be understood.

To rationalize the essential function of a mode of reasoning, but at the same time not to simply reduce it to premises or beliefs, I hypothesize that a mode of reasoning is associated with some hidden universal principles. More precisely, we can imagine that every mode of reasoning is embedded in an underlying cosmology. The essential core of any cosmology consists in some (hidden or explicit) general assumptions. Those general assumptions function just as the universal principles do. Historically, the universal principles of a cosmology were some explicit factual beliefs about the existential state of the world observed by a cultural or scientific tradition. A tradition is best characterized by the specific way of understanding the world. It is likely that, at the early development of a tradition, both the view of the order of things and the place of human beings in it on the one hand and the attitude toward the world on the other were initially basic premises or beliefs about the world. It is reasonable to assume that during the historical development of the tradition, those basic premises or beliefs operated *constantly* within the tradition and eventually become treated as *constants* within it to the extent that they become incorporated into the conception of understanding to generate a unique way of understanding the world. As a result, a unique mode of reasoning evolved from the same process.

If the above hypothesis is right, although there may not be one-to-one correspondence between a mode of reasoning and its underlying cosmology, it is possible to unearth the hidden assumptions or universal principles inherent in the underlying cosmology of a mode of reasoning by means of analyzing the formal features of the mode of reasoning. It is these hidden assumptions of the underlying cosmology of a mode of reasoning—functioning as universal principles—that determine the truth-value status of the sentences embedded with the mode of reasoning. Then to say that a mode of reasoning determines the truth-value status of sentences is actually to say that the hidden assumptions associated with the mode of reasoning determine the truth-value status of the sentences. In this way, we can effectively resolve the internal conflict between two merits of a mode of reasoning: as a form of reasoning and as a truth-value condition.

The Paracelsan sentence (15) 'Mercury salve might be good for syphilis because of associations among the metal mercury, Mercury, the market place, and syphilis' sounds so alien to us that it is hard to assign a truth-value to it. One might say that this is because the sentence is embedded in the Renaissance mode of reasoning, which is totally alien to us. But more precisely, the truth-valuelessness of the Paracelsan sentence is because we cannot comprehend the Renaissance medical, alchemical, and astrological doctrines of resemblance and similitude, which had internalized into the Renaissance mode of reasoning. These doctrines, functioning as the universal principles of the Paracelsan language, assign positive truth-value status to Paracelsan sentences. Failing to identify and comprehend them renders the sentences truth-valueless.

To further illustrate and support the above hypothesis, let us follow the historical path to trace the evolution and development of the pre-modern Chinese mode of reasoning from the corresponding pre-modern Chinese cosmology and its hidden assumptions in the next section.

4. The Pre-Modern Chinese Mode of Reasoning

In our case study of traditional Chinese medical theory, I suspected that the reason why, for Western physicians, a great many of the substantial sentences in Chinese medical theory sound nonsensical and cannot be assigned truth-values is not that they cannot understand the words or cannot translate the sentences of the language, but that the mode of reasoning underlying Chinese medical theory is totally alien to them and is difficult to grasp. I hinted from time to time that within the Chinese cultural/linguistic setting Chinese medical theory was bound up with a specific mode of reasoning in such a way as to make conceptual access for people from Western culture difficult.[3]

To unearth the dominant mode of reasoning of the pre-modern Chinese intellectual tradition, the Yin-Yang doctrine is the best place to start. In Chinese philosophy, the Yin-Yang doctrine sounds very simple but its influence has been extensive. It is not an exaggeration to say that no aspect of Chinese civilization, whether metaphysics, natural sciences, social and political theories, or art, has escaped its influence. In simple terms, the doctrine claims that all things and events in the universe are produced and controlled by two forces or principles, namely, the yin and the yang. The yin, which represents the negative, passive, weak, and destructive side of the universe, is associated with softness, cold, cloud, rain, winter, femaleness, and that which is inside and dark. The yang, which represents the positive, active, strong, and constructive side of the universe, is associated with hardness, heat, sunshine, spring and summer, maleness, and that which is outside and bright. The Yin-Yang doctrine is associated with the doctrine of Five Agents or Elements (wu-hsing): Metal, Wood, Water, Fire, and Earth. They are not so much five sorts of matter but five sorts of processes. The Five Agents are not only an elaboration of the Yin-Yang idea, but actually add the important concept of rotation, i.e., things succeed one another as the Five Agents take their turns.

The concepts of the yin and the yang and Five Agents go far back to the Ch'un Ch'iu period (the Spring and Autumn period, 722 – 4481 BC) when the concepts were used in magic and divination. In their original forms, the concepts were employed to lay stress on the mutual influence supposed to exist between 'the Way of Heaven' (T'ien, both the Heaven of Will and the Heaven of Nature) and human affairs. During the Warring States period (403 – 222 BC), the ancient Yin-Yang School carried on the ancient Chinese tradition in its religious and scientific aspects, further elaborating these religious ideas, developing and transforming them into the first unified system of cosmology in Chinese civilization—a positive systematic explanation of the way of the operation of the world (Y. Fung, 1947, p. 116). The universe was conceived of as a well-coordinated system and a process of self-transformation in which everything is related to everything else.

The Yin-Yang School and its cosmology exerted a profound influence upon the succeeding development of the Chinese intellectual tradition. Turning to the Han dynasty (206 BC – AD 220), the Yin-Yang doctrine came to be almost completely amalgamated with Confucianism, which is called by historians Yin-Yang Confucianism, and became the mainstream of the pre-modern Chinese intellectual tradition (Fung, 1952 and 1953, pp. 7-87). Tung Chung-shu, one of the greatest Confucians, is the representative of Yin-Yang Confucianism. He was chiefly instrumental in making Confucianism the state ideology dominating China until 1905. In Tung Chung-shu's hands, the two basic ideas of the ancient cosmology took a step forward. First, the universe is treated as an organic whole that is composed of ten parts: Heaven, Earth, the yin, the yang, wood, fire, soil, metal, water, and man.

> According to this way of thinking, the universe is an organic structure, and the controlling power in the structure is Heaven. Heaven and earth are the boundary wall, whilst the Yin and Yang and the Five Forces are the framework of the structure. In terms of space, wood belongs to the east, fire to the south, metal to the west and water to the north, whilst soil

occupies the central position. These five forces are very like pillars supporting the universe. In terms of time, four of the five forces control the four seasons, and each is the ch'i of one season, wood being that of spring, fire that of summer, metal that of autumn, and water that of winter. Soil has nothing particular which it controls, but it is the central authority of the four seasons. (Fung, 1947, p. 120)

Within this organic structure, everything undergoes constant transformation. The final cause of these transformations is the yin and the yang. The yin and the yang are of opposing natures. The yang is 'the blessing of Heaven' which is advantageous to birth and growth, while the yin is 'the punishing of Heaven' which is disadvantageous to birth and growth. Hence, if one flourishes, then the other declines. For example, the flourishing and the decline of the yin and the yang cause the changes in the seasons. When the yang flourishes, it helps wood to be dominant. When this happens, it is spring, and all things give birth (Fung, 1947, pp. 118-21).

In addition, according to Tung Chung-shu, not only are things related generally, but they also activate each other. Especially, things of the same kind energize one another. This mutual correlation among things presupposes a pre-established harmony. Actually, the idea that all forces and things in the universe are harmonized has become a typical Chinese traditional conception. The pre-established harmony manifests itself best in the correspondence or mutual influence between human affairs and Nature or in the unity of man and Nature. It is not just because there is a common law governing both man and Nature, but mainly because 'Nature and man form one body' and the same material forces of the yin and the yang create and control both of them (W. Chan, 1963, pp. 246, 271). Among the myriad things in the universe, man is the being imbued with the highest spiritual quality and stands the highest in the scale of values. Man's superiority to ordinary things lies in his intimate association with Heaven and Earth. Actually, man and Heaven belong to the same kind of things. 'Heaven also has a ch'i of pleasure and a ch'i of anger, a heart of sorrow and a heart of joy, just as happens in men' (Fung, 1947, p. 121). On the other hand, man's bodily structure is an edition of Heaven. 'There is a tallying of Heaven and Earth and a reproduction of the Yin and the Yang permanently established in the human body'. More specifically, 'In the body there are 366 small components parts, making the sum total of the days in the year, and twelve major parts, making the sum total of the months in the year. Within, there are five viscera, making the sum total of the Five Forces. Without, there are four limbs, making the sum total of the four seasons' (Fung, 1947, p. 122). Thus, to Tung Chung-shu, man is the universe in miniature: man is the microcosm, Nature the macrocosm, and Heaven is a universe-man.

Man being thus, it follows that man stands along with Heaven and Earth, and the three of them make a perfect trinity. Man's task is to be a complement to Heaven and Earth. This establishes the unique position of man in the universe. The relation between man and the universe being thus, it is natural to expect certain correlations in the yin or the yang operations of Heaven with events in human affairs. 'When Heaven is about to make the yin rain fail, for example, people feel sleepy. The theory is that when the yin force in Heaven and Earth begins to dominate, the yin force in people responds by taking the lead' (D. Wong, 1989, p. 149).

Now, we can identify four essential universal principles[4] that constitute the core of the pre-modern Chinese yin-yang cosmology:

(CM1) The Yin-Yang doctrine: All things and events in the universe are produced and controlled by the yin and the yang.

(CM2) The Five-Element doctrine: Everything is made of five different elements, namely, Metal, Wood, Water, Fire, and Earth.

(CM3) The principle of pre-established harmony: There exist mutual correlations among things in general and the correspondence or mutual influence between human affairs and Nature in particular.

(CM4) The doctrine of constant transformation: Within the organic structure of the universe, everything undergoes constant transformation. These four universal principles are factual beliefs about the existential state of the universe perceived by the Chinese in the pre-modern Chinese intellectual tradition.

In the same historical process in which the pre-modern Chinese yin-yang cosmology rose and developed, the pre-modern Chinese mode of thinking evolved along the way. This mode of thinking can be best characterized by the specific way of understanding the world perceived within the pre-modern Chinese tradition. Shaped by the underlying cosmology, this mode of thinking involving both understanding and attunement has the following three formal features: the understanding of (a) the order of the world, and (b) the place of human beings in it, as well as (c) the attitude toward the world. According to this mode of thinking, in brief, the world is composed of and operated by the yin-yang and the Five Forces. Human beings, holding the most exalted position in it, become attuned to the world to such an extent that they became a shadow in brief of the universe, and are melted into it. Because of such a close relationship with the universe, the ancient Chinese did not treat the surrounding world as the object outside that human beings could love, rebel against, or control. They appreciate the blessing (associated with the yang part of the universe) and accept punishing (associated with the yin part of the universe) from Heaven and Earth. This attitude toward the world and the close correlation between the universe and human affairs shows that for the ancient Chinese, understanding the universe was intertwined with achieving attunement with it. Understanding and attunement are related so closely that attunement eventually became a necessary condition of understanding. Actually, the intertwining of attunement and understanding has been a persistent theme presented in almost all schools of pre-modern Chinese philosophies and recurring throughout the Chinese intellectual tradition (Wong, 1989, pp. 150-51).

Sinologists call the mode of thinking illustrated above 'associated thinking'. J. Needham describes this pre-modern Chinese mode of thinking in the following way:

The symbolic correlations or correspondences all formed part of one colossal pattern. Things behaved in particular ways not necessarily because of prior actions or impulsions of other things, but because their position in the ever-moving cyclical universe was such that they were endowed with intrinsic natures which made that behavior inevitable for them. If they do not behave in those particular ways they would lose their relational positions in the whole (which made them what they were) and turn into something other than themselves. They were thus parts in existential dependence upon the whole world-organism. And they reacted upon one another not so much by mechanical impulsion or causation as by a kind of mysterious resonance. (Needham, 1956, p. 281)

We can see through this mode of 'associated thinking' a distinctive logical form of rational justification or the very form of reasoning. During the historical development of the mode of associated thinking, the association of attunement and understanding had been constantly evaluated highly and gradually incorporated into the mainstream values of the Chinese culture. Consequently, it eventually became institutionalized as a dominant *rule of rational justification*. The rule not only determines what facts count as evidence for justification (in what ways they count and the weight of evidence in support of propositions), but also determines what states of affairs count as permissible facts. For example, according to this form of reasoning, the states of affairs about the interaction between the yin-yang parts in the human body and the yin-yang forces in Heaven count as permissible facts. For the same reason, the interaction provides good evidence for the justification of the claim that sleepiness is caused by rainfall.

It should be clear by now that the pre-modern Chinese mode of reasoning was embedded within traditional Chinese medical theory. In fact, traditional Chinese medical theory was established based on the four basic assumptions of the pre-modern Chinese cosmology (the Yin-Yang doctrine, the Five-Element doctrine, the principle of pre-established harmony, the doctrine of constant transformation). We can see how heavily Chinese medicine has leaned on the Yin-Yang doctrine from a classical Chinese medical work, Huang-ti Nei-ching (Classic of Internal Medicine of the Yellow Emperor). As mentioned earlier, these four initial factual beliefs had operated constantly in the pre-modern Chinese intellectual tradition and were eventually internalized into the pre-modern Chinese way of thinking.

Embedded within this mode of reasoning, many sentences of traditional Chinese medical language are underlain by one or a few of the four universal principles of the corresponding pre-modern Chinese cosmology. For example, sentence (4) ('The association of the yin rain from Heaven and the yin force in the human body makes one sleepy') presupposes the Yin-Yang doctrine and the principle of pre-established harmony (here, the mutual correlation between Heaven and human beings). Since these presuppositions are held to be true within the context of the pre-modern Chinese intellectual tradition and medical community, the sentence is regarded as having a truth-value. Therefore, when I claim from time to time that the pre-modern Chinese mode of reasoning determines the truth-value status of many substantial sentences of Chinese medical language, what I really mean is that it is the basic universal principles of the associated cosmology, which function as state-of-affairs presuppositions, that determine the truth-value status of the sentences. Although I will still adopt the first way of speaking, my intention should not be misunderstood.

Notes

1 Some philosophers see the uncertainty principle as a metaphysical claim about the indeterminacy (nonexistence) of simultaneous values for position and momentum. Others see it as just an expression of epistemic limits of our knowledge. No matter whether we take it as a metaphysical or an epistemic principle, we can treat it as a universal principle as we define here.
2 Anne Hiskes has put her finger on this point in our discussion of a draft of this chapter.
3 The following part is inspired by D. Wong (1989) on the pre-modern Chinese style of reasoning.
4 I do not intend to give a complete list of the central doctrines of the pre-modern Chinese cosmology here.

Chapter 11

Categorical Frameworks

1. Categorical Frameworks

Very briefly put, a categorical framework is a specific category system of a P-language that describes the structure of the world perceived through the language community by categorizing its entities. There is one kind of category system that deserves our special attention: scientific taxonomy. Many scientific category systems are taxonomies, such as animal taxonomy (the Linnaeun system), plant taxonomy, etc. Of course, not all category systems are taxonomies. A category system is a taxonomy if and only if it breaks up into disjointed categories with exclusive relationships between any two categories at the same level. For the purpose of explication, we may conceive of a taxonomy as having both a vertical and a horizontal dimension. The vertical dimension concerns the level of inclusiveness of the category—the dimension along which terms such as collie, dog, mammal, animal, and living thing vary. The horizontal dimension concerns the segmentation of categories at the same level of inclusiveness—the dimension along which terms like dog, car, bus, chair, and sofa vary.

Kuhn worked out a theory of lexical/taxonomic structure of scientific languages[1] in his taxonomic interpretation of incommensurability, some aspects of which we have touched on in chapters 2 and 7. For clarity, let us recapture some essential points of the theory here.

(a) The basic members in a taxonomy are taxonomic categories or kind-terms. For Kuhn, a taxonomic category is nothing other than a kind denoted by a kind-term. Kind-terms refer to a widespread category that includes natural kind-terms (including mundane kinds such as 'tigers', 'lemons', 'water', 'gold', 'metal', 'heat', etc., and cosmic kinds such as 'phosphorous', 'electricity', 'hydrogen', 'H_2O', etc.), scientific kind-terms (the terms used in some branch of sciences as well as some terms from common language when they have a role in the specialty in question, such as 'plutonium', 'the Compton effect', 'apparatus', 'instruments', etc.), and many others. Some kind-terms (such as 'force', 'compound', 'phlogiston', 'planets', 'mass', 'element', etc.) have a greater level of generality than other kind-terms (such as 'alloy', 'metal', 'physical body', 'salts', 'gold', 'water'). Since these more general kind-terms figure importantly in fundamental laws about nature, we have referred to them as 'high-level theoretical kind-terms'; accordingly, these less general terms have been referred as 'low-level empirical kind-terms'. A kind is what is denoted by a kind-term. In the following discussion, we will not distinguish kinds from kind-terms. Since any tree of kinds can be presented as a tree of names of kinds or kind-terms, we can discuss these properties at the lexical level.[2]

(b) There are two principles associated with kind-terms: (i) The Projectibility Principle: Kind-terms are clothed with expectations about the extensions of the terms, since to know any kind-term at all is to know some generalizations satisfied by its tokens and to form expectations about the unexamined events (potential tokens). Some of the expectations about kind-terms (especially for the low-level empirical kind-terms) are normic (admit exceptions), and the others (especially for the high-level theoretical kind-terms) are nomic (exceptionless), which are usually laws of nature. More importantly, these expectations of kind-terms are projectible in the sense that these expectations enable members of a language community who use the kind-terms to project the use of the terms to other unexamined situations. (ii) The No-overlap Principle: No two kind-terms at the same level of a (stable) taxonomic tree may overlap in their extensions. As a result, there are only two possible types of class relationships between two kind-terms, that is, either inclusive when a kind (cats) is included in a higher-level kind (mammals) or exclusive when two kinds (cats and dogs) are at the same level.

(c) A lexicon is a structured conceptual vocabulary of a given language, which is the *mental module* in which *individual members* of a language community store the community's shared kind-terms/kind-concepts (clothed with expectations about their extensions) to describe and analyze the natural and social worlds surrounding them. The lexicons of the various members of a language community may vary in the expectations about the referents of the same kind-term they induce. Indeed, it is the different expectations about the shared kind-terms possessed by each member that distinguish different individual lexicons in a community. However, although different members of a language community may have different lexicons due to difference in expectations (different criteria for determining referents) about the shared kind-terms, these different lexicons are mutually congruent, or have a common structure. It is the shared lexical structure, instead of individual lexicons possessed by each member, that binds the language community together and at the same time isolates it from other communities.

To illustrate the lexical structure of a P-language, imagine, for a moment, that all shared kind-terms in a language community are connected to form a lexical network in which each kind-term is a node from which radiates net lines to tie some terms together and distance them from others, thus building a multidimensional structure within a lexicon. For different individuals, different materials which represent different expectations or criteria of the extension of the nodal term are used as net lines to connect nodes. What such homologous structures preserve, instead of the common materials for the net lines, is both the shared taxonomic kind-terms (the shared nodes) and the similarity relationships between them (the way nodes are connected). This shared structure among different individual lexicons in a language community is what Kuhn calls the lexical structure (lexical taxonomy) of the language. Stating it formally, the lexical structure of a language is the conceptual/vocabulary structure shared by all members of the language community, which provides the community with both shared taxonomic categories (kind-terms) and shared (similarity/dissimilarity) relationships between them.

All scientific languages are P-languages with lexical taxonomies, such as the Copernican taxonomy (in which the extension of 'planets' includes 'earth' but not 'moon' and 'the sun'), the Ptolemaic taxonomy (in which 'moon' and 'the sun' were in the extension of the kind 'planets' but 'earth' was not), the taxonomy of Aristotelian mechanics (in which the kind-term 'motion' not only includes the change of position of a physical body, but also the change of quality such as growth—the transformation of an acorn to an oak and transition from sickness to health), the taxonomy of Newtonian mechanics (in which 'motion' only refers to the change of position, not the change of quality), the taxonomy of the phlogiston theory ('phlogiston', being its primary kind-term), and the taxonomy of traditional Chinese medical theory, etc. The list can go on and on. For a concrete example of the taxonomy of a language, consider the ancient Greeks' taxonomy of the heavens on which Ptolemaic astronomy was based:

> For the Greeks, heavenly objects divided into three categories: stars, planets, and meteors. We have categories with those names, but what the Greeks put into theirs was very different from what we put into ours. The sun and moon went into the same category as Jupiter, Mars, Mercury, Saturn, and Venus. For them these bodies were like each other, and unlike members of the categories 'star' and 'meteor'. On the other hand, they placed the Milky Way, which for us is populated by stars, in the same category as the rainbow, rings round the moon, shooting stars and other meteors. There are other similar classification differences. Things like each other in one system were unlike in the other. Since Greek antiquity, the taxonomy of the heavens, the patterns of celestial similarity and difference, have systematically changed. (Kuhn, 1992, p. 19)

2. The Relativity of Lexical Taxonomies

Since Aristotle, there has been an old epistemological problem about the relation of the categories of mind and the categories of the world. Corresponding to the two extremes of this relation are two extreme positions on the formation of human category systems. At one extreme, environmental determinism claims that human categorization is fully determined by the structure of the world we live in. The categories that we isolate from the world of phenomena we do find there because they stare every observer in the face. The environment does, according to this position, all of the work of categorization. Since the world we perceive has a fixed structure and our categorization is fully determined by it, we would eventually cut the world at its joints.[3] At the other extreme, many others look to the order in the organism, and especially to the form of its cognitive constructs as the basis for the coherence of categorization. According to this intellectually based position, human categorization exclusively depends upon contextual factors embodied in cultures or language communities. The mind does all of the work of categorization.[4]

A better way to take is a middle road that countenances both pure environmental and pure intellectual bases for human categorization. This approach attempts to address the effects on our categories both of the discontinuities in nature and of our cognitive constructs, but with emphasis on the central role of theories or cognitive constructs in human categorization. There has been a developmental trend in cognitive science and psychology on human categorization, a trend away from a perceptual account of categorization, and toward a more theoretical and interest-relative basis for categorization. The issue on categorization has to do with explaining the categories found in a culture or its counterpart, a scientific community, and coded by the language of that culture or community at a particular time. Many studies in cognitive psychology, cognitive science, anthropology, and the history of science in the past three decades have found that the major part of human categorization is neither biologically fixed nor environmentally determined, but is perceived through the lenses of cultures and languages: it is determined by different contextual factors and varies widely across contexts. More specifically, there are three distinguished contextual factors that affect human categorization: cultural factors, linguistic factors, and cognitive factors.

Cultural Factors

Category systems can shift with cultures and traditions. Many types of categories, although perhaps not all categories,[5] are culturally relative. The research in ethnobiological classification has found that many folk classifications of plants and animals are culturally relative. For example, the following is a taxonomy of the animal kingdom attributed to an ancient Chinese encyclopedia entitled the *Celestial Emporium of Benevolent Knowledge*: All animals were divided into (a) those that belong to the Emperor, (b) embalmed ones, (c) those that are trained, (d) sucking pigs, (e) mermaids, (f) fabulous ones, (g) stray dogs, (h) those that are included in this classification, (i) those that tremble as if they were mad, (j) innumerable ones, (k) those drawn with a very fine camel's hair brush, (l) others, (m) those that have just broken a flower vase, and (n) those that resemble flies from a distance (L. Borges, 1966, p. 108).

It is well known that O. Spengler has expanded the group of culture-relative categories to include cognitive categories, and developed his general thesis of the dependence of categories on cultural contexts, based on investigation of a few high cultures of history. According to Spengler, the so-called *a priori* contains, besides a small number of universally human and logically necessary forms of thinking, forms of thinking that are universal and necessary not for humanity as a whole, but only for a particular culture. So there are different 'styles of cognition' characteristic of certain cultures. Even mathematics is relative to certain civilizations. The mathematical formulae as such carry logical necessity; but their visualizable interpretation, which gives them meaning, is an expression of the 'soul' of the civilization that created them. In this sense, our scientific world picture is only of relative validity. The basic scientific concepts or categories, such as infinite space, force, energy, motion, etc., are expressions of our accidental type of mind, and do not necessarily hold for the world picture of other civilizations (L. Bertalanffy, 1955, pp. 251-3).

While Spengler was concerned with a small number of high cultures, American anthropologists took into account the cultures of primitive tribes and reached a similar conclusion. For example, as shown by L. Barsalou and D. Sewell (1984), people who have different cultural backgrounds often have different opinions about how typical of its category a certain instance is. 'Groups of subjects taking different points of view generated substantially different graded structures for the same category. In birds, for example, the robin and eagle were typical from the American point of view, whereas the swan and peacock were typical from the Chinese point of view' (L. Barsalou, 1987, p. 107).

Linguistic Factors

Parallel to Spengler's cultural relativism of the categories, B. Whorf developed his thesis of the dependence of categories on linguistic factors.[6] It was a commonly held belief before Whorf that the cognitive process of all human beings possesses a common logical structure that operates prior to and independently of communication through languages. It follows that different languages are no more than merely different instruments for describing the same states of affairs or saying the same things differently. Whorf challenges this doctrine by proposing his hypothesis of linguistic relativity:

> [T]he background linguistic system ... of each language is not merely a reproducing instrument for voicing ideas but rather is itself the shaper of ideas. ... We dissect nature along lines laid down by our native language. The categories and types that we isolate from the world of phenomena we do not find there because they stare every observer in the face; on the contrary, the world is presented in a kaleidoscopic flux of impressions which has to be organized by our minds—and this means ... by the linguistic system in our mind. (Whorf, 1956, pp. 211-13)

In other words, the structure of the language one habitually uses influences the manner in which one understands reality. Accordingly, the picture of the universe as perceived by users of different languages shifts from tongue to tongue.

According to Whorf, language affects thought by means of the kinds of classifications it 'lays upon' reality. Nature is, in reality, a kaleidoscopic continuum. The units that form the basis of the grammar and vocabulary of each language serve both to classify reality into corresponding units and to define the fundamental nature of those units. There are two basic forms of classification perpetuated by a language. One is overt classification made by language vocabulary at the level of lexicon. Any lexical difference between languages implies a difference in the thought content of the speakers. The other is covert classification at the grammar level. The most basic units of grammar, which Whorf claims formed the basis of the metaphysics of language, are none other than the most general grammatical classes of the language, such as nouns, adjectives, and verbs in English. Whorf speaks of the semantic correlates of grammatical classes as the 'covert categories' of the language that provide the speakers of a language with those covert classifications of reality. For example, in Indo-European languages, substantives, adjectives, and verbs appear as basic grammatical units. A sentence, which is essentially a combination of these basic grammatical units, contains a separable subject and a predicate. This scheme of a persisting entity (represented by the subject) separable from its properties (represented by the predicate) and active or passive behavior had a profound effect on the categories of occidental thinking, from Aristotle's categories of 'substance', 'attributes', and 'action' to the antithesis of matter and force, mass, and energy in physics. By contrast, Indian languages (such as Nootka or Hopi) do not have these parts of speech or separable subject and predicate. Rather they signify an event as a whole. When we say, 'A light flashed', the Hopi use a single term, 'flash (occurred)'. La Barre has vividly summarized this viewpoint:

> Aristotelian Substance and Attribute look remarkably like Indo-European nouns and predicate adjectives. ... More modern science may well raise the question whether Kant's Forms, or twin 'spectacles' of Time and Space are not on the one hand mere Indo-European verbal tense, and on the other hand human stereoscopy and kinaesthesis and life-process—which might be more economically expressed in terms of the c, or light-constant, of Einstein's formula. But we must remember all the time that $E = mc^2$ is also only a grammatical conception of reality in terms of Indo-European morphological categories of speech. A Hopi, Chinese, or Eskimo Einstein might discover via his grammatical habits wholly different mathematical conceptualizations with which to apperceive reality. (La Barre, 1954, p. 301)

Cognitive Factors

The structure of a category system can also vary and shift with the development of human knowledge. The later Kuhn's studies on taxonomic structures of scientific languages reach a similar conclusion as that of Spengler's and Whorf's: the relativity of categories. Kuhn provides us with plenty of case studies from the history of sciences about how lexical taxonomies shift with different scientific theories. In particular, the Kuhnian thesis of the relativity of scientific taxonomies is essentially parallel to the Spenglerian thesis of cultural relativity of categories and the Whorfian thesis of linguistic relativity of categories. The Spenglerian thesis is based upon the few high cultures (civilizations) of history, the Whorfian thesis upon the linguistics of primitive tribes, and the Kuhnian thesis upon scientific development through different historical periods, especially during the periods of so-called scientific revolutions.

According to Kuhn, taxonomization is a process consisting of three interrelated variants: categorizing the domain into taxonomic categories; distributing items into pre-existing categories; and establishing the relationships between two categories in a taxonomy. The change of any one of these variants will change the taxonomic structure of a P-language.

Categorization and recategorization Whorf has shown us in his empirical study of languages that different languages categorize the world in different ways. Sometimes two competing languages may categorize a common domain so differently that there is virtually no substantial overlap between their taxonomies. For instance, Chinese medical theory and Western medical theory classify a common domain in totally different ways and thus have totally disparate systems of medical categories. There is not any major overlap between the taxonomies of these two medical theories.

A new taxonomy may be created from an old one by removing some previous categories completely. For instance, the category 'phlogiston' in phlogiston theory disappeared completely from the oxygen theory. Such recategorization can crucially affect the extensions of interrelated categories, and thus change the lexical structure of the previous language.

Redistribution[7] As Kuhn illustrated below, during revolutionary transitions of scientific theories, a natural family could cease to be natural; its members were redistributed among pre-existing sets by transferring one or more singular objects or subcategories from one category to another, thus changing the membership of the categories.

> The techniques of dubbing and of tracing lifelines permit astronomical individuals— say, the earth and moon, Mars and Venus—to be traced through episodes of theory change, in this case the one due to Copernicus. The lifelines of these four individuals were continuous during the passage from heliocentric to geocentric theory, but the four were differently distributed among natural families as a result of that change. The moon belonged to the family of planets before Copernicus, not afterwards; the earth to the family of planets afterwards, but not before. Eliminating the moon and adding the earth to the list of individuals that could be juxtaposed as paradigms for the term 'planet' changed the list of features salient to determining the referents of that term. Removing the moon to a contrasting family increased the effect. That sort of redistribution of individuals among natural families or kinds, with its consequent alteration of the features salient to reference, is, I now feel, a central (perhaps the central) feature of the episodes I have previously labeled scientific revolutions. (Kuhn, 1979, p. 417)

In the above case, what transferred in redistribution are singular objects, i.e. the moon and the earth, resulting in a change in the category of the high-level theoretical kind-term, 'planet'. There is another kind of redistribution in which the categories of a pair of high-level theoretical kind-terms change due to switching a low-level empirical subcategory between them. For instance, the categories of the kind-terms 'mixture' and 'compound' altered because alloys were compounds before Dalton and were mixtures afterward.

Similarity relationship As discussed in chapter 7, by focusing on the family resemblance relationship, Kuhn shows that two competing scientific communities at some stage of development many adopt different category systems due to holding different networks of similarity relations among objects and situations. For example, the similarity relationship between the sun and the earth was changed during the transition from Ptolemaic to Copernican astronomy. The sun and the earth were no longer put into the same category after Copernicus. Accompanying the shift of similarity relationships, the prototypes within taxonomic categories change. Due to the change of the prototype and the similarity relationship, a natural family or a taxonomic category ceases to be natural since its members are redistributed among pre-existing and newborn sets. Thus, the old category system is replaced by a new one.

Lastly, we have to emphasize that the three contextual factors we have identified above are not independent of one another. Culture and language are not two independent variables in the formation of a category system. It is impossible to separate a language from its cultural background. They are the two sides of the same coin. In some sense, we can say that a culture is reflected, in a concentrated form, in its language. A language is actually a micro-culture. Therefore, we should say that categories and their structures are found or bound in a culture and *coded* by the language of the culture. Furthermore, many taxonomies are not found in a culture all the time but rather found in a culture at a particular point in time. When we talk about a specific taxonomy found in a culture and coded in the language of that culture, we really mean the taxonomy, relative to the culture and its language, at a particular stage of the historical development of the culture and its language. Therefore, a language is not merely an abstract system of symbols outside cultural and historical contexts, but is an organic system of symbols embodied in a specific culture at a specific historical period. Considering a language in this sense, it is not an exaggeration say that each language has its own lexical taxonomy.

To sum up, human category systems do not remain stable across different contexts. Instead, human categorization is shaped by different contextual factors and varies widely across contexts. These findings challenge the traditional absolutistic view of categorization that found its foremost expression in Kantian conceptual absolutism. According to Kant, there are so-called forms of intuition (space and time) and the categories of the intellect (such as substance and causality), which are universally commitments for any rational beings. Accordingly, natural sciences based on these categories are equally universal. Physical science using these *a priori* categories—Euclidean space, Newtonian time, and strict deterministic causality—is essentially classical mechanics, which, therefore, is the absolute system of knowledge, applying to any phenomenon as well as to any mind as the observer. However, the development of natural sciences destroys the dream of absolutism. While categories (such as space) used to appear to be absolute for any rational observer, they now appear as changing with the advance of scientific knowledge (such as non-Euclidean spaces or the many-dimensional configuration spaces of quantum theory). Little is left of Kant's supposedly *a priori* and absolute categories.

3. Categorical Frameworks and Sortal Presuppositions

Sortal Presuppositions

In chapter 8, I have formally clarified and defined logical presupposition, especially existential presupposition. A logical presupposition relation can be defined by logical implication within an uninterpreted language. However, the more interesting presupposition relations, which we encounter frequently in scientific languages, are so-called analytic presuppositions including sortal and state-of-affairs presuppositions. Since analytic presuppositions are meaning/interpretation dependent, they can only be defined by analytic implication within an interpreted language. It is hence hard to formally define analytic presuppositions that can apply to any interpreted language. Below, we will only focus on sortal presuppositions.

Our languages (both natural and scientific) are many-sorted languages. That is to say, all terms (names, descriptions, and variables) are classified into different sorts or categories. With every k-place predicate, P, a k-tuple of sorts or categories is associated: Whenever P $(t_1, \ldots , t_k)$ occurs, it is required that the k-tuple of sorts or categories of $\{t_1, \ldots , t_k\}$ should be the one associated with P. Put in another way, each predicate P in a possible world is assigned a specific set of objects, $<O_i>$, $i = 1, 2, 3, \ldots$ to which the predicate P is applicable. We call this set of objects the category, sort, or significant range of the predicate P in question. For example, for a one-place predicate ' … is red', its category will be all sense-perceivable physical objects in our contemporary English speech community. Violation of this restriction will lead to so-called Ryle's category mistakes, such as

(33) p is tuned to G sharp.
(34) My soul is red.
(35) The barn is grammatical.
(36) The earth is more honest than Mars.

According to Ryle's original doctrine of 'category mistake', all sentences involving category mistakes are meaningless because they are ungrammatical or ill-formed. But the doctrine of literal meaninglessness of category mistakes is highly suspect. It is plain that sentences with category mistakes are different from *gibberish* such as 'Three spadlaps sat on a bazzafrazz' or 'Umph the but g kreeplat blunk'. The sentences like (33)-(36) contain neither any non-word nor any other illicit surface-grammatical concatenation. Whether they are meaningful appears to be language dependent. For example, sentence (34) is truth-valueless for contemporary English speakers (call it language community E). However, the same sentence may be truth-valueful or actually describe a state of affairs when considered within another language community. Imagine a primitive tribe (call it language community T) whose people believe that each person's soul is colored and that red color signifies courage. Then it is perfectly meaningful for a tribe's warrior to declare proudly, 'My soul is red'. The real problem with sentences involving category mistakes is not that they are meaningless or make no sense in general, but that they are truth-valueless or pointless for the user of a specific language community. For this reason, I side with J. Martin in attributing category mistakes to be the failure of sortal presuppositions.[8]

In fact, both (33) and (34) and their negations

(~33) p is not tuned to G sharp.
(~34) My soul is not red.

presuppose the following sentences, respectively:

(33a) p is capable of producing a musical tone.
(34a) My soul is capable of being colored.

Since these sortal presuppositions are held to be false for language community E, two presupposing sentences (33) and (34) are neither true nor false. More specifically, if an atomic sentence is thought to be made up of a subject and a predicate, such as

(23) S is P,

we can determine its truth-values in the following way. The extension of the predicate S must fall in the category of the predicate P. We define the extension of a predicate as a subset of its category. We can further assign truth-value status and truth-values to an atomic sentence (23) according to where its subject S falls: If S falls in the extension of P, then (23) is true; if S falls inside the category of P but outside the extension of P, then (23) is false; if S falls outside the category of P, then (23) is neither true nor false. Values for molecular sentences are then calculated according to Kleene's strong matrix (refer to Table 8.3 in chapter 8). For example, since the subject 'my soul' of (34) is outside the category 'is colored' (all sense-perceivable objects), (34) is neither true nor false for language community E.

Like other types of presuppositions, a sentence may presuppose (necessarily) many different sortal presuppositions. For example, both sentence (34) and its negation (~34) presuppose (34a). Since a soul is a non-sense-perceivable entity (let us take this for granted for the sake of argument) for language community E, sentence (34a) in turn presupposes the sentence,

(34b) Some non-sense-perceivable entities are capable of being colored.

Based on rules of presuppositions CP2 and CP3 given in chapter 9, sentences (34a), (34b), and (34a and 34b) are all the presuppositions of (34). Of these different presuppositions, some of them are more fundamental than others in the sense that some presuppositions presuppose others without being presupposed by them. A sortal presupposition SP will be regarded as being more fundamental than another sortal presupposition SP*, if SP is presupposed by SP*. For instance, for sentence (34), presupposition (34b) is more fundamental than presupposition (34a) since (34a) presupposes (34b). Furthermore, if a sortal presupposition is so fundamental to a language that it sets (fully or partially) the boundary for the category (sort, or significant range) of a predicate in the language, then we will call it an *absolute sortal presupposition* for the language. For example, the category of the predicate, ' ... is red', is the collection of all sense-perceivable physical objects in language community E. This category is determined by an absolute sortal presupposition in this language, namely,

(34c)   Only sense-perceivable physical objects are capable of being colored.

This is the reason why (34) is neither true nor false when considered within community E, since a sortal presupposition of the sentence, i.e., (34b), directly contradicts the absolute sortal presupposition, (34c), of language community E. By contrast, the same sentence (34) would be either true or false when considered within language community T since the fundamental presupposition of the sentence, namely (34b), partially sets the category of the predicate '... is red' that includes some immaterial entities like souls.

If the essential function of an absolute sortal presupposition is to determine (fully or partially) the category of a predicate in a language, then we can further *assume* that each language has its own system of shared absolute sortal presuppositions that set the boundaries for the categories of most predicates in the language (since the categories of most predicates in a language are relatively fixed). The existence of the system of shared absolute sortal presuppositions for each language is predicted by our formal theory of presupposition, although which specific absolute sortal presuppositions are held in a language cannot be identified by such a formal theory. Unlike logical presuppositions, which are defined by (formally) logical implication in an uninterpreted language, sortal presuppositions are analytic, and thus are defined by (formally) analytic implication in an interpreted language. While what a sentence logically presupposes can be determined by referring to the grammatical form of the sentence and is hence independent of the meaning of any particular descriptive term, what a sentence analytically presupposes cannot be determined by inspection of the grammar alone, but depends on the meaning of a particular descriptive term. The fact that 'S is a cat' and 'S is not a cat' presupposes 'S is an animal' depends on the meaning of 'cat' in a specific interpreted language. Therefore, it is the task of empirical semantics to identify the particular sortal presuppositions, including the absolute sortal presuppositions, of a specific language.

A Categorical Framework as a System of Sortal Presuppositions

There is one effective way to identify the system of shared absolute sortal presuppositions of a language without appealing to a formal theory of presupposition. Consider the Ptolemaic sentence (19),

(19)   Some planets travel around the earth.

(19) presupposes at least two assertions about the categorical status of the earth:

(19a)   The earth is not a planet.
(19b)   The earth is a star.

Here we take 'revolve' to mean a relation between two separate objects in that one object revolves around the center of gravity of the other. So (19a) is analytically implied by the verb 'revolve' in (19). (19b) can be derived from (19) based on the definition of 'star' (All the planets revolve around a star). (19a) is one absolute sortal presupposition of the Ptolemaic language about the category of the predicate, '… is a planet', which excludes 'the earth' from the category of the kind-term 'planets'. Similarly, (19b) is another absolute sortal presupposition of the Ptolemaic language about the category of the predicate '… is a star' that includes 'the earth' in the category of the kind-term 'stars'. In general, according to the category framework of Ptolemaic astronomy, the sun and moon were planets, like Mars, Venus, and Jupiter; the earth was not, but belonged to the natural family of stars. In contrast, after the transition from Ptolemaic to Copernican astronomy, the earth was a planet; the sun was a star; and the moon was a satellite.

The above case suggests that the categorical framework of each P-language actually functions as the system of shared absolute sortal presuppositions that sets the boundaries for the categories of most predicates in the language. Since each language has its own unique lexical taxonomy, it also has its own system of shared absolute sortal presuppositions. This conclusion is further supported by one key feature of the lexical taxonomy of a language, namely, the closure of lexical taxonomy: The lexical taxonomy of a language involves something like a 'closure'—there are things that cannot be said without violating the lexical taxonomy in question. The closure of a taxonomy is implied by the Whorfian thesis. According to Whorf, the system of classifications of a language, especially the covert classifications, creates '*patterned resistance* to widely divergent points of view' (Whorf, 1956, p. 247) (because of their subterranean nature they are 'sensed rather than comprehended—awareness of [them] has an intuitive quality' (Whorf, 1956, p. 70)). This patterned resistance may oppose not just the truth of the resisted alternatives but often the absolute sortal presuppositions underlying the system of classifications that an alternative has been presented.

Consider a familiar example. The Copernican system of classifications of celestial bodies (The earth is put into the category of planets) creates a 'patterned resistance' to the sortal presupposition (19a) or (19b) that is presupposed by the Ptolemaic sentence (19). The same is true for the Ptolemaic system of classification that creates a 'patterned resistance' to a sortal presupposition, say,

   (37a) The sun is not a planet,

of many Copernican sentences, say,

   (37)   The planets revolve about the sun.

As a result, the Ptolemaic sentence (19), when considered within the Copernican language, is neither true nor false, since its sortal presupposition (19b) is held to be false within the Copernican language. Accordingly, the Copernican sentence (37), when considered within the Ptolemaic language, is neither true nor false, since its sortal presupposition (37a) is not held to be true within the language of Ptolemaic astronomy. It is clear that the Copernican (or the Ptolemaic) taxonomy not only sets the boundaries for the categories of predicates such as '… is a planet' and ' … is a star', but actually determines the truth-value status of the sentences involved with these predicates. In this sense, we can say that the Copernican (the Ptolemaic) taxonomy constitutes its own unique system of absolute sortal presuppositions.

   In general, if two different language communities have two different category systems that offer 'patterned resistance' to one another in the sense of suspending each other's absolute sortal presuppositions, then the members of one community can use one sentence to make an assertion while the apparently same sentence cannot be used to make an assertion within the other community. Similarly,

> with the Aristotelian lexicon in place it does make sense to speak of the truth or falsity of Aristotelian assertions in which terms like 'force' or 'void' play an essential role, but the truth-values arrived at need have no bearing on the truth or falsity of apparently similar assertions made with the Newtonian lexicon. (Kuhn, 1993a, p. 331)

This is because there is no way, even in an enriched Newtonian vocabulary, to convey Aristotelian assertions regularly misconstrued as asserting the proportionality of force and motion or the impossibility of void. In the Newtonian lexicon, these assertions cannot be expressed fully since to express them one has to invoke the Aristotelian lexicon, which is 'patternedly' resisted by the Newtonian lexicon.

4. Compatible versus Incompatible Metaphysical Presuppositions

Suppose that $<MP_1>$ and $<MP_2>$ are two distinct sets of M-presuppositions of two disparate P-languages PL_1 and PL_2. We can define a pair of compatible M-presuppositions as:

   $<MP_1>$ and $<MP_2>$ are compatible iff $<MP_1>$ and $<MP_2>$ are or could both be held to be true by the advocate of PL_1 or PL_2.

Consequently, the members of two language communities with compatible sets of M-presuppositions agree on the truth-value status of sentences of the other language, although they may differ in assigning truth-values to the same sentences. In contrast, two sets of M-presuppositions of two different P-languages are incompatible if the M-presuppositions of one language are categorically rejected by the advocate of the other. Two incompatible M-presuppositions assign opposite truth-value status to numerous core sentences of one language under consideration.

Accordingly, two P-languages PL_1 and PL_2 are compatible if their M-presuppositions $<MP_1>$ and $<MP_2>$ are compatible with one another; otherwise the two languages are incompatible.

In the case in which one language is a subset of the other competing language, whether the two sets of M-presuppositions are compatible depends on the specific relation between them. Phlogiston theory was formed by adding the vocabulary and the universal principle of the existence of phlogiston to Aristotelian metaphysics such that the former was embedded within the Aristotelian language. Since the advocate of phlogiston theory accepts Aristotelian metaphysics and the Aristotelians could accept the existence of phlogiston, the two languages are compatible. In contrast, as shown in chapter 5, although the Newtonian language of space L_N includes the Leibnizian language of space L_Z as its sublanguage, the M-presuppositions of the two languages are still incompatible. This is because the Leibnizians are obliged to reject the M-presupposition of the Newtonian language, namely, the existence of Newtonian absolute space. Consequently, both languages assign different truth-value status to the sentences in the set S_N-S_L.

Among three types of M-presuppositions identified, our definition of the compatibility of M-presuppositions can be applied directly to existential presumptions and universal principles since they can be formulated as a set of countable statements. For example, the core sentences of Newtonian mechanics are underlain by the principle of construction. That is, properties such as shapes, masses, and periods are internal properties inherent in objects themselves and change only by direct physical interactions. But according to the theory of relativity, there are no such inherent properties (at least by the interpretation of Einstein and Bohr), for shapes, masses, and periods are relational properties between physical objects and co-ordinate systems. These properties could change even without any physical interaction between the objects in which those properties are supposed to inhere if the original co-ordinate system is replaced by another. From the viewpoint of the theory of relativity, the Newtonian principle of construction is simply false. Therefore, the universal principle of the Newtonian language and the corresponding universal principle of Einstein's language (no matter what it is) are incompatible. A similar analysis can be easily applied to other P-languages with incompatible universal principles or existential presumptions. We will not belabor them here.

However, it seems difficult to apply our definition directly to categorical frameworks. Intuitively we cannot ask the question of whether a category system is true or false. We had better come up with a different criterion to test the compatibility of two competing category systems.

Consider the following hypothetical case study about color classification.[9] Different languages may (and actually do) divide the spectrum in different ways and thereby have different color predicates. Imagine three sets, C_1, C_2, and C_3, of color predicates in three different languages L_1, L_2, and L_3. C_1, C_2, and C_3 divide the spectrum in such a way that no color predicates of one set match up with the color predicates of any other set. Thus, we have three different category systems of color. Let us further suppose that C_1 has 'red', 'orange', 'yellow', 'green', 'blue', and 'purple' as its finest-grained color predicates; C_2 has 'rorange', 'ygreen', and 'bpurple'; C_3 has 'red*', 'orange*', 'green*', and 'blue*' at the corresponding level of discrimination. The extensions of C_2 can be imagined to be distributed along the spectrum so that 'rorange' matches up with the shades of both 'red' and 'orange', 'ygreen' with both 'yellow' and 'green', and 'bpurple' with both 'blue' and 'purple' in C_1. The extension of C_3 can be imagined to be shifted along the spectrum to such a degree that 'red*' applies to some shades that C_1 would call red and some shades that C_1 would call orange; 'orange*' applies to some shades that C_1 would call orange and the shade that C_1 would call yellow, as well as others that C_1 would call green; 'green*' applies to some shades that C_1 would call green and some shades that C_1 would call blue; 'blue*' applies to some shades that C_1 would call blue and some that C_1 would call purple.

The categorical mismatch between C_1 and C_2 is adjustable since each color predicate of C_2 (namely, each color category) can be defined in C_1. The predicate 'x is rorange' in C_2 can be expressed as 'x is either red or orange but not both' (take 'or' in exclusive sense) in C_1, 'x is ygreen' as 'x is either yellow or green but not both', and so on. In this way, C_2 can be incorporated fully (without loss) into C_1 although there are no shared categories between them. In this case, we say that C_1 and C_2 are two compatible category systems. By contrast, the mismatch between C_1 and C_3 is so disparate that it is impossible to define each color predicate of C_3 in C_1, and vice versa.

Perhaps one might suggest a possible formulation as follows: The speaker of L_1 can interpret the speaker of L_3 as having the concept 'reddish-orange', or 'shades which overlap with red and orange' for the concept 'red*' in C_3. Similar formulations can be given for the concepts 'orange*', 'green*', and 'blue*' in C_3. But this proposal faces at least one crucial objection. Since we have supposed that the color predicates in both C_1 and C_3 are the finest-grained in L_1 and L_3, the speaker of L_1 has no way to identify the extension of the concept 'reddish-orange', and thereby the extension of the concept 'red*' is undetermined for the speaker of L_1. Without the identification of the extension of the concept 'red*', any formulation of 'red*' (in L_1) cannot be done. Therefore, it is clearly not the case that C_3 can be *incorporated without remainder* into C_1. In this case, we say that C_1 and C_3 are two incompatible category systems.

Based on the above illustration, we can formulate the following distinction between compatible and incompatible category systems:

> Two category systems (or taxonomies) are compatible iff one overlaps *substantially* with the other[10] to the extent to permit *incorporating* one into the other.

Of course, the relationship between C_1 and C_2 is not the only kind of possible compatibility relations between two category systems. It is, for instance, possible that taxonomy T_1 may be included fully within taxonomy T_2 because T_2 is extended from T_1. Recall the Newton–Leibniz debate on the absoluteness of space. Although Newton's and Leibniz's theories of space present different conceptions of space, both theories have 'overlapping' or 'common parts'. We have shown that Newton and Leibniz have no disagreement about the truth-value status of the classical Euclidean language L_E, which is a common language shared by both sides. Actually, Leibniz's language of space L_Z is an extension of L_E. The taxonomy of L_E is included fully within L_Z. Therefore, the taxonomies of L_E and L_Z are compatible. As to the relation between Newton's language L_N and Leibniz's language L_Z, the situation is a little different. L_Z is a sublanguage of L_N. Within the overlapping part (namely, within L_Z), since both languages have identical taxonomies (more precisely, part of Newton's taxonomy is identical with Leibniz's), both have compatible taxonomies. The controversy arises when the speaker of Leibniz's language goes beyond L_Z and asks whether the sentences in the set S_N-S_L have factual meanings. For example, from Leibniz's point of view, core Newtonian sentences (9) ('The body b at time t could have located in a different place') and (10) ('Spatial location of the body b at time t_1 is different from its location at time t_2) are actually truth-valueless or factually meaningless. But such a controversy about the truth-value status of sentences in question is caused by an underlying universal principle (whether there exists Newtonian absolute space), not caused by incompatible taxonomies of the two languages.

However, one might think that the central expressions used in the above definition, i.e., '*incorporated*' and 'overlap *substantially* with', are too vague to be useful to make an effective distinction. Fortunately, Kuhn has provided us with another criterion to identify two incompatible taxonomies based on the projectibility principle and the no-overlap principle of kind-terms: Two category systems or taxonomies are incompatible if the extensions of some shared theoretical kind-terms in two taxonomies overlap (but are not co-extensive) in some local area to such an extent that incorporating one into the other will directly violate the no-overlap principle. In addition, two category systems or taxonomies can be incompatible if they are mismatched to such an extent that they are either totally disjointed or lack any major overlap. Those two criteria have been presented earlier. We will not belabor them here.

Notes

1   See Kuhn, 1983b, pp. 682-3; 1988, p. 11; 1991, pp. 5, 11-12; 1993a, pp. 315, 325, 329.

2   In his discussion of kinds, Kuhn does not explicitly discuss the ontological status of kinds that concerns, for example, the problem of natural kinds (cosmic vs mundane natural kinds). However, from Kuhn's position on the ontological status of similarity relations and his attitude toward metaphysical realism, it is reasonable to regard Kuhn as taking a conceptualist position on kinds: There are not any sets, kinds, universals, classes out there in the world that cut the nature at its joints. Kinds or universals can exist in particulars, but there are none *priori* to particulars. There are real things out there, and we divide them into kinds according to both nature's way (the things in nature distinguishing themselves into various segmentation) and our conceptions (the lexicon, i.e., the module in which members of a speech community store the community's kind-terms). See Kuhn, 1993a, pp. 315-16 and Hacking, 1993, pp. 277, 291.

3   There are a number of problems with this position. One is that such environmental factors are ultimately insufficient to account for the richness and diversity of human categorization.

4   The real problem with this position consists in its assumption that segmentations of the world are originally arbitrary. Taking this view, human categorization may become the arbitrary product of historical accident or even of whimsy.

5   Perceptual categories, such as the categories of color, may be universal. Even so, they still leave room for cultural differences.

6   In some sense, we can say that the Whorfian thesis of linguistic relativity is part of a general conception of cultural relativism developed in the first half of the twentieth century.

7   Kuhn, 1970b, pp. 269, 275-6; 1979, p. 417; 1987, pp. 8, 10.

8   See J. Martin, 1975, 1979 for a formal definition of sortal presupposition.

9   The following hypothetical case study is a modified version of a similar case study given by M. Khalidi (1991, pp. 73-8). I adopt this case for a very different purpose.

10 An overlap between two taxonomies is different from an overlap between two kind-terms (or two categories). The no-overlap principle only prohibits the overlap between two kind-terms or categories at the same level of a taxonomy, but does not prohibit the overlapping of two taxonomies.

Chapter 12

The Failure of Cross-Language Understanding

As we have demonstrated repeatedly, incommensurability is a semantic phenomenon closely related to the problem of cross-language communication: Whether successful linguistic communication is possible between two sufficiently disparate P-language communities. At the extreme end of the spectrum of different degrees of communication breakdown is the failure of mutual understanding. Understanding the members of one's own language group, even within one's own family, can be taxing. Understanding cross-differences of classes, races, genders, religions, cultures, and traditions can seem all but impossible. However, we find that understanding an unknown P-language with incompatible M-presuppositions turns out to be more challenging and problematic. It is not just the vocabulary and grammar of the language that is unknown, but also the way of thinking and justification, the mode of reasoning, the categorical framework, and the cosmology, all of which constitute the linguistic context from which the language derives its life, that are incomprehensible to us. The failure of mutual understanding between two incompatible P-language communities suffices to break down the cross-language communication between them.

We have contented ourselves so far with vague and pragmatic pronouncements on the notions of understanding and communication. To give a sufficient interpretation of the concept of incommensurability, the time has come to clarify the notions of the failure of cross-language understanding and cross-language communication breakdown. Presumably, to do this we need a theory of cross-language understanding, which would specify both necessary and sufficient conditions of cross-language understanding, and a theory of cross-language communication. However, considering the vagueness and difficulties that go with the notions of understanding and communication, to produce a general account of cross-language understanding and communication would be much more than we can achieve here. Fortunately, such a general account is not needed either considering our limited purpose here. What concerns us are not the notions of understanding and communication in general, but specific senses of understanding and communication: effective cross-language understanding and successful cross-language communication. Furthermore, to explain the *failure* of cross-language understanding or cross-language communication *breakdown*, it would be sufficient if we are able to locate some essential semantic and/or conceptual obstructions between two P-languages and thus to identify a significant *necessary condition* of effective cross-language understanding and that of successful communication between two language communities. We will deal with the failure of cross-language understanding in this and the next chapter (chapter 13) and leave the topic of communication breakdown for chapters 14, 15, and 16.

1. The Propositional Understanding

The Notion of Propositional Understanding

The notion of effective cross-language understanding will become clear in due course. Nevertheless, a few preliminary remarks about the concept, the subject, and the object of effective cross-language understanding should be helpful here. First, what do we mean by '*to understand* a (declarative) sentence or a language'? 'To know or comprehend its meaning', of course, one may say. But what is supposed to be included in its meaning? What is the meaning of a sentence anyway? To answer these questions, we need to know the basic functions of linguistic expressions to be understood. German psychologist K. Bühler (1934) put forward a tripartite schema of language functions that places the linguistic expression in relation to the speaker, the world, and the hearer/interpreter: The speaker comes to an understanding with the hearer about something in the world. According to Bühler's semiotic model of linguistic expression, a linguistic expression functions simultaneously as symbol (correlated with states of affairs), as symptom (depending upon the speaker's intention), and as signal (its appeal to the hearer). J. Habermas[1], following K. Bühler, J. Austin, and J. Searle, identifies those three aspects of linguistic expressions as the three structural components of speech acts or the three dimensions of the meanings of utterances:

(a) The propositional content: *What is said* literally and explicitly, not implicitly, with a linguistic expression that is supposed to represent states of affairs and can be defined in terms of its truth conditions in the case of a sentence.
(b) The expressive content: What is *intended* with a linguistic expression by the speaker.
(c) The illocutionary content: What is *used* in a speech act to enter into a relationship with the hearer/interpreter.

Presumably, a comprehensive understanding of a linguistic expression should deal with all these aspects of meaning. More specifically, understanding a written declarative sentence usually involves not only a comprehension of its propositional content, but also, if not always, appreciation of the intention of the author, the 'illocutionary force' with which it is issued, and the conversational implicatures of what is said in some specific contexts, an appreciation that goes far beyond knowing its propositional content. Following the convention, we can separate the propositional content of a declarative sentence from the other two contents, i.e., the expressive and the illocutionary content, and call the former the semantic content and the latter the pragmatic content.

Accordingly, we can identify at least two notions of understanding. On the one hand, to understand a sentence could mean to grasp its propositional content (not propositional attitude), the thought expressed, or its literal meaning (conventional meaning, namely, the meaning determined by the linguistic conventions of a language), which we can call *propositional understanding*.[2] Following the dominant view these days in the philosophy of language, we need to distinguish propositional *attitude*, including the content of a belief, expressed by a speaker in terms of a sentence from the propositional *content* asserted by the sentence. The content of beliefs and other kinds of propositional attitude are characteristically *intensional*: The content of a speaker's belief that '*A* is *F*' may differ from the content of another of his belief that '*B* is *F*', even if '*A* is *B*'. This is because belief is taken by many to be a relation between believers and propositions or 'modes of acquaintance with propositions or ways in which a believer may be familiar with propositions' (N. Salmon, 1986, 441). Hence, the content of beliefs might be essentially the property of particular persons, so that different interpreters could apprehend the same state of affairs in similar ways, though never in the same way. In contrast, the propositional content and the literal meaning are characteristically *extensional*: If *A* and *B* are co-referential, then the literal meaning or the propositional content of the sentence '*A* is *F*' and '*B* is *F*' is the same. Subsequently, (literal) meaning and propositional content are objective, or at least intersubjective, in the sense that they are not in the head of an individual believer, but are properties of a language that are publicly accessible and communicable.

On the other hand, to fully understand a sentence could mean, besides to grasp its propositional content, to capture the illocutionary force that the sentence induces on the hearer, the speaker's particular intention in uttering the sentence on a particular occasion, or the speaker's particular beliefs and other propositional attitudes associated with the utterance. Clearly, since the propositional content of a sentence is the semantic foundation on which its other dimensions of meaning depend, propositional understanding constitutes the central core of any notion of understanding.

Since we are mainly dealing with scientific languages, which are public languages composed of declarative statements, we can, for the sake of clarity, focus on their semantic dimension only. In fact, recall that Kuhn and others discover that a failure of cross-language understanding between two scientific languages cannot be simply taken as evidence of the interpreter's limitation of knowledge or lack of interpretative skills. It is a failure of mutual understanding between two P-language communities as a whole, rather than between some individual speakers with different dialects, intentions, attitudes, or conflicting interests. It clearly involves some deep *semantic and/or conceptual obstructions* between two substantially different languages. Therefore, in our following discussion, we will be only concerned about the essential core of any notion of understanding, i.e., propositional understanding. More precisely, corresponding to the speaker's act of expressing a thought via a sentence to the interpreter about a state of affairs, there stands the interpreter's apprehension of the thought expressed:

> The interpreter is able to understand a sentence S if and only if he or she can grasp/apprehend its *propositional content*.

Second, who is supposed to be the subject of propositional understanding? We are not interested here in how each individual member of a language community understands another language, but rather how a language community as a whole understands it. Thus, *a* speaker's (as an individual speaker of a language) meaning or intention on a particular occasion and *an* interpreter's (as an individual interpreter of a language) personal apprehension of a sentence have no place in our discussion. Whenever I mention '*the* speaker" or '*the* interpreter' of a language, I mean all the speakers or interpreters as a whole or the representative speaker or interpreter, which roughly corresponds to U. Eco's 'Model Reader' who agrees to abide by the rules and conventions of a linguistic game in order to reach a coherent understanding, rather than some specific individual speaker or interpreter with different transient psychological states, such as Eco's empirical reader with a specific personal *encyclopedia* or world knowledge. Based on a commonly accepted assumption about linguistic competence—the competent speakers of a language can and do effectively understand the language and fully communicate with one another—the so-called literal meanings of sentences of a language are actually the meanings understood and used by the speaker of the language. In this sense, 'the literal meanings' of the sentences of a language are actually what I might call, more precisely, 'the speaker's language-meanings' (not an individual speaker's meaning), or 'language-meanings' for short to emphasize the fact that the meanings are *conventionally* established by the language. Although I will still, following the convention, talk about the literal meaning in the discussion below, please do bear in mind that I am actually referring to the speaker's language-meaning.

Finally, 'what to be understood' is a language as a whole, not some isolated words or sentences of a language. Of course, the language to be understood is not just any kind of language, but rather a P-language as we have identified.

Some Accounts of Propositional Understanding

A variety of accounts of cross-language propositional understanding have been proposed. None of them, however, are able to identify a *significant necessary* condition of understanding. I have discussed one account extensively in chapter 2, that is, the translational account of understanding, according to which the interpreter can understand an alien language by translating it into the interpreter's own language. I argued that translation is neither a necessary nor (necessarily) a sufficient condition for understanding. Understanding an alien language is always possible through language learning no matter whether the interpreter can translate it into his or her own home language.

Others contend that one can understand a language by understanding its components (the compositional account of understanding). But understanding a language is obviously different from understanding its words or sentences. In fact, in any given case of cross-language understanding, there would always be a failure to grasp or understand some of its specific parts. The experience of understanding tells us that sometimes the interpreter can understand a sentence of another language without understanding some of the terms involved. For example, for a sentence of phlogiston theory, 'phlogiston is phlogiston', failing to understand the term 'phlogiston' does not keep the interpreter from understanding the sentence itself. To understand each word of a sentence is not necessary for understanding the sentence. More to the point, it is quite uninteresting to learn that understanding a language requires a comprehending of its parts. Even if understanding the parts of a language is necessary for understanding it, it is too trivial a position to be useful as an explication of the failure of cross-language understanding. In addition, as I will argue later, understanding each word of a sentence is not sufficient to understand the sentence as a whole.

One might argue that these problems with the compositional account of understanding lie in an ill-formed view of meaning and understanding, namely, that a word or a sentence could have its meaning in isolation. But to understand one part, words or sentences, of a language we have to understand the language as a whole, since the meaning of a part is determined by its place in the whole language. Therefore, others who follow the holistic view on meaning and understanding correctly insist that to understand a sentence of a language one needs to identify the unique role or the function of the sentence in the language. However, how can we identify the unique role or the function of every sentence in a language? Presumably, we need a recursive procedure to identify the roles of infinite sentences from some limited subset if language is learnable. As D. Davidson points out, the role of a sentence in a language can be determined by constructing a theory of truth for the language based on Tarski's semantic notion of truth. Then, the above functional account of understanding can be reduced to the truth-conditional account of understanding.

Among many other accounts of propositional understanding that I will not discuss here, Davidson's truth-conditional account of understanding is by far the most appealing one (Davidson, 1984). According to it, the Tarskian semantic notion of truth plays the most essential role in linguistic understanding due to a necessary conceptual connection between the knowledge of (Tarskian) truth conditions and understanding. To understand a sentence, it is necessary and sufficient to know its truth conditions; to understand a language is to know the truth conditions of any sentences of the language.

Many have challenged Davidson's basic doctrine that Tarski's semantic notion of truth is central to a theory of understanding by casting doubt on the effectiveness of the knowledge of Tarskian truth conditions to understanding. Some raise doubt about the thesis that the knowledge of truth conditions is *necessary* for cross-language understanding. They claim that the requirement is too strict since Tarskian truth conditions of many sentences are either unknowable or unavailable to the interpreter. Other alternatives are offered that do not appeal to truth conditions, such as M. Dummett's (1993) verificationist conditional account, which focuses on assertability conditions, instead of truth conditions, and W. Sellars' and G. Harman's (1984) conceptual role account. On the other hand, for some, the knowledge of truth conditions is too trivial to be sufficient for understanding. In my opinion, whether the knowledge of Tarskian truth conditions is sufficient for understanding depends on what counts as qualified knowledge of the truth conditions. If we strengthen the knowledge of the truth conditions strictly enough— for example, if 'to know the truth conditions of a sentence S in a language L' means 'to know the proposition expressed by the T-sentence of S in terms of knowing the T-sentences for all other sentences in L'—I fail to see why one cannot understand a language by knowing the truth conditions of all its sentences.

However, the real trouble with the truth-conditional account is that even if we were able to achieve understanding by strengthening the knowledge of the truth conditions, we would have to give up the hope that *the same knowledge* could serve as a necessary condition for understanding. This can be shown briefly as follows. Suppose that an ordinary English speaker Jenny understands a sentence S in English,

(S)   Smith is cowardly.

That means that Jenny grasps the proposition expressed by S in English. But for Jenny to know the truth conditions of S, she has to know the proposition expressed by the T-sentence of S,

(T_S)   'Smith is cowardly' is true in English if and only if Smith is cowardly.

To know what is expressed by T_S, Jenny has to know that it is verified as true in terms of a theory of truth for English that gives all T-sentences to other sentences in English. Jenny needs a few strong assumptions to derive the propositional knowledge of T_S from the propositional knowledge of S, such as the assumptions that Jenny tacitly knows Tarski's semantic notion of truth and that she tacitly knows how to construct a Tarski-type theory of truth for English. I do not think an ordinary competent English speaker can and should master these assumptions in order to understand English. A speaker may understand a sentence well without knowing its truth conditions. Without specifying a significant necessary condition for understanding, the truth-conditional account cannot explain the possible failure of understanding between two substantially disparate languages.

Please notice that I do not intend to argue here that the knowledge of truth conditions *cannot* be a necessary condition for understanding. As I have mentioned above, it depends upon what counts as the *knowledge* of truth conditions. If we interpret the knowledge of truth conditions weakly enough, it could be a necessary condition for understanding. For example, if we interpret the content of the 'know that' clause as a propositional attitude of Jenny's, then for Jenny to understand an utterance of 'Smith is cowardly', it is at least necessary for her to know that it is true if and only if Smith is cowardly. For her to know that theorem, it is not necessary for her to have any ideas about Tarski's semantic notion of truth and of how to construct a Tarski-type theory of truth for English.

More importantly, as to be argued later, it is the knowledge of *truth-value* conditions, not the knowledge of *truth* conditions, that plays an essential role in cross-language understanding.

2. The Essential Role of Metaphysical Presuppositions in Cross-Language Understanding

A P-language is underlain by a particular cosmology loaded with a set of existential presumptions and universal principles, embodies a unique mode of reasoning, and is embedded with an exclusive lexical structure. It is not too much of an exaggeration to claim that a P-language as such reflects a form of life. The conceptual richness of a P-language determines the depth and inclusiveness of understanding it. A P-language is fully intelligible, and its purported justification is adequately understood by an interpreter only if its underlying M-presuppositions, including its cosmology and its mode of reasoning, are fully comprehended. Any interpreter who fails to comprehend the underlying M-presuppositions cannot see the 'point' of the language, and thus cannot become an engaged communicator. Consequently, effective cross-language understanding cannot be achieved. This is the reason why in so-called 'abnormal discourse' (to borrow R. Rorty's terminology) effective mutual understanding across two conceptually disparate P-languages is problematic and difficult.

Consider the following three core sentences from three different P-languages:

(4)   The association of the yin and rain makes people sleepy.
(30)   Element *a* contains more phlogiston than element *b*.
(31)   Electrons have eternally hidden nuclei.

Recall that (4) is a sentence from the language of traditional Chinese medical theory (CMT); (30) from the language of phlogiston theory; and (31) from the language of an imaginary physical theory. Now consider Dr Smith, an interpreter who is educated within contemporary Western scientific tradition. Suppose that he is familiar neither with phlogiston theory nor with CMT and its underlying pre-modern Chinese mode of reasoning. Is Dr Smith able to effectively understand (4), (30), and (31)?

Dr Smith can understand each word of (31). After all, he knows what an electron is, what it means for something to be eternally hidden, and what it means to say that a particle possesses a nucleus. Certainly one may be forgiven for thinking that the hypothesis about the eternally hidden nuclei no more involves 'an empty play of words' than claims about permanently confined quarks seriously discussed and defended in contemporary physics. Besides, Dr Smith realizes that (31) is in good syntactic and semantic order. Contrary to M. Schlick and others who believe that (31), like claims of a pseudoscience, is senseless, (31) does make sense to Dr Smith. Presumably, as long as a sentence does not contain any meaningless words and is not ill-formed, it *makes sense to* the interpreter. In fact, it is logically possible to have a context in which (31) could be used to say something true or false.

However, even if Dr Smith can make sense of (31), he is still unable to understand what is being *said* by it. The trouble is not that (31) involves meaningless words or combines meaningful words in an illegitimate way, but rather that it is odd to the point of being unintelligible (in the sense of failing to have a point) in *contexts* that are conceptually recognizable to him. For Dr Smith, there is no context, as far as he can tell, in which such a claim can be deemed true or false. He can, as M. Schlick (1991) argues, convince himself that if he asked a question of the speaker, 'What do you actually *mean* by the presence of this nucleus?' then the speaker would have to admit, 'everything would be exactly as before'. Thus he can justifiably conclude that the speaker 'had not succeeded in conveying to us the meaning of the hypothesis that electrons have eternally hidden nuclei'. In addition, a sentence can be used to make a point only within a suitable context. (31) could be said to have a point only when it is put forward within the context that Dr Smith is obliged to reject as conceptually unsuitable or impossible. Thus, (31) fails to make a point for him as an interpreter. Consequently, Dr Smith can neither understand *what is being said* nor grasp *the thought* expressed by the sentence.

The importance of the above observation is that although the interpreter from another language can know the meaning of each word of a sentence of an alien P-language, he or she might still not be able to effectively understand it if he or she is not aware of its *point* or is unable to grasp *its propositional content*. So declaring that one can *make sense of* a sentence is altogether different from declaring that one can understand *what the sentence is saying*.

Our second observation is more significant. We find that although both (4) and (30) are unintelligible to Dr Smith, it is important to make a distinction between the two. An old P-language may be forgotten, but can still be made intelligible to the modern reader who is willing to spend the time relearning it. In contrast, some P-language—especially when embodied within a substantially disparate intellectual or cultural tradition—indicates such a radically disparate mode of thinking and/or categorical framework as to require something far more complicated than mere learning of the language itself. In order to understand it, one has to learn the whole *form of life* behind it (with the mode of reasoning and the category system as its cores). Two distinct languages of phlogiston theory and CMT are good examples of this contrast.

The primary cause for Dr Smith failing to understand (30) lies in the meaningless term 'phlogiston'. As long as he learns the meaning of the term and the corresponding existential assumption (the existence of phlogiston), he is able to understand (30). Although Dr Smith does not believe there is such a substance as phlogiston, he can work it out and understand *the point* of what Priestley is saying when presenting his phlogiston theory. This is mainly because Priestley's phlogiston theory, lying within the same intellectual tradition as modern science, is *conceptually recognizable* to Dr Smith. After being given the meaning of 'phlogiston', he is able to identify and comprehend a metaphysical presupposition of the language, i.e., the existence of phlogiston. He can thus fully recognize the truth-value conditions of (30).

For contrast, turn to CMT. As we have pointed out in chapter 5, Dr Smith's failure to understand CMT is not due to the difficulty of translation of the language of CMT into modern English or a corresponding modern scientific language. For example, his failure of understanding (4) does not just lie in the meaningless term 'yin'. He can understand the meaning of 'yin' by being given a plain definition. But even if he could make sense of (4) (in the sense that it does not involve any meaningless term and the terms are not combined in any illegitimate way), he would still be left in a fog. What is *the point* of what is being presented or argued for by (4)? Or what is *the thought* that (4) expresses? To know the point of what is being said by (4), Dr Smith has to know the proper contexts in which (4) can be used to say something true or false, namely, to know its truth-value conditions. But such contexts are not *conceptually recognizable* to him, for he does not comprehend the conceptual framework (consisting of the pre-modern Chinese mode of reasoning, its underlying yin-yang cosmology, and the related medical category system) within which these possible contexts are constructed. To grasp the thought expressed by (4), Dr Smith needs to comprehend the pre-modern Chinese mode of reasoning and the related category system that are central to the thought and presupposed by the proposition expressed by (4).

Unfortunately, it is the pre-modern Chinese mode of reasoning and the related category system that are totally alien and scarcely comprehensible to Dr Smith. The goal of attunement to nature was so highly valued in the pre-modern Chinese culture that it became a necessary condition of understanding and rational justification. Then it is possible that what ancient Chinese thought to be rationally justifiable and perfectly intelligible is not at all rationally justifiable and intelligible in an intellectual tradition that severs the connection between understanding and attunement. This is what actually happens when Dr Smith encounters CMT. As C. Taylor points out, the world for the European intellectual tradition ceased to be a possible object of attunement after the rise of modern science. Instead, the world became alienated from human beings and became the object of investigation, experiment, and control. The original connection between understanding and attunement was severed, dismissed as mere projection onto the world order of things human beings find meaningful (Taylor, 1982). Therefore, the pre-modern Chinese mode of reasoning, which values attunement so highly, is totally alien to the modern Westerner and thus hard to understand. This explains why many substantial sentences of the language of CMT such as (4) sound so strange to Dr Smith that he cannot fully grasp them.

It seems to be clear that the understanding of CMT is, for Dr Smith, an entirely different exercise from relearning phlogiston theory. Of course, it does not mean that Dr Smith cannot learn the language of CMT. However, for Dr Smith, the language acquisition process involved in learning CMT is different from that involved in learning phlogiston theory. Using a metaphor, we can say that the former is a 'wholesale' learning process while the latter is a 'retail' learning process. CMT reflects a unique belief system and embodies a specific form of life. By studying the Chinese intellectual tradition, Dr Smith should be able to comprehend the pre-modern Chinese mode of reasoning, the underlying yin-yang cosmology, and the categorical system presupposed by the language of CMT. After such a 'wholesale' learning, he is able to understand effectively the theory and even talk in the pre-modern Chinese way. However, Dr Smith at best can start the pre-modern Chinese way of speaking only if he becomes alienated or dissociated from the thought and the way of speaking used in the modern Western intellectual tradition.

The essential role of M-presuppositions in cross-language understanding is observed by many others. Recall that Hacking emphasizes the essential role of a style of reasoning in understanding a scientific tradition, such as Paracelsan medical theory: 'Understanding is learning how to reason' (Hacking, 1982, p. 60). Until one has learned how to reason in the speaker's way one cannot understand effectively what the speaker is saying. This is especially true for the understanding between two P-languages within two substantially disparate traditions or cultures. 'Understanding the sufficiently strange is a matter of recognizing new possibilities for truth-or-falsehood, and of learning how to conduct other styles of reasoning that bear on those new possibilities' (Hacking, 1982, p. 60). While Hacking emphasizes the necessary role of modes of reasoning in effective cross-language understanding, Kuhn, in a similar spirit, explores the role of a categorical framework in effective understanding. To understand effectively a scientific language, one has to familiarize oneself with its lexical structure. For instance, the Aristotelian assertions in which terms like 'force' and 'void' play an essential role are perfectly intelligible with the Aristotelian lexicon in place. But apparently similar assertions are barely intelligible to a Newtonian who is unfamiliar with the Aristotelian lexicon. This was why Kuhn became lost when he first encountered Aristotelian physics.

For A. MacIntyre, 'a language may be so used ... that to share in its use is to presuppose one cosmology rather than another ...' (MacIntyre, 1985, p. 7). A language reflects a way of life. Choice between two languages is to choose between two alternatives and, sometimes, incompatible sets of beliefs and ways of life. 'Moreover each of these sets of beliefs and ways of life will have internal to it its own specific modes of rational justification in key areas and its own correspondingly specific warrants for claims to truth' (MacIntyre, 1985, p. 8). In this way, we are unable to find application for the concepts of truth and justification that are independent of the standards of one language community or the other. Therefore, the language of a culture or tradition is fully intelligible and its purported justification is adequately understood for the interpreter from another radically different culture or tradition only if its underlying cosmology, its way of life, and its scheme of beliefs embodied in the language are fully comprehended (MacIntyre, 1985, p. 13). Similarly, J. Habermas contends that understanding a language is sharing a form of life (a lifeworld). Forms of life cannot be described fully in another incompatible language. One who does not share the form of life cannot see the 'point' of the language and cannot put oneself in the position of being an engaged communicator. Consequently, effective understanding cannot be achieved.

Let us collect all these insights that we have revealed so far about the role of M-presuppositions in cross-language understanding. Theoretically, 'what a sentence of an alien language means' can be used to refer either to the meanings of the *words* used in the sentence, or to the thought expressed by it. To understand what a sentence of an alien language means is not just to know the meanings of its words. A good dictionary can help us with that. But it cannot help us understand the thought expressed by the sentence. To effectively understand a sentence of an alien language is not just simply to make sense of it, but rather to grasp the thought expressed by it. To know the thought expressed by a sentence, it is necessary to know that it is assertable or that it has a point, and to know what it asserts or what its point is. As S. Cavell notes, 'we can understand what the *words* mean apart from understanding why you say them; but apart from understanding the point of your saying them we cannot understand what you mean' (1979, p. 206). If a sentence is comprehensible to the interpreter, he or she has to understand the point of what is being said, being presented, or being argued.

Whether or not a sentence of a P-language, when considered within the context of the interpreter's language, can be used to make an assertion (has a truth-value) is language dependent. More precisely, as I have argued in chapter 9, the assertability of a sentence of a language is determined by the M-presuppositions of the language. A sentence that is apparently the same could be used to assert something or have a point within the context of one language but without a point in a rival one. This establishes the fact that in order to capture the point of what is being said by a sentence of an alien P-language, it is necessary to comprehend its M-presuppositions. There is a conceptual bridge that connects the two referents of the expression of 'what a sentence of an alien language means', namely, 'the *meanings* of the words used by the sentence' on the one hand and 'the *thought* expressed by the sentence' on the other. The connection is not established by universals, propositions, or rules, but rather by the M-presuppositions that make certain syntactical utterances become *assertions*. In conclusion,

A P-language is fully intelligible to the interpreter *only if* he or she can conceptually recognize and comprehend the M-presuppositions underlying the language, including the mode of reasoning with its underlying cosmology, the categorical framework, as well as other universal principles and existential assumptions.

Thus, we have so far identified *a significant necessary condition* of effective understanding of a P-language, i.e., the knowledge of its M-presuppositions.

It seems to be questionable for some whether one needs to share M-presuppositions of an alien language in order to conduct some ordinary linguistic acts. For example, one could order a bowl of gavagai stew from a native whether or not one shares the M-presuppositions of the native language in which 'gavagai' could mean rabbit, undetached rabbit parts, rabbit time slice, etc. To respond, I think we should distinguish how to effectively understand a language from how to conduct an ordinary linguistic act. If one wants to understand the native language in which 'gavagai' plays an essential role (as 'phlogiston' does in the phlogiston theory), then one needs to recognize and comprehend the specific lexical taxonomy about 'gavagai'. In contrast, to conduct a rather ordinary linguistic act of ordering a bowl of gavagai stew from the natives, it is not necessary for one to share the lexical taxonomy. Within a concrete linguistic context, there are some other more direct ways to perform such an act. One could simply point to the stew that looks like rabbit stew and say, 'I want this'. However, suppose that 'gavagai' means different things to the natives at different times (it means 'rabbit' in the morning, but 'rabbit time slice' in the afternoon). If one says 'I want gavagai stew' in the afternoon, the native would be quite confused.

Returning to our initial question, Dr Smith is able to make sense of, but is unable to effectively understand (4), (30), and (31) because he cannot comprehend the respective M-presuppositions underlying these sentences. After relearning the meaning of 'phlogiston', he can quite well understand (30). However, to fully understand (4), he has to learn the pre-modern Chinese form of life—with the pre-modern Chinese mode of reasoning, the yin-yang cosmology, and its category system as its core—together with Chinese medical theory itself.

3. A Truth-Value Conditional Account of Understanding

It has been argued above that effective understanding of a sentence in an alien language should be distinguished from making sense of the sentence. But what exactly does it mean 'to make sense of' a sentence on the one hand and 'to effectively understand' it on the other? It seems safe to say that one understands a sentence if one knows what it means. According to this everyday manner of speaking, the phrase 'to understand' is an abbreviation of the phrase 'to know the meaning of'. But to make sense of a sentence seems no more than to make it meaningful. If so, the distinction between effective understanding and making sense seems to be blurred.

Suppose someone was now to say sentence (16), 'The present king of France is bald'. No one would deny that (16) is cognitively significant. The problem is how to explain the significance of such an obviously vacuous sentence. For B. Russell, only sentences with truth-values could be significant. A sentence with a nondenoting subject like (16) has a truth-value (it is false) and thus is significant (Russell, 1905). By contrast, for P. Strawson, a sentence could be both significant and truth-valueless. For example, a vacuous sentence is truth-valueless (due to the failure of its presupposition) although it is obviously significant. Strawson thinks that the alleged connection between significance and bivalence should be severed based on his distinction between a sentence and the use of a sentence. The same sentence can be used, by different persons in different linguistic contexts, to make different assertions; different sentences can be used to make the same assertion. Significance is a semantic property of sentences while truth-value is the function of the use of sentences (or a semantic property of assertions).

> To give a meaning of a sentence is to give *general directions* for its use in making true or false assertions. … For to talk about the meaning of an expression or sentence is not to talk about its use on a particular occasion, but about the rules, habits, conventions governing its correct use, on all occasions, to refer or to assert. So the question of whether a sentence or expression is *significant or not* has nothing whatever to do with the question of whether the sentence, *uttered on a particular occasion*, is, on that occasion, being used to make a true-or-false assertion or not. (Strawson, 1950, p. 220; italics as original)

For example, sentence (16), uttered by someone today, is certainly significant, since every word of it is meaningful and it follows correct grammatical conventions. Nevertheless, this does not mean that any particular use of (16) has to be true or false. If it were uttered by someone today, it is truth-valueless (since one of its presuppositions, (16a), is false). But if the same sentence were uttered by someone in the reign of Louis XV, it had a truth-value. Therefore, whether a sentence is significant should be separated from whether it has a truth-value in some specific context. Specifically, according to Strawson, to say that a sentence is significant is to say that it has a *possible* truth-value. Or, more precisely, as long as the sentence satisfies linguistic conventions governing its correct use, it *could be used* in certain contexts to say something true or false.

Sensefulness and Common-Sense Understanding

Strawson's above distinction tells us that meaningfulness and sensefulness are conceptually connected with the *possible* truth-value, instead of with the *actual* truth-value of a sentence. 'Sensefulness' may be defined based on logically possible truth-value:

> A sentence S of an alien language L is senseful to the interpreter who speaks language L_1 if and only if S when considered within the context of L_1 has a *logically possible* truth-value.

Recall that, as we have clarified in chapter 9, a sentence has a logically possible truth-value if it could be used to say something true or false in certain logically possible contexts. Usually, as long as a sentence is in good semantic and syntactic order, it could be used to say something true or false in certain logically possible contexts. Most of the so-called pseudo-science assertions identified by logical positivism belong to this category. For instance, sentence (31) is in a good semantic and syntactic order such that there are some logically possible contexts in which (31) could be used to say something true or false. Hence, (31) has a sense. In contrast, the following sentences are senseless:

(38)   Loves Jenny falling Bob in.
(39)   Three sapdlaps sat on a bazzafrazz.
(40)   The number two is red.

(38) is not a well-formed sentence. (39) contains meaningless terms. (40) is little different from the first two. Having fixed on the natural language meanings of the subject and predicate, we find that (40) is neither true nor false since its sortal presupposition, i.e., 'A number can be colored', is false. It is not just truth-valueless in some possible worlds, it is rather neither true nor false in any possible worlds in which we continue to use these words with their ordinary meanings. In other words, sentence (40) could not be used, in any context, to make an assertion.

Accordingly, the notion of 'making sense of' or common-sense understanding (i.e., 'understanding$_c$') may be defined as follows:

The interpreter can make sense of or understand$_c$ S if and only if (a) S is senseful to him or her and (b) he or she knows the sense of S.

Meaningfulness and Effective Understanding

However, to say that a sentence has a *logically* possible truth-value does not mean that the truth-value is *conceptually* recognizable to the interpreter, since the interpreter may not be able to recognize its truth-value conditions. For example, our Dr Smith could not specify the truth-value conditions of (4). The sentence has no conceptually possible truth-value to him. To distinguish making sense from effective understanding, it is useful to distinguish meaningfulness from sensefulness based on conceptually possible truth-values defined in chapter 9:

A sentence S in an alien language L is meaningful to the interpreter who speaks language L$_1$ if and only if S when considered within the context of L$_1$ has a *conceptually possible* truth-value.

Recall that S has a conceptually possible truth-value for one if and only if one can recognize and comprehend its truth-value conditions. According to Convention P, one can know the truth-value conditions of S only if one is able to comprehend its presuppositions. One can know the truth-value conditions of core sentences of a P-language only if one can know their shared absolute presuppositions. Those presuppositions are nothing but what we call the M-presuppositions of the language. As we have argued in the previous section, to comprehend the M-presuppositions of a P-language is necessary for one to effectively understand the language. Thus, we can define the notion of effective understanding as follows:

> The interpreter who speaks language L_i can *effectively* understand a sentence S of an alien language L if and only if (a) when considered within the context of L_i, S is meaningful to the interpreter, and (b) the interpreter knows the literal meaning (the propositional content) of S.

The notion of effective understanding so defined suggests that the understanding of a sentence S of a language is actually a two-staged cognitive process. First, is S meaningful (or does S have a conceptually possible truth-value) to the interpreter? S has a conceptually possible truth-value to the interpreter if and only if the interpreter can *recognize and comprehend* the truth-value conditions of S. Second, if S is meaningful to the interpreter, then what is its meaning? According to Davidson's truth-conditional theory of meaning, to know the truth conditions of S is sufficient to know its literal meaning. Notice that the truth conditions of concern here are not Davidson's truth conditions of a sentence with an *actual* truth-value, but rather the truth conditions of a sentence with a *conceptually possible* truth-value. To indicate this distinction, we may call the latter the *possible* truth conditions of S. The interpreter knows the literal meaning of S if he or she knows its *possible* truth conditions.

Knowing the truth-value conditions of a sentence is a prerequisite to knowing its truth conditions. For this reason, the account presented above may be referred to as *the truth-value conditional theory of understanding*. According to this theory, the interpreter can effectively understand a sentence only if he or she knows its truth-value conditions. Accordingly, the interpreter is able to effectively understand a P-language only if he or she knows the truth-value conditions of its core sentences. Furthermore, as was argued earlier, the interpreter can know the truth-value conditions of the core sentences of a P-language only if he or she is able to comprehend its M-presuppositions. This brings us back to the same conclusion drawn from the previous section: It is necessary for the interpreter to be able to identify and comprehend the M-presuppositions of a P-language in order to understand the language effectively.

It might be asked what counts as to comprehension of the M-presuppositions of a P-language? Or what are the conditions for comprehending them? An M-presupposition of a P-language (i.e., 'phlogiston exists') is usually a statement or a set of statements that are accepted as either true or false by both sides of the communication. Based on Davidson's truth-conditional theory of understanding, it can be comprehended by knowing its Tarskian truth conditions.

Factual Meaningfulness

At last, we notice that some sentences of an alien language are meaningful to the interpreter only when they are put forward within certain possible contexts that the interpreter is obliged to reject as unsuitable. This is because one may be able to identify and comprehend a conceptually possible context in which a sentence says something true or false but does not regard the context as a genuine one. For instance, although the sentence (30) is meaningful for modern chemists who happen to know the existential assumption behind it, they simply deny the truth of the assumption. Similarly, we can understand the story of *Snow White* very well by imagining a world in which the story tells us something true or false. But we deny the actual reality of such an imaginable world. In this case, sentences can have conceptually possible truth-values (be meaningful for the interpreter), but lack actual truth-values. We say that those sentences are not factually meaningful to the interpreter. By definition,

> A sentence S of an alien language is factually meaningful to the interpreter who speaks language L if and only if S when considered within the context of L has an *actual* truth-value.

According to the notion of actual truth-value defined in chapter 9, a sentence of an alien language has an actual truth-value for the interpreter if and only if the interpreter holds its underlying presuppositions to be true. That means that S is factually meaningful to the interpreter if and only if S's presuppositions are held to be true. For the advocate of phlogiston theory, (30) is factually meaningful, no matter whether it is actually true or not, since the existential presupposition of (30) is held to be true by the advocate. When the interpreter regards the core sentences of a language as factually meaningful, he or she commits him or herself to the truth of its M-presuppositions.

A Semantic Indicator of the Failure of Effective Understanding

We have concluded that a P-language is fully intelligible to the interpreter *only if* he or she can conceptually recognize and comprehend the M-presuppositions underlying the language. The interpreter who fails to grasp the M-presuppositions of the language could not effectively understand it. However, M-presuppositions, such as the mode of reasoning or the lexical taxonomy of a P-language, are not easily identifiable. This poses a problem in knowing whether the interpreter is able to identify and comprehend the M-presuppositions of a P-language. It would be helpful if we could locate a clearly identifiable semantic indicator for the failure of cross-language understanding.

As observed earlier, for the Western physician, due to the inability to identify and comprehend the underlying pre-modern Chinese mode of reasoning and its associated yin-yang cosmology, the core sentences of traditional Chinese medical theory are simply not candidates for truth or falsity. In other words, to the Western physician, they do not possess conceptually possible truth-values. Similarly, Hacking has noticed a strong linguistic correlate of our failure in understanding Paracelsus. Numerous Paracelsan sentences, when considered within the context of modern scientific theories, do not have any conceptually possible truth-values, because we cannot comprehend the Renaissance mode of reasoning underlying the language. Kuhn concurs. When the modern interpreters find Aristotelian sentences hard to understand, the trouble is not that they think Aristotle wrote *falsely*, but that they cannot attach *truth or falsity* to many Aristotelian core sentences since the Aristotelian lexical taxonomy presupposed by the sentences is totally alien to them.

Our case study of the Newton–Leibniz debate on the absoluteness of space (chapter 5) tells us a different story. We found that the confrontation between the Newtonian and the Leibnizian languages of space is different from the confrontation between Chinese and Western medical languages. Not like the Western physician who cannot identify and comprehend the M-presuppositions of the Chinese medical theory, the Leibnizian is able to identify and understand the underlying M-presupposition of the Newtonian language, namely, the existence of Newtonian absolute space. Hence, for the Leibnizian, the set of sentences S_N-S_L has conceptually possible truth-values (i.e., is meaningful), and thus the Newtonian language is perfectly intelligible. Nevertheless, the Leibnizian categorically rejects the assumption of absolute space, which makes the sentences in S_N-S_L *actually* truth-valueless (lack of actual truth-values). Therefore, it is the lack of conceptually possible truth-values, not the lack of actual truth-values, that is semantically correlated with the failure of cross-language understanding.

In fact, we could derive such a semantic indicator from our truth-value conditional account of understanding. When the interpreter fails to comprehend the underlying M-presuppositions of an alien P-language, based on Convention P, the core sentences of the language would be rendered conceptually truth-valueless to the interpreter since he or she cannot identify and comprehend the truth-value conditions of those sentences. Lack of conceptually possible truth-values makes those sentences meaningless to the interpreter, and results in his or her failure of effective understanding of the alien language. We can hence use the lack of *conceptually possible* (not actual) truth-values as a strong semantic indicator of the failure of effective understanding on the interpreter's part. In general, if the speaker of a P-language PL_1 is unable to recognize and comprehend the M-presuppositions of an alien P-language PL_2, then the core sentences of PL_2 when considered within the context of PL_1 will lack conceptually possible truth-values. Suppose the speaker of PL_2 faces the same situation when encountering PL_1. Then a conceptually possible truth-value gap occurs between PL_1 and PL_2, which signifies semantically the failure of mutual cross-language understanding between the two language communities.

4. Complete Communication Breakdown

It has been argued that the interpreter can effectively understand a P-language only if he or she is able to identify and comprehend its M-presuppositions. M-presuppositions are language dependent. Hence, effective understanding is language dependent. Just as it is not useful to ask whether a sentence itself is true or false, but only whether its specific use within a linguistic context is true or false, so it is not useful to ask whether or not a sentence in isolation is meaningful. We can only ask whether, when considered within the context of a specific language, it is meaningful, and what its meaning is. Therefore, the core sentences of a P-language that are meaningful and can be understood in the contexts of its own or some other compatible P-languages (languages with compatible M-presuppositions) might not be fully understood when considered within the context of some incompatible P-languages (languages with incompatible M-presuppositions).

We need to emphasize that the essential role of contexts in effective understanding is not that the same sentence is meaningful in one context, but meaningless in the other. Rather, it is the difference between using a sentence to say something true or false in one context when the same sentence says nothing at all in the other context. It is not a contrast between affirmation and counter-affirmation, but a contrast between affirmation and silence. When a sentence is considered within an unsuitable context, it could sound unintelligible to the point where what is being said by the sentence totally eludes the interpreter.

The Projective Approach to Cross-Language Understanding

The language-dependent feature of effective understanding has a significant impact on cross-language understanding. Presumably, the degrees of difficulty and the types of understanding involved when the interpreter encounters an alien P-language depend upon the relation between the alien language and the interpreter's own language, especially the familiarity of the interpreter to the M-presuppositions of the alien language. First, there will be cases in which the M-presuppositions of the alien P-language are compatible with those of the interpreter's own language, which we will call *normal discourse*. Within normal discourse, we may make a further rough-and-ready distinction based on whether the interpreter is familiar with the alien language. When the interpreter is familiar with the language, understanding happens without a glitch. When the language is initially unfamiliar to the interpreter but either lies in the recent past of the interpreter's own cultural or research tradition or comes from a closely related cultural or research tradition, the interpreter could still have a measure of tacit familiarity with its M-presuppositions by virtue of his or her own historical, cultural, or research tradition. Thus, the interpreter can still comprehend the M-presuppositions of the alien language based on his or her historically formed tacit 'pre-understanding'. The understanding of the language could be readily obtained.

Second, by contrast, there will be cases in which the interpreter encounters an alien P-language whose M-presuppositions are incompatible with those of his or her own language, which we will call *abnormal discourse*. Even within an abnormal discourse in which the M-presuppositions of an alien language are substantially distinct from those of the interpreter's own, the interpreter could still be familiar with the language if the interpreter can gain access to it through interpretation already essayed by historians or anthropologists (such as the language of Aristotelian physics) or if two languages coexist within a broad intellectual tradition (such as the Newtonian language and the Leibnizian language of space). However, the real challenge arises when the interpreter faces an alien P-language with which he or she is not familiar and has not even a tacit familiarity because it belongs to a cultural, research tradition other than his or her own (such as traditional Chinese medical theory for a Westerner), or because it exists in the distant past and lacks a continuous history of interpretation and appreciation within the interpreter's own cultural tradition (such as the Paracelsan medical language for a contemporary Westerner).

Obviously, normal discourse scarcely concerns us here; our central concern in explication of incommensurability is with abnormal discourse. When two disparate P-language communities confront one another, each with its own body of M-presuppositions, but lacking a knowledge of each other, the speaker of one P-language often falls into the temptation of approaching the other unknown P-language by imposing, reading into, or projecting the categories, beliefs, or the mode of reasoning embodied in the speaker's own language upon the other. Each community will usually represent the beliefs of the other within its own tradition, in abstraction from the relevant tradition of the other. This was the strategy used by Kuhn in that hot summer of 1947 when he attempted to 'approach Aristotle's texts with the Newtonian mechanics' (Kuhn, 1987, p. 9). This is a phenomenon frequently encountered by historians or anthropologists. For lack of an alternative historians or anthropologists are tempted to understand an old or alien text as they would if it had occurred in either contemporary discourse or in their own culture or tradition. As we have discussed, Davidson defends such a projective understanding by presenting his version of the principle of charity, which requires, in interpretation, to put the speaker in general agreement with the interpreter according to the interpreter's view. The principle is obviously powerful and could at least be utilized to jump-start the interpretation process. However, the principle is *too* 'powerful' to the extent that it is vulnerable to being abused for ethnocentric ends. The principle tells us that to understand others, we have to render them so that as much as possible they speak the truth, make valid inferences, and so forth according to our notion of truth, truth conditions, and logical rules. We have to make them think like us, attribute our own beliefs to them, and interpret their utterances within the system of our own language.

There is a hidden assumption behind the above projective approach to understanding: Others are basically like us, sharing the same linguistic conventions, belief systems, and, most importantly, M-presuppositions. This assumption is a manifestation of absolutism in cross-language understanding, which is based on a basic conviction that there is or must be some permanent, ahistorical, culture-transcendent matrix or framework (R. Rorty) to which one can ultimately appeal in determining the nature of rationality, intelligibility, truth, reality, and morality. To make the discourse of others intelligible and rational one needs to be able to find some area of agreement. Within the analytic tradition, the desired agreement has often been imagined to lie in some common language. Specifically, in the discussion of cross-language understanding the agreement manifests itself as shared or compatible M-presuppositions. Absolutists believe that there is or should be a common or third language between any two languages if understanding between them is possible. Such a third language would have two central features: (a) It does not presuppose allegiance to either of the two sets of M-presuppositions associated respectively with the two languages in question, nor presuppose any other set of M-presuppositions that might compete for allegiance with those two; (b) It must be able to provide semantic and ontological resources for an accurate representation of the two languages, including the M-presuppositions embodied in each language.

Complete Communication Breakdown in Abnormal Discourse

The projective approach to understanding is justifiable only when the above absolutist assumption is sound. In normal discourse in which the M-presuppositions of an alien P-language are compatible with that of the interpreter's own language, the projective approach can proceed without much difficulty. The interpreter is able to understand the other language since he or she can recognize its M-presuppositions by way of analogy or by simply taking his or her own as the other's. Normal discourse is supposed to be contrasted with abnormal discourse in which both sides speak P-languages with incompatible M-presuppositions. Roughly, normal discourse proceeds at the stage of Kuhnian normal science while abnormal discourse happens during Kuhnian scientific revolution.[3] However, the projective approach would ensure the failure of cross-language understanding in abnormal discourse. Projecting the M-presuppositions of the interpreter's own language upon an alien P-language would suspend or distort the M-presuppositions of the latter. Suspending the alien P-language's M-presuppositions would make its core sentences conceptually truth-valueless and hence suspend all empirical contents of meaningful statements of the language. By distorting them, the interpreter puts the original meaningful statements out of their appropriate contexts and hence causes them to lose their original meanings. Either way prevents recognition and comprehension of the M-presuppositions of the alien language. Lack of knowledge of the M-presuppositions of an alien P-language is sufficient to preclude effective understanding of the language. From each point of view, the key concepts and core statements of the other—just because they are presented apart from the linguistic context constituted by its own M-presuppositions from which they draw their conceptual life—will necessarily appear without context, lack justification, and hence become meaningless and unintelligible. Therefore, in abnormal discourse, the projective approach to understanding is doomed to failure.

This is the real source of the failure of cross-language understanding that we have often experienced between two conceptually disparate languages.

When the speakers of two P-languages cannot understand one another, no communication acts can be carried out. The communication between the two P-language communities breaks down *completely*, as is often experienced by the Chinese medical community and the Western medical community. I call this kind of communication breakdown between two P-language communities caused by lack of effective understanding *complete* communication breakdown.

Notes

1   Habermas, 1992, p. 57; 1987, p. 62.
2   In the following discussion, I will treat the literal and the propositional content of a declarative sentence and the (objective) thought expressed by it roughly as the same conception and ignore the subtle distinction between them.
3   The terms 'normal discourse' and 'abnormal discourse' are borrowed from R. Rorty (1979, Ch. 7), but I use them in a different way here.

Chapter 13

Hermeneutic Understanding in Abnormal Discourse

A complete communication breakdown between two P-language communities due to lack of mutual understanding is a rare phenomenon. Within the same cultural or intellectual tradition, such a breakdown can only exist between two disparate P-languages separated by time and where one has undergone a radical conceptual shift; for example, the communication breakdown between the speakers of Aristotelian physics and Newtonian physics (as Kuhn experienced in 1947) or between the speakers of Paracelsan medical theory and contemporary Western medical theory. A complete communication breakdown can also happen between two P-language communities existing in two radically disparate intellectual traditions or cultures, for example, the complete communication breakdown experienced by Chinese medical and Western medical communities.

More significantly, complete communication breakdowns are not only rare, but also are *contextual* and *temporary* phenomena, which in principle can be overcome. We have found that mutual *effective* understanding, which is *propositional* understanding in character, between the speakers of two P-languages with disparate M-presuppositions is problematic and difficult, for it is hard for one side to identify and comprehend the M-presuppositions of the other's language. Worse still, a complete communication breakdown between them occurs when the speakers of one language attempt to approach the other language by projecting the categories, beliefs, and mode of reasoning of their own language onto the other language. However, the failure of the projective approach to understanding in abnormal discourse by no means implies the impossibility of mutual understanding with one another. It certainly does not mean that mutual understanding between two P-language communities with incompatible M-presuppositions is *unattainable in principle*. It would be a mistake to think that one cannot understand an alien P-language *per se*. Any human-made P-language (thought, theory, ideology, conceptual scheme), no matter how remote it is from our own, is an open linguistic system, open to possible modification, expansion, and evolution (both syntactically and semantically), and open to human understanding. One can at least make an alien language intelligible, no matter how strange and remote it is, by language learning. 'Shared neurology and overlapping environments make it extremely likely that any speaker of one human language can, with sufficient effort, always learn another' (Kuhn, 1999, p. 34).

This conclusion is not just our sheer conviction, but it is actually implied by my truth-value conditional theory of cross-language understanding. Since understanding and intelligibility of any P-language is linguistic-context dependent, anything which can be said in one P-language[1] can, with imagination and effort, be understood by the interpreter of another P-language as long as he or she is willing to immerge him or herself into the linguistic context. No matter how sedimented a language, culture, or tradition might become, it is never resistant to unfolding itself and disclosing something new. In fact, if we take understandability or interpretability (not translatability as Quine and Davidson do) as the standard of languagehood, then the idea of a language being categorically unintelligible or unlearnable in principle is incoherent.

The occurrence of a complete communication breakdown between two P-language communities indicates that cross-language understanding occurs in abnormal discourse. The failure of the projective approach in abnormal discourse only shows that the process of understanding is not yet finished, which needs to be completed by approaching the aliens in a different way. It reminds us that *the kind of understanding* involved in abnormal discourse—which is, as we will show, hermeneutic, not propositional in nature—is more subtle, more elusive, and more dialectic than that involved in normal discourse. We had better proceed with care.

I have suggested that the as-up-to-now complete communication breakdown due to lack of effective understanding in abnormal discourse can be overcome by somehow reaching understanding. The question faced by would-be communicators, when experiencing a complete communication breakdown, is this: How can we do justice to an unfamiliar P-language without falsifying or distorting it? To ask the same question with a Kantian twist: How is understanding in abnormal discourse *possible*? What *kind of understanding* is more effective in abnormal discourse? In the following sections, we will discuss three ways to achieve mutual understanding in abnormal discourse: the adoptive approach to understanding, language learning, and Gadamer's hermeneutic understanding.

1. The Adoptive Approach to Cross-Language Understanding

Put Oneself into the Other's Skin

The failure of the projective approach tempts many into the opposite approach: 'going native' by putting themselves into the alien's skin, or jumping into the stream of the other's consciousness. Compared with the projective approach that is a third-person, objectivist approach to understanding, this is a first-person, phenomenological approach to understanding. This temptation is very strong in the field of anthropology, which is a wide-ranging discipline where many issues touched upon in discussions of incommensurability come into sharp focus. One general problem that has been exercised in methodological discussion in anthropology is this: How is anthropological knowledge of the way natives think, feel, and perceive possible? As many believe, anthropological understanding stems from some sort of extraordinary sensibility, an almost preternatural capacity to think, feel, and perceive like natives. It is believed that rather than attempt to place the experience of natives within the framework of our own, we must, if we are to achieve understanding, set our own framework aside and view their experiences within their own framework. For example, according to anthropologist B. Malinowski (1965, pp. 11-15, 21-2), the task of anthropological linguists is not to show the deficiencies of 'primitive' languages or to locate common roots, if any, between the alien language and their home language. Instead, the meanings of words and sentences must be understood and described within 'contexts of cultural reality' often radically different from interpreters' home language. Practical and ritual activities, facial expressions and gestures, story telling and conversation, social interactions and ceremonial interchanges, all contribute to the social contexts in which the language must be understood.

Against Kantian and Enlightenment efforts to portray knowledge and understanding in universal terms, the romantic and historical hermeneutists of the nineteenth century contend that the task of understanding is to get beyond our own perspectives and to think in terms of the context of the author's intention or 'the true meaning' of the text. For F. Schleiermacher, the understanding of unfamiliar texts is essentially a matter of psychological interpretation, a placing of oneself in the mind of the author, a recreation of the author's creative act. Thus, the art of understanding becomes the reconstruction of the original production. While Schleiermacher emphasizes the subjective aspect of understanding, W. Dilthey focuses on the historical context of understanding. For Dilthey, the development of historical consciousness is its liberation from every dogmatic bias. Historical consciousness means understanding things in terms of the values and standards of their own time, putting the interpreter's own values and standards out of action. Dilthey seeks to overcome the historical conditionedness of the interpreter and achieve objective knowledge of the historical conditioned. Historians must place themselves within the spirit of the age they are studying and think with its ideas and its thoughts, not with their own. In fact, historical understanding requires the self-extinguishing of the interpreter.

Perhaps the two most provocative versions of the adoptive approach to understanding are P. Winch's social-cultural relativism and B. Whorf's linguistic-cultural relativism. Winch is occupied with the problem of determining what is the best way to understand and interpret different cultures and societies so that we can learn from them. He finds that many influential studies of primitive societies have gone astray because they are frequently insensitive to the different 'points' and 'meaning' of the other's activities and beliefs that would reveal to us 'different possibilities of making sense of human life'. Winch is protesting against the pervasive ethnocentricism at the time, whereby we measure and judge what is initially alien to us by our present standards of rationality, righteousness, and truth, as if those standards were the sole and exclusive measures of those 'universal' human values. To understand a society or culture is, for Winch, to make it intelligible. But the notion of intelligibility is context-dependent. Out of context, the notion is systematically ambiguous. 'A single use of language does not stand alone; it is intelligible only within the general context in which the language is used' (Winch, 1958, p. 39). Each society or culture has its own criterion of rationality and intelligibility. We cannot say which one is better. The criterion is formed and defined within a specific form of life. A form of life, as Wittgenstein defines and argues, is the given that we have to accept. Forms of life endorsed in different societies or cultures may be so radically different from each other that in order to understand alien or primitive societies we not only have to *bracket* our prejudices and biases, but also have to *suspend* our own standards and criteria of rationality and intelligibility. Therefore, to understand an alien or a primitive society it is necessary to understand it somehow from within, on its own terms and by its own standards (Winch, 1958, pp. 18, 35, 39, 40, 107).

Based on his extensive research on the language of the American Hopi, Whorf argues that the Hopi language is embedded with (or, simply *is*) a metaphysics and a cosmology radically different from those found in European languages. When a linguist 'imposes his own ontology and linguistic pattern on the native', he fails to understand the Hopi speaker. The Hopi grammar, with no verb tenses as we know them and with no reference to kinetic rather than dynamic motion, is not an alternative way to express Western Newtonian time and space. It is an alternative way of thinking of and understanding the basic elements of reality. Although pragmatically equivalent English expressions can be 'recast' in Hopi terms, this should not be confused with understanding. To truly understand Hopi, one has to try to think in the Hopi way, or at least to achieve relative consonance with the system underlying the Hopi view of the universe. In Whorf's notion of understanding, not-native expressions are mapped approximately onto the usage of a home language, but the concepts of the home language are reshaped to communicate a different and alien view of the world.

You Do Not Have to Be One to Know One

As a solution to the failure of understanding in abnormal discourse, the adoptive approach to understanding faces many problems. H.-G. Gadamer puts it best:

> In reading a text, in wishing to understand it, what we always expect is that it will inform us of something. A consciousness formed by the authentic hermeneutic attitude will be receptive to the origin and entirely foreign features of that which comes to it from outside its own horizon. Yet this receptivity is not acquired with an objectivist 'neutrality': it is neither *possible, necessary*, nor *desirable* that we put ourselves within brackets. (1979, pp. 151-2; my italics)

Just how can one jump into the stream of the other's experience? Gadamer reminds us that an interpreter always belongs to a certain tradition, culture, and language before a tradition, culture, or language belongs to the interpreter. One is ontologically grounded in one's own tradition, culture, and language. It is impossible for one to leave behind one's own tradition and to place oneself in other traditions. This is because one's tradition, culture, and language are not contingent properties of an abstract self but are constitutive of one's being (though one's own tradition and horizon can evolve and be modified and expanded). One simply cannot place oneself in the other tradition without losing oneself. 'We become fools of history if we think that by an act of will we can escape the prejudgments, practices, and traditions that are constitutive of what we are' (R. Bernstein, 1983, p. 167). Anthropologist C. Geertz concurs:

> The ethnographer does not, and in my opinion, largely cannot, perceive what his informants perceive. What he perceives—and that uncertainly enough—is what they perceive 'with', or 'by means of', or 'through' or whatever word one may choose. (Geertz, 1979, p. 228)

No matter how accurate or half-accurate one can be about what one's informants are 'really like', it comes from the ethnographer's own experience, not from the informants' experience. 'In the country of the blind, who are not as unobservant as they appear, the one-eyed is not a king but spectator' (Geertz, 1979, p. 228).

The champion of the adoptive approach could maintain that even if one accepts as an empirical fact that one cannot sufficiently attain the ideal of leaving oneself behind, this would still be a legitimate ideal that one should strive to reach as far as possible. For Gadamer, what is wrong with the adoptive approach to understanding is not just that it is impossible, but rather that it is undesirable, and even detrimental, for achieving cross-language understanding. As I will detail later, according to Gadamer's hermeneutic understanding, not only do all understandings inevitably involve some prejudices, but also the understanding of the alien tradition, culture, and language is only possible based on prejudgments provided by one's own tradition. The past can only be revealed to one in one's present situation; the other can only speak to one in one's own tradition. Therefore, to 'try to escape from one's own concepts in interpretation is not only impossible but manifestly absurd. To interpret means precisely to bring one's own preconceptions into play so that the text's meaning can really be made to speak for us' (Gadamer, 1989, p. 397).

Moreover, the adoptive approach to understanding is founded on a monologue model of meaning, according to which the meaning of a text is inherent in the text to be understood, as it is given by the author and co-determined by his or her historical situation. The text is self-enclosed and its meaning is self-contained—simply there to be discovered—independent of the interpreter. The aim of understanding is to grasp the subjective intentions of the author or the original meaning of a text. The interpreter, therefore, has to purify him or herself of any prejudices and immerse himself or herself in the original context in order to understand the other. As we will see shortly, this model of meaning is challenged and rejected by Gadamer's dialogue model of meaning of his philosophical hermeneutics.

Lastly, the adoptive approach is not necessary because 'you do not have to be one to know one' (Geertz, 1979, p. 227). For example, you do not have to blind yourself in order to understand the language of the blind. An 'account of other people's subjectivities can be built up without recourse to pretensions to more-than-normal capacities for ego-effacement and fellow feeling' (Geertz, 1979, p. 240). There is actually a better way to achieve cross-language understanding without falling into the temptations of imprisoning oneself either within the alien's mental horizon or within one's own mental horizon. What I have in mind is the hermeneutic approach to understanding to be discussed below.

2. Gadamer on Hermeneutic Understanding

We have argued so far that one can fail to understand an alien P-language (tradition, culture, language, texts, the works of art) in two contrastive ways: by 'going native' through becoming one with members of the alien culture, or by maintaining an imperial distance and projecting one's M-presuppositions onto the others. In either case, one cannot achieve the goal of effective cross-language understanding, learning of insights, and reaching agreement. If interpreters simply abandon whatever they think about the world, then they neither understand the alien P-language, nor are they capable of learning from others; 'going native' is imitation, not insight. Fortunately, Gadamer's hermeneutic understanding seems to point a way out of such a dilemma.

Hermeneutics is concerned with all situations in which we encounter meanings that are not immediately understandable. The task of hermeneutics in general is to render intelligible a text that was previously considered alien and strange. The nineteenth century romanticist and historicist concept of hermeneutics, represented by F. Schleiermacher and W. Dilthey, was transformed by M. Heidegger and H.-G. Gadamer in the twentieth century. For Heidegger (1962), understanding is ontological in the sense that it is not one human faculty among others, but the primordial mode of being of *Dasein*. For *Dasein*, to be is to understand. Gadamer emphasizes further that understanding is universal: it underlies all human activities; nothing is, in principle, beyond understanding. Gadamer's abiding concern has been to understand understanding itself, or to answer the question, 'How is understanding possible?' (Gadamer, 1989, p. xxx). To do so, he sets out to expose, in his philosophical hermeneutics, the fundamental conditions or presuppositions of human understanding.

In the following discussion, we will focus only on those aspects of Gadamer's philosophic hermeneutics relevant to our topic, that is, how to reach cross-language understanding in abnormal discourse.

The Dialogue Model of Meaning

To understand a text is to grasp its meaning. Presumably, a theory of understanding bears a close relation to a theory of meaning. As far as the determination of the meaning of a text is concerned, there are three crucial issues in the theory of meaning subject to contentious debate: (i) Who determines the meaning of a text? Is it the author's intention alone? Does a reader's participation in the process of understanding also play a role in the constitution of meaning? If it does, How much does a reader contribute to the meaning?[2] (ii) Is apprehension of meaning essentially a private psychological state or a public linguistic activity? (iii) Is the meaning of a text—no matter what counts as the bearer of meaning, either the intention of the author or the propositional content expressed by the text—a fixed, self-contained mental or linguistic entity, or the product of a dynamic, open-ended process?

As we have mentioned, the adoptive approach to understanding is underwritten by what I refer to as the monologue model of meaning.[3] According to this classical model, it is the author's intention alone, along with the necessary linguistic conventions of the author's language, that determines the meaning of a text. The meaning is a self-closed linguistic entity waiting to be discovered. Hence, to understand the text is to recapture the author's intention through decoding the meaning of the text. In contrast, Gadamer's notion of hermeneutic understanding subscribes to what I will call the dialogue model of meaning. According to this more recent model, neither meaning nor understanding should be identified with psychological states or processes of mind (whether the author's or the reader's). Meaning and understanding are for Gadamer an objective, nonnatural feature of the world, not private psychological data, but rather essentially and intrinsically linguistic entities that are publicly accessible and communicable. Meaning is not in the author's head, but in the text itself. However, to say that the meaning is in the text itself does not mean, for Gadamer, that the text itself contains the complete meaning given by the author alone, independent of the interpreter's participation.

To see what is wrong with the monological notion of meaning, Gadamer asks us to reflect upon our experience with the understanding of the works of performing art. Art is, for Gadamer, by nature presentational, which means that a work of art should not be thought of as a self-enclosed object or thing detached from and indifferent to a spectator. Instead, a work of art is a dynamic event or happening of being, which 'is merely repeated each time in the mind of the viewer' when the viewer interacts with and participates in it. 'The ontological significance of representation lies in fact that "reproduction" is the original mode of being of the original art work itself. ... The specific mode of the work of art's presence is the coming-to-presentation of being' (Gadamer, 1989, p. 159). 'The being of a work of art is play and it must be perceived by the spectator in order to be actualized' or be made complete (Gadamer, 1989, p. 164). Therefore, the meaning of a work of art can only be determined and contextualized through a dynamic interaction and transaction between the work of art and the spectator who shares in it. Then, Gadamer asks, 'Is this true also of the understanding of any text. Is the meaning of all texts actualized only when they are understood?' (Gadamer, 1989, p. 164) Gadamer's answer is a categorical 'yes'.

> Understanding must be conceived as a part of the process of *the coming into being of meaning*, in which the significance of all statements—those of art and those of

everything else that has been transmitted—is formed and made complete. (Gadamer, 1975, p. 146, my italics)[4]

In fact, such a dynamic, procedural notion of understanding is best conveyed by the German term for 'understanding', *Verstehen*. To use '*Verstehen*', Gadamer intends to stress its close affinity to *Verstandigung*, 'coming-to-an-understanding' or 'reaching-an-understanding' *with* someone *about* something. This tripartite model of understanding reminds us of K. Bühler's tripartite schema of language functions introduced in chapter 12.

Since meaning is always coming into being through the 'happening' of understanding and understanding always happens within certain contexts, the hermeneutic understanding is an open-ended process, which can never (ontologically) achieve finality. It is always open and anticipatory. The meaning of a text is not simply there waiting to be discovered. It makes no sense to talk about *the* single correct interpretation. 'We understand in a *different way, if we understand at all*' (Gadamer, 1989, p. 297; italics as original).

The Role of Prejudices in Understanding

Gadamer's dialogue model of meaning/understanding has a direct impact on the role of the interpreter's own historical situation, prejudgments, or tradition in understanding. The adoptive approach to understanding is justified only under the assumption that a text to be understood has its own fixed meaning that can be isolated from the interpreter's participation. If one's participation in understanding is part of the making of the meaning, it is an illusion to think that one can eliminate one's anticipatory prejudgments or prejudices, to somehow abstract oneself from one's own historical context or cultural/intellectual tradition in order to enter 'into' the minds of the others that are to be understood.

But such an explication of the role of the interpreter's prejudices and prejudgments in understanding appears to bring us back to either the projective approach to understanding or some form of more sophisticated relativism—similar to the relativist stand underlying the adoptive approach to understanding—which Gadamer's hermeneutic understanding is supposed to refute. On the one hand, since one always takes oneself along whenever one understands and there is no way out of one's own perspective, one seems to be justified to project one's prejudgments into understanding others—a version of absolutism lurks in the background. On the other hand, if the meanings of what is to be understood are conditioned by the interpreter's coming-to-understanding, and this subjective understanding is different for different interpreters within different historical contexts and cultural settings, then it seems that we lose the very target of understanding since there is no objective, integrated meaning to understand at all. If so, does it not mean that all understandings are equally good, or even that 'anything goes'? To see Gadamer's responses, we need to appreciate fully the basic functions of prejudices and prejudgments in understanding others.

Ontologically, our prejudices or fore-structures determine our being. 'It is not so much our judgments as it is for our prejudices that constitute our being. ... They are simply conditions whereby we experience something—whereby what we encounter says something to us' (Gadamer, 1976, p. 9). These prejudgments, prejudices, and fore-conceptions that determine who we are now, are not of our own making, are not an act of subjectivity, but are determined by what Gadamer calls the 'commonality which binds us to the tradition' (1989, p. 293). As Gadamer sees it, we belong to a tradition before it belongs to us. In one sense, a tradition is handed down to us, which we are thrown into without choice. In another sense, a tradition does not just stand over us; it is always part of us at the same time. Belonging to a tradition is standing in a happening of tradition. It is our tradition, through its sedimentations as prejudgments, that constitutes what we are in the process of becoming.

As a precondition of understanding, our prejudices and fore-understanding are what make our initial understanding possible. 'All understanding inevitably involves some prejudice' (Gadamer, 1989, p. 270). In fact, 'a person who is trying to understand a text is always projecting' (Gadamer, 1989, p. 267). Whatever is understood is understood on the basis of a pre-understanding or anticipation of meaning (fore-meaning). Since our prejudices are determined by our tradition, wherever there is understanding there is pre-understanding determined by the happening of tradition in which we stand. Then, what is the mechanism through which tradition informs understanding? It is through language. A tradition is sedimented in a language—in its lexicon, in its grammar, in its underlying presuppositions, and in its way of justification and reasoning—and is transmitted and handed down linguistically. Moreover, understanding is a linguistic activity and is discursive, and as such is subject to the traditions sedimented in language use. Since we are 'thrown into' a tradition, the projections of meaning that make understanding possible are 'thrown' projections. Furthermore, the projection of meaning is not so much the act of an individual interpreter but the act of a member of a historically conditioned language community. Thus, as determined by tradition through language, understanding is not and cannot be arbitrary. On the contrary, 'understanding reaches its full potential only when the fore-meanings that it begins with are not arbitrary' (Gadamer, 1989, p. 267). Therefore, *understanding is to be thought of less as subjective act than as participating in an event of tradition*, a process transmission in which past and present are constantly mediated' (Gadamer, 1989, p. 290; italics as original).

To claim that our being and our understanding are conditioned and shaped by our tradition in terms of our prejudices and fore-structures does not necessarily commit us to R. Carnap's 'Myth of Framework'—that we are enclosed within a wall of fixed prejudices which isolate us from others. For Gadamer, 'prejudices are biases of *openness* to the world' and to others from different traditions. They are by nature an open system—always anticipatory, always subject to transformation, and always future-oriented. To see how prejudices can evolve during understanding and how we can reach out to others in terms of our prejudices, we have to see what happens when we encounter others.

Understanding Others—the Hermeneutic Circle and Fusion of Horizons

Gadamer fully realizes that although fore-understanding and prejudices are indispensable for understanding, not all prejudices are conductive in fostering understanding. We have 'to distinguish the true prejudices, by which we *understand*, from the *false* ones, by which we *misunderstand*' (Gadamer, 1989, pp. 298-9; italics as original). Unfortunately, 'he [the interpreter] cannot separate in advance the productive prejudices that enable understanding from the prejudices that hinder it and lead to misunderstanding' (Gadamer, 1989, p. 295). Rather, this separation must take place in the very process of understanding, namely, through the dialogical encounter with the others who are apparently distinctly alien to us. 'Only through others do we gain true knowledge of ourselves'. Self-knowledge—in this case, the knowledge of our false, unjustified prejudices—is achieved only with the dialectical interplay with the 'other'.

When we initially approach the aliens or alien texts, we always (we 'have to', to be more precise) understand them through the lens of our fore-meanings and other prejudices conditioned by our tradition. But our projection of meanings often hinders our understanding. We all have 'the experience of being pulled up short by the text. Either it does not yield any meaning at all or its meaning is not compatible with what we had expected' (Gadamer, 1989, p. 268). This is especially the case when we attempt to understand texts belonging to other traditions that are significantly distinct from ours. The hermeneutic task is to find the necessary resources *in our tradition* to enable us to understand what initially sounds alien to us without distorting it. However, no matter how alien a text is to us, we should still be able to find some degree of affinity or familiarity between the alien text and our language and experience; otherwise, it would no longer be intelligible to talk about understanding. Such 'a tension between alienness and familiarity', using R. Bernstein's phrase, indicates that our initial fore-meanings and prejudices are not justified.

The hermeneutic circle We learn from the failure of initial understanding that:

> We cannot stick blindly to our own fore-meaning about the thing if we want to understand the meaning of another. Of course this does not mean that when we listen to someone or read a book, we must forget all our fore-meanings concerning the content and all our own ideas. All that is asked is that we remain open to the meaning of the other person or text. But this openness always includes our situating the other meanings in relation to the whole of our own meanings or ourselves in relation to it. (Gadamer, 1989, p. 268)

It is here that Gadamer formulates his own version of the hermeneutic circle. The circle in the traditional sense, referring to the dialectic interaction between the understanding of part and whole of a text, does not take the fore-structures of the interpreter into account. For Gadamer (following Heidegger), hermeneutic understanding is the dialectical interplay between the interpreter's fore-structures and the text to be understood, not just between part and whole of a text. On the one hand, although interpreters have to rely on their fore-structures and prejudices in any understanding, they must be 'on guard against arbitrary fancies and the limitations imposed by imperceptible habits of thought' which lead to misunderstanding (Gadamer, 1989, p. 266). Interpreters must be guided by the 'things in themselves' and keep their gaze fixed on them, i.e., the texts to be understood. Interpreters must be open minded, to listen to, to share, to participate with them so the texts can 'speak to' them; they must be receptive to the claims to truth that a text makes upon them. On the other hand, openness and receptiveness to the 'things in themselves' are possible only because of the fore-structures and prejudices that are constitutive of interpreters' being and only in terms of '*justified* prejudices' that open and guide them to other's languages, experiences, and traditions. This requires that interpreters be able to identify unjustified prejudices, revise them, and replace them with 'more suitable ones'. But such a separation of justified ('true' or 'enabling') prejudices from unjustified ('false' or 'blind') ones can be carried out only because of the interplay or clash between the unjustified prejudices and the 'things in themselves'. Therefore, hermeneutic understanding involves constant movement from less suitable prejudices to more suitable ones. 'This constant process of new projection constitutes the movement of understanding and interpretation' (Gadamer, 1989, p. 267). The issue is not whether we need prejudices in order to understand, but what kind of prejudices we need. 'The important thing is to be aware of one's own bias, so that the text can present itself in all its otherness and thus assert its own truth against one's own fore-meanings' (Gadamer, 1989, p. 269).

True prejudices and correct understanding The question left is how we can know whether we have reached a correct understanding (if any at all). Another closely related question is what exactly counts as *true* or legitimate prejudices that promote a correct understanding? Those two notions are obviously related: A correct understanding is the one that lets the *true* meaning reveal itself or lets the 'things in themselves' speak for themselves, while true prejudices are those to be confirmed by the 'things in themselves' (Gadamer, 1989, p. 267). Obviously, these two definitions do not help at all unless we are told what 'true meaning', 'real meaning', or the 'things in themselves' to be understood are. Gadamer is rather vague about these terms and uses them rather elusively. But we know what Gadamer does *not* mean by the terms. By '*real* meaning', Gadamer certainly does not mean the self-contained unit inherent in the text, independent of the interpreter's participation. Recall that, for Gadamer, meaning is always coming into being through the 'happening' of understanding. Similarly, he surely does not use the term, the 'things in themselves', in the Kantian sense, since what the 'things themselves' say could be different to interpreters situated in different traditions.

I think the real answer may be revealed by Gadamer's notion of truth (one of his most elusive notions) used in 'true meaning'. It is clear that Gadamer does not treat the truth as correspondence, nor the propositional truth either. Instead, he seems to adopt a mixture of holistic, coherent, and pragmatic notions of truth. R.

Bernstein suggests that Gadamer is in fact appealing to a concept of discursive truth 'that amounts to what can be argumentatively validated by the community of interpreters who open themselves to what tradition "says to us"' (Bernstein, 1983, p. 154). Such a reading seems to be confirmed by Gadamer's formulation of his hermeneutic circle, which is similar to the traditional formulation of the circle as the movement between part and whole:

> The anticipation of meaning in which the whole is envisaged becomes actual understanding when the part that is determined by the whole themselves also determines this whole. ... Thus the movement of understanding is constantly from the whole to the part and back to the whole. Our task is to expand the unity of the understood meaning centrifugally. The harmony of all the details with the whole is the criterion of correct understanding. The failure to achieve this harmony means that understanding has failed. (1989, p. 291)

Gadamer here speaks of what he calls the 'anticipation of completeness' through projection, according to which only what really constitutes a coherent unity of meanings is acceptable and should be our goal of understanding (Gadamer, 1989, pp. 293-4). In seeking to understand a text, we first project tentative fore-meanings for the text as a whole (the whole). But those fore-meanings have to be revised in light of attention to the details of the text (the part). Through constant negotiation and argumentation between our evolving fore-meanings (the whole) and the text itself (the part), we move toward 'the harmony of all the details with the whole'. Our projected fore-meanings are confirmed by the 'thing in itself' of the text when a unity of meaning (the real, true meaning) is borne out. 'The only "objectivity" here is the confirmation of a fore-meaning in its being worked out. Indeed, what characterizes the arbitrariness of inappropriate fore-meanings if not that they come to nothing in being worked out?' (Gadamer, 1989, p. 267) A true or legitimate prejudice is one that leads to a coherent understanding of the text when we attend to the details of the text in light of it, or which survives engagement with the text by resulting in a coherent lifeworld—a coherent interconnection of meanings, beliefs, values, and norms. Accordingly, an understanding is correct or warranted when it reveals the true meaning of what is to be understood.

We have to be cautious here. For Gadamer, to say that an understanding is correct actually means that it is warranted under the current situation; it does not imply that there is one and only one correct understanding. This is because

> [the] discovery of the true meaning of a text or a work of art is never finished; it is in fact an infinite process. Not only are fresh sources of error constantly excluded ... but new sources of understanding are continually emerging that reveal unsuspected elements of meaning. (Gadamer, 1989, p. 298)

Of course, this does not lead to the relativist position that all understandings are equally good. Many understandings are not warranted by appropriate forms of prejudices that can yield true meanings.

Fusion of horizons We have seen so far that coming-to-an-understanding is a dynamic process, involving the dialectical interaction between the prejudices conditioned by one's tradition and what is to be understood. Such a hermeneutic circle applies to both the intra-tradition and inter-tradition understanding. What we are interested in here is primarily inter-tradition understanding. What happens to the two distinct traditions, cultures, and languages involved in cross-language, cross-tradition understanding: Are both intact, is one replaced by the other, or are both fused into one? To answer this, Gadamer introduces the notion of horizon.

Gadamer's notion of horizon is an extension of the Husserlian notion from perception to understanding. He defines the horizon as 'the range of vision that includes everything that can be seen from a particular vantage point' (1989, p. 302). One's particular horizon or viewpoint is formed by one's particular tradition, culture, language, historical past, and situation, which embraces not just immediate context of fore-meanings that one is currently engaged with, but the broader context which conditions them. Ultimately it is a whole historical lifeworld—the world in which one lives out one's life, a web of meanings, beliefs, values, and norms—the world from a particular historical standpoint or a form of life. In this sense, the notion of horizon is as broad as the notion of tradition.

A horizon is, by definition, limited and finite; but it also is essentially an open system, open to self-transformation and open to being understood. To have a horizon means not being limited to what is nearby but being able to see beyond it. 'Tradition is not simply a permanent preconception; rather, we produce it ourselves inasmuch as we understand, participate in the evolution of tradition, and hence further determine it ourselves' (Gadamer, 1989, p. 293). 'So too the closed horizon that is supposed to enclose a culture is an abstraction. ... The horizon is, rather, something into which we move and that moves with us' (Gadamer, 1989, p. 304).

When one tries to understand a horizon other than one's own, especially an alien horizon from an alien tradition, one cannot simply *replace* one's own horizon with the alien's for the reasons we have given in the discussion of the adoptive approach to understanding. It is neither possible nor desirable. The placing of oneself in an alien horizon is not and cannot be a case of leaving one's own horizon behind. One has to place *oneself* in the other horizon, and this self is an ontological being with its essence determined by its own tradition and horizon. One can experience the genuine otherness only by placing *oneself situated within one's horizon*, as a filter to separate the oneness from the otherness, in the other's horizon. Just as one can only gain knowledge of oneself through the eyes of the other, the genuine otherness can be revealed to one only against the background of one's oneness. Therefore, for Gadamer, the understanding of the other can be achieved by a 'fusion of horizons', a fusion between the horizons of the interpreter's and the interpreted, whereby the interpreter's horizon is enlarged and enriched in terms of the engagement with the horizon of the interpreted. This fusion of horizons is not a technique to be employed in inter-tradition/culture understanding, but what happens whenever there is understanding, whether it is of something from the past, something from another culture, something in another language, or even another person within the same tradition and speaking the same language. We have more to say about a 'fusion of horizons' later in chapter 15.

3. Language Learning as Hermeneutic Interpretation

If to understand an alien P-language is to grasp accurately its propositional contents within its linguistic context, can there be a better way to do so than through language learning by immersing ourselves in the context? We know that the most efficient way for the anthropologist to understand the native encountered is to learn the native's language by living in the community. The same should be true for our would-be communicator who faces a complete communication breakdown due to lack of understanding of an alien P-language. Against the translational approach to understanding, Feyerabend remarks that 'we can learn a language or a culture from scratch, as a child learns them, without detour through our native language' (1987, p. 76). As Kuhn emphasizes repeatedly, although two scientific languages with unmatchable lexical taxonomies cannot be translated into one another, access to the other incommensurable linguistic grid is still possible by learning the other language together with its taxonomy.[5] Although to understand some radically disparate P-language, one has to learn it in a wholesale manner (learn the language together with the underlying mode of reasoning, the cosmology, the categorical framework, and other factual commitments), it only makes understanding difficult, not impossible in principle. After learning, the would-be communicator could become a bilingual, a person who can effectively understand both P-languages in play. For example, for a Western physician (call her Ms Jones), after 'whole-sale' learning of Chinese medical theory, the pre-modern Chinese mode of reasoning and the yin-yang cosmology, its category system, and even the Chinese language, she can understand the language of the traditional Chinese medical theory very well.

As an extreme version of the adoptive approach, the language-learning approach of cross-language understanding faces many similar problems. For example, our adult language learner (for instance, Ms Jones) who tries to learn an alien P-language (the traditional Chinese medical language) is not a child with a mind of 'blank slate'. When she approaches the alien language, she is already equipped with a set of incompatible M-presuppositions brought with her own home P-language (the contemporary Western medical language). Experience and observations are theory-laden. It is almost impossible for our language learner with a prejudged mind to enter into the others' stream of experiences and truly 'go native'. Like Geertz's ethnographer, the adult language learner does not, and largely cannot, perceive what the native speaker of the P-language perceives.

The language-learning approach of cross-language understanding still faces some other unique difficulties. Language learning is not universally accessible for the interpreter. Many language communities and their linguistic contexts that the interpreter tries to understand no longer exist, for example, the Paracelsian medical language. All that the interpreter has access to are the written texts and historical background through the eyes of historians and anthropologists. It is simply impossible for us to learn them as a child does. Besides, some researchers—historians of science, for example, such as our Ms Jones—can, by suitably immersing themselves in some living alien tradition to learn its language (such as the Chinese medical language), come to understand the language well. But it does not mean that this particular specialist's understanding is what underwrites the communicative capacities of the overwhelming number of participants in the tradition (such as the Western medical community as a whole).

I have to admit that the above two concerns are only pragmatic and do not touch the root of language learning. Here is the real problem of language learning. To learn a second language, one cannot simply leave one's own language behind. Instead, one has to approach the language from the horizon of one's own language, or let the language speak to one who is already grounded within one's own linguistic tradition. Seeing from the perspective of Gadamer's hermeneutics, language learning in general, and the language acquisition of an alien P-language in particular, is actually a process of hermeneutic interpretation. As Gadamer puts it:

> It is impossible to understand what the work has to say if it does not speak into a familiar world that can be found *a point of contact* with what the text said. Thus to learn a language is to increase the extent of what one can learn. ... To have learned a foreign language and be able to understand it means nothing else than to be in a position to accept what is said in it *as said to oneself.* The exercise of this capacity for understanding always means that what is said *has a claim over one*, and this is impossible if one's own 'worldview and language-view' is not also involved. (1989, p. 442; italics as original).

Kuhn seems to realize this also. For the same reason, Kuhn prefers to call the understanding acquired by language learning 'interpretation'.

> It [referring to interpretation] is an enterprise practiced by historians and anthropologists, among others. Unlike the translator, the interpreter may initially command only a single language. At the start, the text on which he or she works consists in whole or in part of unintelligible noises or inscriptions. ... If the interpreter succeeds, what he or she has in the first instance done is learn a new language. (Kuhn, 1983b, pp. 672-3)

> Anything which can be said in one language can, with imagination and effort, be *understood* by a speaker of another. What is prerequisite to such understanding, however, is not translation but language learning. Quine's radical translator is, in fact, a language learner. (Kuhn, 1988, p. 11)

Here Kuhn associates understanding with language learning, and further treats the language learning process as the process of Quine's radical translation. By doing this, Kuhn reveals one central feature of his notion of understanding through language learning, namely, its openness and indeterminacy. In general, the process of learning an alien language may well involve a complex process of reinterpretative reconstruction that is not just purely reproductive, but also a *productive* activity. This is especially true when the language to be learnt embodies a set of M-presuppositions incompatible with those of the interpreter's own language.

Kuhn actually has the notion of hermeneutic understanding in mind when he talks about interpretation and language learning. The language-learning approach is different from both the adoptive approach (which is usually attributed to Kuhn's relativism) and the projective approach to understanding (which Kuhn is supposed to argue against). While both the projective and the adoptive approaches are *propositional* in nature (both are based on the notion of propositional understanding), understanding through language learning is hermeneutic in character, which necessarily involves the hermeneutic circle between the interpreter's language and the alien language. In fact, Kuhn explicitly formulates his own version of the hermeneutic circle in a very sketchy way in the following passage:

> When reading the works of an important thinker, look first for the apparent absurdities in the text and ask yourself how a sensitive person could have written them. When you find an answer, I continue, when those passages make sense, then you may find that more central passages ones you previously thought you understood, have changed their meaning. (Kuhn, 1977a, p. xiii)

Moreover, to be able to understand a P-language is different from being able to *use* the language effectively. Recall that, according to Kuhn's projectibility principle, kind-terms of a scientific theory are clothed with expectations about their referents. These expectations are projectible so that they enable members of the language community to project the use of these terms to other unexamined situations. Although the interpreter can learn and understand the kind-terms in an alien language, it does not mean that he or she can use them *projectibly* in different situations. The same problem faces our language learner Ms Jones. She can understand Chinese medical theory through the learning process. But it does not ensure that she is able to use *productively* the kind-terms, such as 'the yin' and 'the yang', employed by the theory (to extrapolate the terms to new cases).

4. The Hermeneutic Dimension of Incommensurability [6]

The relevance of Gadamer's discussion of hermeneutic understanding to our question at hand is palpable, namely, how cross-language understanding is possible in abnormal discourse,. In fact, as I see it, Gadamer's philosophic hermeneutics provides a further support for our presuppositional interpretation of incommensurability as cross-language communication breakdown. Let me be more specific.

For Gadamer, our basic 'hermeneutic experience is verbal in nature', that is, both our being-in-the-world and the-world-for-human-beings (the lifeworld or the world we live in) are primordially linguistic in nature. Human language is not simply one of the distinctively human capacities that enable us to deal with the world, 'language was human from its very beginning'. As we have seen, for Gadamer, understanding is both ontological and universal: it is the primordial mode of human existence (our being-in-the-world) and underlies all human activities. Besides, understanding is essentially linguistic and discursive. Thus, 'language is originally human means at the same time that man's being-in-the-world is primordially linguistic' (Gadamer, 1989, p. 443). In addition, 'language is not just one of man's possessions in the world, rather, on it depends the fact that man has a *world* at all. ... But this world is verbal in nature'. It is not as though, on the one hand, there is language, and, on the other hand, the world. 'Not only is the world world only insofar as it comes into language, but language, too, has its real being only in the fact that the world is presented in it' (Gadamer, 1989, p. 443). Following M. Scheler, Gadamer contrasts the concept of world with that of environment. All living creatures have their environment, but only humans have a world. 'Man's relationship to the world is characterized by freedom from the environment. This freedom implies that linguistic constitution of the world. Both belong together. To rise above the pressure of what impinges on us from the world means to have language and to have "world"' (Gadamer, 1989, p. 444).

Gadamer contends that the thesis of the integration between language and world or the linguistic constitution of the world is the real thrust of W.V. Humboldt's idea that 'languages are worldviews' or 'a language-view is a worldview'. However, Humboldt's concept of language is the linguist's notion of language, namely, particular types of natural languages as lexicon or grammar. For Gadamer, it is not by virtue of its form that a language is a worldview, but rather by virtue of what is embodied in it. 'If every language is a view of the world, it is so not primarily because it is a particular type of language (in the way that linguists view language) but because of what is said or handed down in this language' (Gadamer, 1989, p. 441). 'What constitutes the hermeneutical event proper is not language as language, whether as grammar or as lexicon; it consists in the coming into language of what has been said in the tradition' (Gadamer, 1989, p. 463). For example, it is not Chinese language *per se*, as a natural language with its unique grammatical structure and lexicon, but rather the Chinese cultural tradition embodied in it, as handed down linguistically by the Chinese language, that constitutes the worldview of the Chinese. In other words, it is as embodying the happening of tradition that a language is a worldview. Moreover, for Gadamer, 'tradition is essentially verbal in character' (Gadamer, 1989, p. 389). Tradition is embodied or sedimented in language, transmitted and handed down linguistically.

The contact point between Gadamer's linguistically formulated and transmitted tradition as worldview and our concept of P-language has become evident. It would not be too much of an exaggeration to claim that our concept of P-language is, roughly, a linguistic counterpart of Gadamer's notion of tradition, which itself is linguistically formed and transmitted. Just as there are multiple P-languages, there exist different traditions, each of which corresponds to a linguistically constituted world or worldview and has its own unique path of development. A language community situated in a particular cultural or linguistic tradition sees the world differently from another language community belonging to a different tradition. Distinct traditions or language-worlds might be in conflict with one another, which puts mutual understanding between them at risk. The comparison between P-language and tradition becomes more striking if we consider the similar roles of the M-presuppositions of a P-language and the prejudices of a tradition. *Pre*-judices and *pre*-judgments conditioned by a tradition are actually *pre*-suppositions of a P-language. Of course, Gadamer's notion of prejudice is a broader notion than that of M-presupposition. But I think Gadamer would agree that the three major types of M-presuppositions I have identified are clearly prejudices that shape one's worldview and are preconditions for cross-tradition/cross-language understanding.

Now, let us see what happens when we try to understand an alien P-language or language-world from the perspective of Gadamerian hermeneutics. To begin with, we are always ontologically situated in a history, a culture, a form of life, or a tradition articulated in our own home P-language. From this given situation—the existential state of the language community—the language community forms a specific mode of reasoning, a unique categorical framework, and universal principles about the world surrounding it. When confronting an unfamiliar alien P-language or language-world, it is natural for us to project these prejudgments or M-presuppositions from our home P-language onto the other. The projection is not in the first place a matter of choice. Rather, we are 'thrown' into it. All understanding is projective. Contrary to the adoptive approach to understanding,

> [t]he fact that our experience of the world is bound to language does not imply an exclusiveness of perspectives. If, by entering foreign language-worlds, we overcome the prejudices and limitations of our previous experience of the world, this does not mean that we leave and negate our own world. Like travelers, we return home with new experiences. Even if we emigrate and never return, we still can never wholly forget. (Gadamer, 1989, p. 448)

Instead, the prejudgments or presuppositions from our own P-language should be considered almost like transcendental 'conditions' or the springboard of understanding from which we can reach out to the other language. In this sense, our own language is not a restriction but the very principle of understanding. Therefore, we do not need to (and cannot either) bracket or forget our own language in order to understand the others.

However, this by no means restricts us to the absolutistic projective understanding. On the one hand, we cannot rely blindly on the prejudgments or M-presuppositions of our own P-language by taking the projective approach from our own P-language as the proper one. In order to understand an alien P-language, we must make our own situation (here, the M-presuppositions of our own home P-language) transparent so that we can appreciate precisely the otherness without concealing the proper meaning of the other language by allowing our unelucidated prejudices to distort it. To reveal self-reflectively the M-presuppositions of our own

P-language, it is necessary to allow otherness of an alien language to be disclosed. Only after we are conscious of the M-presuppositions embedded within our own P-language, can we allow the other language to speak for itself, to reveal its proper meaning (the speaker-language meaning). On the other hand, only in confrontation with an alien P-language and its M-presuppositions, can we hope to get beyond the limits of our present horizon. It is precisely in and through the understanding of an alien P-language (culture, tradition, or form of life) and a realization of how different others are from our own that we can come to a more sensitive and critical understanding of our own P-language (culture, tradition, or form of life) and of those M-presuppositions or prejudices that may lie hidden from us. Self-consciousness is always relative to consciousness of otherness and alienness. To understand an alien P-language, we must participate or share in it, listen to it, open ourselves to what the language is saying, and allow it to speak to us.

The above dialectical interplay between understanding our own P-language and understanding an alien P-language is Gadamer's hermeneutic circle of understanding. Whilst we must not allow ourselves to be blinded by the prejudgments or M-presuppositions coming with our own language; otherwise we will be unable to open ourselves to the other and allow the other to speak to us. Neither should we bracket all the prejudgments or M-presuppositions of our own P-language or jump into the stream of the other's experience. It is only the play of these M-presuppositions that allows us to engage in the process of understanding, to reach out to the other, and to capture the speaker-language meaning. Therefore, mutual understanding in abnormal discourse necessarily involves a constant movement back and forth between our own language and the other language. The speaker-language meaning can only be grasped through this circle of understanding.

In fact, both Kuhn and Feyerabend have somehow experienced hermeneutic understanding in their encounter with incommensurable texts. As we have discussed at the very beginning of the book, Kuhn was extremely perplexed by Aristotle's assertions about motion, which, for Kuhn, are not even false, but senseless, absurd, and having no truth-value. Even so, Kuhn was able to comprehend the basic subject matter that Aristotle discussed, i.e., mechanics. What really perplexes Kuhn is not the total incomprehension of the texts, but rather the tension he felt between alienness and familiarity of the text to him. Such a tension indicates Kuhn's initial failure of understanding of Aristotle's mechanics. In Gadamerian terms, Kuhn's failure of understanding was the result of his initial projecting of the prejudices (M-presuppositions) of Newtonian mechanics onto Aristotelian mechanics. But Kuhn did not blindly stick to his own fore-meanings and prejudgments. More careful attention to the details of the text made Kuhn suspect that the fault might be his own approach of reading Aristotle. He revised and readjusted his assumptions and continued to puzzle over the details of the text by approaching Aristotle from a different horizon. The effort paid off and his perplexities suddenly vanished in that memorable hot summer day when 'suddenly the fragments in my head sorted themselves out in a new way, and fell into place together' (Kuhn, 1987, p. 9).

What Kuhn has described here is actually his personal encounter with Gadamer's hermeneutic circle, the dialectical interplay between his own fore-structures, fore-meanings (the M-presuppositions of Newtonian mechanics) and the text (Aristotelian mechanics) to be understood. Kuhn was trained in the research tradition of modern physics with Newtonian mechanics as one paradigm. In seeking to understand Aristotle, Kuhn did not leave his prejudices behind. He carried his prejudices from the Newtonian tradition along when he first approached Aristotle. He tried to project his tentative fore-meanings for the text as a whole (the whole). But those fore-meanings were revised in light of attention to the details of the text (the part). Through constant negotiation and argumentation between his evolving fore-meanings (the whole) and the text itself (the part), Kuhn eventually moved towards 'the harmony of all the details with the whole'. At last, his projected fore-meanings were confirmed by the text when the unity of meaning was borne out. The text started to speak its own truth to him.

Feyerabend touches on the hermeneutic circle when he discusses the application of the anthropological method in his classical case study on the shift from 'the archaic style' to 'the classical style' in ancient Greek art, which we discussed at length in chapter 10. Recall that Feyerabend finds out that every formal feature of a style (a theory or a language) corresponds to assumptions (hidden or explicit) inherent in its underlying cosmology. To understand the shift between the two incommensurable art styles, Feyerabend needs to identify and understand the cosmology behind the archaic style, i.e., the cosmology of the archaic Greeks. In contrast with the method of logical reconstruction, which he takes to be the misguided obsession of many philosophers of science of the time, Feyerabend has a great appreciation of the anthropological method used by social anthropologists, such as Evans-Pritchard, who attempt to understand the worldview or cosmology of an alien tribe. For Feyerabend, 'the anthropological method is the correct method for studying the structure of science (and, for that matter, of any other form of life)' (Feyerabend, 1978, p. 252).

To begin with, the anthropologist tries to discover the way in which the cosmology is mirrored in the tribe's language, arts, and daily life. To do so, he has to learn the language and the basic social activities, and to identify key ideas. Whether some ideas are key ideas not only depends upon the roles that the ideas play in the native's cosmology, but also can only be identified against the interpreter's belief system. 'His attention to minutiae is not the result of a misguided urge for completeness but of the realization that what looks insignificant to one way of thinking (and perceiving) may play a most important role in another' (Feyerabend, 1978, p. 250). To understand those key ideas, the anthropologist cannot simply bracket his own knowledge and assumptions and leave them behind to 'go native'. 'The anthropologist carries within himself both the native society and his own background' (Feyerabend, 1978, p. 250). But more importantly, on no account must he attempt a 'logical reconstruction', i.e., to project his own fore-meanings and prejudices into the native, which is the method used by many philosophers of science in their understanding of past scientific practices. 'Such a procedure would tie him to the known, or to what is preferred by certain groups, and would forever prevent him from grasping the unknown ideology he is examining' (Feyerabend, 1978, p. 250).

Clearly, Feyerabend's anthropological method is hermeneutic in character. In fact, he actually presents a version of the hermeneutic circle in the following passage:

> The examination of key ideas passes through various stages, none of which leads to a complete clarification. Here the researcher must exercise firm control over his urge for instant clarity and logical perfection. ... Each item of information is a building block of understanding, which means that it is to be clarified by the discovery of further blocks from the language and ideology of the tribe rather than by premature definitions. ... They [referring to the fore-meanings projected] are preliminary attempts to anticipate the arrangement of the totality of all blocks. They are then to be tested and elucidated by the discovery of further blocks rather than by logical clarification. (Feyerabend, 1978, pp. 251-2)

The parallel between Feyerabend's version of the hermeneutic circle and Gadamerian hermeneutic circle is evident. Both presuppose that what is to be understood (for Gadamer, the alien texts defying our initial comprehension while for Feyerabend, incommensurable scientific research traditions or paradigms), no matter how distant and alien it is to the interpreter, can be understood. Both Feyerabend's and Gadamer's hermeneutic circles are characterized by the constant movement back and forth between 'parts', the detail of the text to be understood, and 'whole', the harmony of all the details as a whole. Both provide us with a way of avoiding two extremes, the adoptive and the projective approaches to understanding.

It is time to summarize what we have learnt from our discussion of the hermeneutic experience of understanding an alien P-language in abnormal discourse, that is, *openness of the incommensurables*. Gadamer, Kuhn, and Feyerabend all believe that 'everything that is intelligible must be accessible to understanding and to interpretation' (Gadamer, 1989, p. 404). It is true that those who are brought up in a particular P-language or a particular linguistic and cultural tradition see the world in a different way from those who belong to other P-languages or traditions. It is true that two P-languages or two historical 'worlds' that succeed one another in the course of the same cultural or linguistic tradition are different from one another; but in whatever P-language or tradition we consider it, it is always a human (i.e., verbally constituted) language-world that presents itself to us. As verbally constituted, every such P-language or language-world is of itself always open to every possible insight and hence to every expansion of its own world picture, and is accordingly open to being understood by others (Gadamer, 1989. p. 447).

As such, we have reached the same conclusion as does R. Bernstein:

> The core of the incommensurability thesis, as we have seen, is not closure and being encapsulated in self-contained frameworks but the openness of experience, language, and understanding. (Bernstein, 1983, p. 108)

This conclusion is the opposite of a prevailing interpretation of the incommensurability thesis: The thesis leads to some form of extreme relativism, either Carnap's Myth of the Framework—what is incommensurable is enclosed within the walls of its own prison house built up by its unique tradition, language, and form of life—or Davidson's Myth of Unintelligibility—an alien conceptual scheme could be extremely remote from ours to the extent of being 'mutually unintelligible' or 'forever beyond our grasp'. On the contrary, if we follow Heidegger to give understanding an ontological orientation by interpreting it as an existential mode of existence of *Dasein* (human existence), we start 'to recognize temporal distance as a positive and productive condition enabling understanding' (Gadamer, 1989, p. 297). Similarly, although the conceptual distance between two incommensurables (P-language, cultural or research traditions, paradigms, forms of life) does cause some difficulty in mutual understanding, it by no means sets insurmountable barriers between both sides. Instead, we should take this challenge as an invitation to genuine understanding, both understanding others and understanding ourselves. We can only truly understand ourselves by comparison with others and by testing our own prejudices during the process of understanding others.

However, we have to end such an optimistic, spirit-lifting journey along the hermeneutic path with a cautious note about mutual understanding in abnormal discourse. Be aware that what we have concluded so far is that it is *possible* to achieve a better, more effective mutual understanding if we follow Gadamer's suggestions. To put the point more precisely, mutual hermeneutic understanding is *possible* in abnormal discourse. Another's horizon is *open* for us to understand. But this does not entail that a *complete, full* (hermeneutic) understanding between two incompatible P-languages is *feasible*. Remember that Gadamer's hermeneutic understanding is a process in which a speaker is coming-to-an-understanding and reaching-an-agreement with someone or with a text about something. It is an open-ended process that can never achieve finality. We will see later in chapter 15 when we discuss Gadamer's conversation model of communication that such an understanding can only be achieved through conversation, which is in essence a process of communication. I will argue that an *undistorted full* cross-language communication between two incompatible P-languages cannot be achieved.

Notes

1   What we are concerned with here is the intelligibility of human linguistic systems (or ideas, thoughts, theories, conceptual schemes), not animal languages or the languages of extraterrestrial aliens, should they exist.

2   A primary contrast is between those who hold that it is the author's intention alone that determines the meaning of a text (literary theorists as E. Hirsch, P. Juhl and historians Q. Skinner and J. Pocock) and those who acknowledge the reader's participation in the constitution of the meaning. For the latter view, there is wide divergence as to how much the reader can contribute to the meaning of a text. Some (such as W. Iser) believe

that the author imposes definite limits on the meaning although it still leaves ample room within those limits for competent readers to form different interpretations. At one extreme stands S. Fish, for whom a text's meaning is so radically indeterminate that readers can create a variety of meanings at will.

3   The monologue mode of meaning identified here roughly corresponds to what M. Dummett calls the code theory of meaning.

4   As my bibliography indicates, I refer to two different translations of Gadamer's *Truth and Method* for more accuracy with the original text.

5   Kuhn, 1983b, pp. 672-3; 1988, p. 11; 1991, p. 5.

6   The basic idea presented below is essentially R. Bernstein's. See his 1983.

Chapter 14

Informative Communication Breakdown: The Transmission Model

We have concluded that mutual understanding in abnormal discourse should be approached through hermeneutic understanding or language learning. However, the real possibility of restoring mutual understanding in abnormal discourse seems to make our incommensurability thesis lose its revolutionary edge and philosophical significance. Let me explain. Since communication breakdown between two incompatible P-languages is caused by lack of mutual understanding, the restoration of mutual understanding appears to render the doctrine of communication breakdown between two P-languages unsubstantiated. To paraphrase Davidson's mock on conceptual relativism (1984, p. 183), 'incommensurability as communication breakdown is a heady and exotic doctrine, or would be if we could make good sense of it. The trouble is, as so often in philosophy, it is hard to improve intelligibility and tenability—by removing the extreme relativistic flavor, i.e., the failure of mutual understanding, off the thesis of incommensurability—while retaining the excitement'. Thus, we seem to face a dilemma: We can either defend the doctrine of cross-language understanding failure in order to keep the sharp edge of the thesis of incommensurability as communication breakdown, but at the same time we have to bite the bullet of extreme relativism; or we have to give up the doctrine of communication breakdown altogether. But if there were no communication breakdown *per se*, then there would be no incommensurability as communication breakdown either.

Fortunately, we do not have to accept the above 'either/or offer'. We can accept the universal understandability of any languages and traditions, and, at the same time, maintain the doctrine of incommensurability as communication breakdown. This is certainly achievable so long as we realize that cross-language understanding (in the propositional sense, as comprehension) should be distinguished from cross-language communication. I will argue that cross-language understanding (in the propositional sense) is a necessary but not sufficient condition for cross-language communication.[1] Although the failure of mutual understanding surely puts communication at risk as I have argued in chapter 12, the success of cross-language understanding does not guarantee successful communication. Successful cross-language communication requires much more than understanding. Even if complete communication breakdown can be overcome in terms of hermeneutic understanding or language learning, there still exist some much more interesting cases of communication breakdowns that are inevitable in abnormal discourse, namely, partial communication breakdown. It is this kind of communication breakdown *per se* in abnormal discourse that substantiates our doctrine of incommensurability as communication breakdown.

1. Hermeneutic Understanding versus Propositional Understanding

Hermeneutic understanding is supposed to be an alternative to avoid antithetical poles of the projective and adoptive approaches to understanding in order to achieve authentic understanding in abnormal discourse. As attractive as it is, what concerns us here is not whether Gadamer's hermeneutic understanding could provide us with *a sort of* understanding in abnormal discourse, which it does, but rather whether it could restore full communication by achieving effective understanding in abnormal discourse, which it does not.

Based on our truth-value conditional account of understanding, effective understanding involved in cross-language communication is *propositional* understanding by nature. According to it, interpreters understand effectively the sentences of an alien P-language if and only if they can grasp the speaker's language-meaning or its propositional contents. In normal discourse, propositional understanding can usually take place smoothly since both sides have shared or compatible M-presuppositions as a common ground to establish necessary logical relations between the propositions under discussion. In contrast, in abnormal discourse the would-be communicators do not even agree on the truth-value status of the sentences under discussion. One sentence that clearly expresses a proposition when considered within the context of one P-language may lack any propositional content when considered within the context of the other. It makes propositional understanding (effective understanding in particular) in abnormal discourse problematic and difficult (although it is still possible).

Gadamer drives home the point that propositional understanding is doomed to failure in abnormal discourse in the following passage:

> There are no propositions which can be understood exclusively with respect to the content that they present, if one wants to understand them in their truth. ... Every proposition has *presuppositions* that it does not express. Only those who think with these presuppositions can really assess the truth of a proposition. I maintain, then, that the ultimate logical form of the presuppositions that motivate every proposition is the *question*. (Gadamer, 1986, p. 226; first italics are my own)

Put into our terminology: Gadamer is actually arguing that proper propositional understanding of an alien language depends on the grasp of the M-presuppositions of the language. In abnormal discourse, if interpreters cannot comprehend the M-presuppositions of an alien language ('do not think with these presuppositions'), they cannot grasp the propositions expressed by the sentences of the language. Consequently, the propositional understanding of the alien fails. The best alternative in this situation is to work out the M-presuppositions of the alien language by engaging in a dialogue to ask questions, to hypothesize the alien way of thinking, and to make comparison between their own language and the alien's. To do this, interpreters inevitably become involved in the hermeneutic circle between their own language and the alien's. It is when propositional understanding in abnormal discourse fails that hermeneutic understanding takes over. This is where Gadamer's notion of hermeneutic understanding comes into play.

However, to say that propositional understanding begins where hermeneutic understanding ends seems to give the reader an impression that hermeneutic understanding is a continuation of propositional understanding. This is not the case. Here we are actually dealing with two very different kinds of conceptions of understanding. Heidegger's and Gadamer's notion of hermeneutic understanding is not intended to be a solution to the traditional epistemological problem of understanding, i.e., propositional understanding. Instead, it is supposed to offer an alternative revolutionary notion of understanding. In fact, Heidegger's and Gadamer's notions of hermeneutic understanding arise as a rebellion against the traditional notion of propositional understanding. The notion of propositional understanding presupposes the monologue model of meaning. According to the model, a sentence possesses a self-sufficient unit of linguistic meaning or a self-contained propositional content that is given by the author and is determined the conventions of his or her language, and thus has nothing to do with the interpreter's involvement (the interpreter's language, tradition, or horizon). It is, therefore, monological.

Heidegger considers propositional understanding to be secondary and derivative from universal hermeneutic understanding. For Heidegger, to understand is to 'be at home with something'. In this sense to understand something is to master it. This is actually what we usually call practical understanding or 'to understand how' in contrast with propositional understanding, i.e., 'to understand what'. Understanding in this sense means less a 'kind of knowledge' than 'knowing one's way around' (Heidegger, 1975, p. 286).

Like Heidegger, Gadamer considers the 'construction of logic on the basis of the proposition' to be one of the 'most fateful decisions of Western culture' (1986, p. 195). For Gadamer, 'Language is most itself not in propositions but in dialogue' (1985, p. 98). Against the primacy of the notion of propositional understanding, which assumes a self-contained propositional content for each sentence, the hermeneutic approach adopts the dialogue model of meaning. It reminds us that the literal meaning or the propositional content of a sentence can never be prescinded from the context in which it is embedded when it is understood, and therefore can never be detached from the process of understanding itself. There is no fixed, self-closed propositional content for each sentence that is waiting to be discovered by the interpreter. Understanding of what a sentence is saying thus becomes an activity of participation, engagement, assimilation, and dialogue, that is, the interpreter's participation in the reformation and enrichment of its meaning, the engagement between the speaker's and the interpreter's traditions or the fusion of horizons, assimilating what is said to the point that it becomes the interpreter's own, and ultimately a dialogue between the interpreter and the text. 'For language is by nature the language of conversation; it fully realizes itself only in the process of coming to an understanding' (Gadamer, 1989, p. 446). In a dialogue, there are no fixed propositions, only questions and answers, which in turn call forth new questions.

In general, the process of hermeneutic understanding well involves a complex process of reinterpretative reconstruction on the part of the interpreter, which is not just purely *reproductive*, but also a *productive* activity. The advocate of propositional understanding may suspect that the so-called hermeneutic understanding is actually what we usually call the process of interpretation. Just as to know an object, to understand a text is a monological or unilateral process in which the subjectivity of the interpreter (the fore-meanings and prejudices coming with his or her own tradition and language) is not supposed to be part of understanding. This is why we always strive for a value-free, objective understanding (at least in scientific understanding). The goal is to attain some finally adequate understanding so that we can obtain a full intellectual control over what is to be understood. But interpretation is supposed to be a different process from understanding. The former is less stringent compared with the latter. To interpret is to see things through the eyes of the interpreter. It is a bilateral and dialogical process in which the interpreter comes to an interpretation through the interplay between his or her own fore-knowledge and what is to be interpreted. '[A]n interpreter's task is not simply to repeat what one of the partners says in the discussion he is translating, but to express *what is said* in the way that seems *most appropriate to him*, considering the real situation of the dialogue' (Gadamer, 1989, p. 308; my italics). Therefore, 'to interpret means precisely to bring one's own preconceptions into play so that the text's meaning can really be made to speak for us (Gadamer, 1989, p. 397). Since both the fore-knowledge of the interpreter and the context in which an interpretation takes place can change, we can never have a final interpretation. Clearly, the advocate of propositional understanding continues, Gadamer confuses understanding with interpretation.

To this charge, Gadamer accepts its conclusion but rejects its premise: The critic is right on the point that hermeneutic understanding is interpretation, more precisely, interpretative understanding; but the critic is wrong to separate understanding from interpretation as two independent processes. In the early tradition of hermeneutics, which seeks to draw a rigorous distinction between understanding (*subtilitas intelligendi*) and interpretation (*subtilitas explicandi*), there is first a pure, objective understanding of the meaning existing in the text, which is free from all prejudices and not 'contaminated' by 'subjective' interpretation. Propositional understanding is supposed to be such an objective understanding. Interpretation functions as a means to the end of or a supplement to understanding. In bold new contrast to this tradition, Heidegger and Gadamer argue that there are no essential differences between understanding and interpretation.

> Interpretation is not an occasional, post facto supplement to understanding; rather, understanding is always interpretation, and hence interpretation is the explicit form of understanding. In accordance with this insight, interpretative language and concepts were recognized as belonging to the inner structure of understanding. (Gadamer, 1989, p. 307)

Understanding and interpretation are internally related or fused, both of which compose one unified process. Every act of understanding involves interpretation, and every act of interpretation involves understanding. Understanding is not just reproductive but, because it involves interpretation, is also a productive process. Understanding always involves something like the application of the text to be understood to the interpreter's present situation. Since we are always understanding in light of our anticipatory prejudgments determined by the effective historical situation in which we live, we always take ourselves along whenever we understand. Meaning is always coming into being through the 'happening' of understanding. 'The interpretative process of understanding ... is simply *the concretion of the meaning itself*' (Gadamer, 1989, p. 397; italics as original). C. Geertz describes the indeterminacy of hermeneutic understanding vividly as follows:

> Understanding the form and pressure of, to use the dangerous word one more time, natives' inner lives is more like grasping a proverb, catching an allusion, seeing a joke—or, as I have suggested, reading a poem—than it is like achieving communion. (Geertz, 1979, p. 241)

Therefore, there is always room for different understandings or interpretations. 'There cannot, therefore, be any single interpretation that is correct "in itself". ... Every interpretation has to adapt itself to the hermeneutical situation to which it belongs' (Gadamer, 1989, p. 397). Hermeneutic understanding is interpretative understanding.

To sum up, although the would-be communicators in abnormal discourse can still understand one another to a certain extent by utilizing the process of hermeneutic understanding, effective understanding (in the sense of propositional understanding) is inevitably partial. When effective understanding in abnormal discourse is incomplete, cross-language communication between two P-language communities is inevitably also partial.

'Hold on', a critic may object, 'your argument starts with a very suspicious premise that understanding used in cross-language communication has to be *propositional*, which you call effective understanding. Why does it have to be so? Why could we not have a different notion of communication that is itself hermeneutic by nature and thereby does not presuppose propositional understanding at all?' I have to admit that our imaginary critic has raised a very good point. Indeed, whether propositional understanding should be an essential part, a necessary condition, of cross-language communication depends upon the notion of cross-language communication to which we subscribe. In fact, as we will clarify shortly, there are at least two competing models of communication: the standard transmission model, and the hermeneutic dialogue model. The transmission model of communication, which is propositional by nature, does presuppose propositional understanding while the dialogue model does not. Considering my specific purpose, there is no need for us to engage in a touchy debate as to which model of communication is better (although I personally favor the dialogue model). My strategy is to show that both models lead to the same conclusion: Partial communication breakdown is inevitable in abnormal discourse even if understanding (especially propositional understanding as comprehension) is no longer an obstacle. For the sake of argument, I will assume that either full effective understanding or (partial) hermeneutic understanding is in place, and see whether full communication can be established between two incompatible P-languages in abnormal discourse.

2. A Bilingual in Cross-Language Communication

Before we tackle the two communication models directly, let us first consider what a bilingual would face in the context of abnormal discourse. By definition, a bilingual of two P-languages should master each language equally as well as any competent native speaker does. A bilingual should be able to understand, speak, use each language, as well as to think, reason, and act in terms of each language. Obviously, understanding is no longer a problem for a bilingual in cross-language communication. We do not have to debate whether such an understanding is propositional or hermeneutic either. A bilingual of two P-languages can achieve sufficient understanding of both languages and thus overcome complete communication breakdown between the language communities in abnormal discourse. Nevertheless, the question remains: Is successful *full* communication possible?

A Bilingual on the Boundary

One may argue that although it is difficult to put oneself into the other's skin, it does not mean that it is impossible. A member from one P-language community can become a member of another P-language community by immersing oneself in the community and living their way of life. With tremendous effort and sufficient time, one could eventually become a bilingual of both P-languages. For example, our bilingual Ms Jones can become a bi-medical ending up talking both medical languages.

Unfortunately, the life of our bilingual who walks on the thin line between two incompatible P-language communities is not a 'happy' one. A *bi*-lingual—who has mastered two P-languages separately—is not a *meta*-lingual—who can speak a meta-language or common language with the two P-languages as its sublanguages. Lack of a common measure, especially shared M-presuppositions, between two substantially different P-languages (such as the Chinese medical language and the Western medical language) often puts our bilingual (Ms Jones) who dwells on the boundary between the two language communities in an awkward situation. It is impossible for Ms Jones to commit herself fully to the two incompatible P-languages since one can neither be incorporated into the other, nor can they coexist peacefully. She cannot think in terms of both P-languages at the same time. She might not be able to live the pre-modern Chinese way of thinking. If she could, then her present Western way of thinking would not survive. In any case, at best Ms

Jones can start to think and talk in terms of the pre-modern Chinese way of thinking only when she becomes *alienated or dissociated* from the Western medical language. By doing so, she will eventually drop out of the language community of Western medical theory. At some point, our bilingual eventually has to decide to which language she wants to commit herself unless she can 'happily' live such a 'double life'. Nevertheless, the decision is not going to be an easy one since she cannot hold both P-languages side-by-side in mind to make a point-to-point rational comparison.

One might ask, why does our bilingual have to commit herself to one of the two competing languages? Could she suspend her judgment and hold onto both languages without committing to either? This is fine if Ms Jones does not need to communicate with the other side. The trouble emerges when she attempts to communicate the ideas of the Western medical language to the Chinese, or vice versa. These ideas are intelligible for the Chinese only if the underlying M-presuppositions are comprehensible to them. This requires that Ms Jones emerge from the way of thinking embedded in the Western medical language and interpret the ideas in the way that the Chinese can understand. Therefore, a bilingual like Ms Jones who inhabits a certain type of frontier situation between two rival P-languages always faces a choice between immerging and emerging: to immerge into an adopted alien language to effectively understand it; to emerge from one's own native language to successfully communicate with the speaker of the alien language. Such frequent switches between immerging into and emerging from a P-language very often put our bilingual in a predicament if she is confused about in which language community the discussion is occurring. The use of one mode of reasoning or lexical taxonomy to make assertions to the speaker who uses the other incompatible one makes understanding problematic, and thus places communication at risk. For example, as Kuhn points out, one cannot speak an alien language (such as the language of Aristotelian dynamics) while using the projectible kind-terms of the other competing language (such as the Newtonian notion of 'motion'). Bilinguals are forced to remind themselves at all times which language is in play and within which language community the discourse is occurring to avoid improper use of a kind-term of one language in the other language community. The inability to use one way of thinking embodied in one language to understand the other language makes full communication between two substantially different languages problematic.

Of course, it is possible for our bilingual to understand the Chinese medical language but not commit herself to it. She might identify and comprehend very well the pre-modern Chinese way of thinking and corresponding categorical framework while rejecting them as either illegitimate or unsuitable. In this case, she actually becomes a *spectator*, not a participant, therefore not an *engaged* conversation partner and communicator. We have more to say about this when we discuss Gadamer's conversation model of communication later.

A Bilingual in Communication

More significantly, let us put those worries about bilinguals surviving on the boundary between two incompatible P-language communities aside and turn our attention to cross-language communication. The achievement of a bilingual is understanding; but understanding is only part of communication. The real issue is this: Even if our bilingual can understand perfectly an alien P-language, it does not mean that she can communicate successfully with its speakers. This is because cross-language understanding between two P-languages itself does not guarantee successful communication between the language communities. Cross-language understanding is only *necessary*, but not *sufficient* for cross-language communication.

For our bilingual (Ms Jones) to be able to communicate with another P-language community (such as the Chinese medical community), she needs first to master both languages (the Chinese and Western medical languages), that is, to be able to *understand* and *use* effectively each language involved. But the fact that our bilingual can understand an alien language does not automatically make her *an engaged communicator* with the speaker of the alien language. Communication is not assimilation. A constructive dialogue between two distinct P-languages requires that each side commit to their own language and manage to communicate their thoughts to the other side. It is not an easy task as Gadamer has realized:

> It is well known that nothing is more difficult than a dialogue in two different languages in which one person speaks one and the other person the other, each understanding the other's language but not speaking it. As if impelled by a higher force, one of the languages always tries to establish itself over the other as the medium of understanding. (Gadamer, 1989, p. 384)

To be an engaged communicator, in addition to understanding the other, our bilingual also has to make herself understood well by the other or to *convey* effectively the information from one language (the Western medical language) to the other,[2] to respond properly to the other's requests, and to engage in constructive dialogue with the other. It is exactly this aspect of cross-language communication that the bilingual of two incompatible P-languages cannot fulfill.

For an illustration, consider the following situation faced by our bilingual, Ms Jones, and our old friend, the Chinese physician Mr Wong. Suppose that both of them work together in a clinic, and both speak Chinese (or English if you wish) as the working language. Suppose that Mr Wong, like our Ms Jones, knows both medical theories inside out. For some unknown reason, Mr Wong practices the Chinese medicine only and Ms Jones only practices the Western medicine. One day, a patient by the name of Jennifer, who is a friend of both physicians, comes to visit the two physicians separately for her painful spleen. Mr Wong diagnoses her illness as an excess of the yin within her spleen (an asthenic spleen). According to Ms Jones, however, Jennifer's illness is caused by a bacterial infection. Jennifer, perplexed with two very different diagnoses, asks the two physicians to meet and explain the difference between them to her. As a friend, she requests that each physician convince her of the correctness of their own diagnosis by presenting arguments to the other. You can imagine what would happen under such a situation. Although both physicians understand the other's diagnosis well, it is hard for them to become *engaged* communicators. They cannot engage in a productive conversation, not because of misunderstanding, but because there is no common ground for them to engage in a constructive conversation and argumentation. There is no way to match what Mr Wong wants to say against what Ms Jones wants to say. Each side cannot make a point-to-point comparison with the other side. Neither can even deny directly the correctness of the other's diagnosis by claiming that it is *false*. In fact, the question of whether the counterpart's diagnosis is *true or false* simply does not arise. Although both can understand one another very well, the communication between them is inevitably partial.

The point of the above thought experiment can be driven home when we compare the previous situation with the following one. Suppose that Jennifer visits two Chinese physicians. One physician diagnoses her illness as an asthenic spleen while the other diagnoses it as a sthenic spleen (the excess of the yang within her spleen). When asked to present their own case, each physician can effectively present his or her argument based on the evidence shared by both (such as symptoms of abdominal distention, loose stool, inappetence, phlegm-retention, oedema, diarrhea, blood in stool, and so on). In this case, both sides can not only understand one another well, but also engage in successful communication by exchanging and comparing their thoughts as well as engaging in constructive rational argumentation.

What we learn from the above case is not only that cross-language understanding is only necessary, not sufficient, for cross-language communication in abnormal discourse; but also that cross-language understanding is actually necessary for our realization of partial cross-language communication breakdown. Communication breakdown between two language communities with incompatible M-presuppositions arises not just from one side's inability to understand the other side (a *complete* communication breakdown), but also precisely from those discourses in which one side is able to understand and see how different the other's beliefs, categorical framework, and mode of reasoning are from their own (a *partial* communication breakdown). Ms Jones and Mr Wong cannot engage in a constructive dialogue with one another, not because the other's beliefs appear bizarre, but because each can understand how the other is tied to a form of life and intellectual tradition manifested as the M-presuppositions of the language to which they are so committed themselves. Ironically, it is precisely our ability to understand the other language that reveals the inability to communicate successfully with the other. In this sense, a bilingual can only make one aware of and appreciate the occurrence of partial communication breakdowns in abnormal discourse, but cannot provide a solution.

3. The Transmission Model of Cross-Language Communication

We have contented ourselves so far with a vague and common sense notion of cross-language communication. To distinguish cross-language communication from cross-language understanding and to argue for the thesis of inevitability of communication breakdown in abnormal discourse, the time has come for us to be more specific about the notion of cross-language communication.

The Varied Senses of 'Communication'

Although humans were anciently dubbed both the 'rational animal' and the 'speaking animal' by Aristotle, only since the late nineteenth century, especially after the linguistic turn in the early twentieth century, have we realized that it is our ability of using language to communicate with ourselves and others, more than our ability to reason (we cannot reason without language), that defines who we are. We might simply define humans as the linguistic communicator. Indeed, there is nothing people do more often than communicate with one another by language. However, the act of linguistic communication is easily accomplished, but not so easily explained conceptually.

As other notions hailed as unmixed goods, the notion of communication suffers from the misfortune of conceptual confusion. Through tracing the historical development of the idea of communication, J. Peters (1999, pp. 6-10) finds that there are at least four clusters of meanings associated with the notion of communication. (a) The word 'communication' originated from Latin word '*communicare*', meaning to impart, share, or to make common. Accordingly, one dominant cluster of meaning in 'communication' has to do with imparting, a *unilateral* process quite apart from the *bilateral* process of exchange, dialogue, interaction. It could mean 'partaking' (partaking of Holy Communion), an act of receiving, not of sending; it could mean a message or a notice to a certain audience, an act of sending, not receiving or exchanging; or it could mean physical connection or linkage. We can call this *the communion model of communication*. (b) Another dominant cluster of meanings involves transmission of physical objects, such as heat, light, electronic signals, or of information, which could be psychological entities, such as ideas, thoughts, or speaker's meanings, or linguistic entities, such as propositional contents, literal meanings. As we will see later, like the communion model, this *transmission model* is not necessarily a bilateral process of exchange and interaction between the sender and the receiver. The transmission process could be one-way, from the sender to the receiver without any feedback from the receiver. (c) If we fold up the linear transmission model to connect its one end with the other so as to make it a two-way, bilateral process of exchange between the sender and the receiver, we have *the exchange model of communication*. It is supposed to involve interchange, mutuality, reciprocity, and engagement, such as exchange of ideas in dialogue, psychosemantic sharing, even fusion of consciousness. (d) Last, 'communication' can mean something very general, as a blanket term for various modes of communication identified above.

Very roughly, any attempt at manifestly having a specific impact on somebody else's cognitive attitudes may be called an act of communication.

The above four models of communication cover linguistic and non-linguistic communication. Linguistic communication stands out from any other forms of non-linguistic communication in that the vehicles or mediums employed in communication belong to a system of semantically interdependent meaning items, which are conventionally established, such as a natural language. Although there is no consensus on what linguistic communication is, especially, what counts as an engaged successful linguistic communication, we can still identify the two most popular philosophical models of linguistic communication directly pertaining to cross-language communication. One is the transmission model, which is the standard model of linguistic communication; and the other is what I shall call the dialogue model, roughly corresponding to the exchange model identified above. We will focus on the transmission model in this chapter and discuss the dialogical model in the next chapter.

The Standard Transmission Model

The transmission view of linguistic communication is the most prevalent conception of communication in our culture, and perhaps in all industrialized cultures (J. Carey, 1992). The philosophical foundation of the transmission model was established by John Locke's empirical epistemology. It is John Locke, more than anyone else, who provides articulate defense of the two doctrines foundational to the classical transmission model: the private mind filled with ideas and linguistic signs as empty vessel to be filled with ideational content. Locke treats the meaning of words as a sort of private property in the individual's interior, that is, the internal stream of ideas derived either from reflection or from sensations which are neither social nor linguistic. For Locke, language does not play any of the significant roles that many philosophers attribute to language, neither as a source of knowledge, nor as a shaper of thinking (B. Whorf), nor as the way we articulate our being in the world (Heidegger), not even as a mode of performing actions (T. Austin); it is rather 'the great instrument' that makes the inner life of ideas publicly accessible and a means of transporting ideas from one speaker to another. Words are at best conventions; they refer to meanings inside people's minds and to objects in the world. When we communicate with others, we trust our private ideas to public symbol proxies by virtue of *encoding* them as linguistic signs. An act of communication is successful if the hearer can replicate the speaker's ideas without distortion in terms of *decoding* the linguistic signs. 'When a man speaks to another, it is that he may be understood; and the end of speech is, that those sounds, as marks, may make known his ideas to the hearer' (Locke, 1996, 3.2.2).

Locke's notion of private ideas and his commitment to the individual as sovereign in meaning-making make linguistic communication both necessary and impossible. It is fundamentally impossible to communicate accurately the ideas in the mind of one person to stimulate the same ideas in the mind of another, unless we can read each other's minds. Therefore, for Locke, all communication, both intra- and inter-language communication, is inherently imperfect. Communication breakdowns loom large in Locke's scenario.

Could we remove the mentalistic elements of the Lockean classical transmission model (meanings as private ideas or mental images given by individuals), but preserve its basic spirit—briefly, linguistic communication as the process of transferring messages from one speaker to another by means of language? The Lockean classical transmission model hinges on the premise that communication is a process whereby an *idea*, as a private mental entity, transfers from *the mind* of the speaker to *the mind* of the receiver. If we substitute a neutral term 'message' for Lockean 'ideas' as what is to be transferred and accordingly de-emphasize the role of *the private mind* of the speaker/receiver in the transmission process, we can identify four basic components involved in the process of linguistic communication: message, speaker, receiver, and linguistic expression.[3] Like our treatment of cross-language (propositional) understanding, 'the speaker' still refers to the representative speaker of a language community; the same is true for 'the receiver'. By 'the linguistic expressions', I mean declarative sentences of a P-language, a system of semantically interdependent meaning items that can be conventionally established by the language. Especially, I will identify 'message' with thoughts, propositional contents, or literal meanings expressed by declarative sentences. To repeat what we have emphasized in chapter 12, I treat the notion of thought here as *extensional*: A thought is not the *intensional content* of the mind, such as Lockean ideas or mental images, or the content of beliefs and other propositional attitudes, which characteristically differ from one individual to another, but is rather taken to be what is expressed by a declarative sentence and corresponds to the propositional content asserted by the sentence. What distinguishes (objective) thoughts from (purely subjective) mental contents is that thoughts can be, or are at least capable of being, true or false while mental contents, such as Lockean ideas, cannot. As such, thoughts are neither inside the head nor outside language. They are publicly accessible to all the competent speakers of a language and are communicable among them.

In contrast, it is common among many to treat linguistic communication as the branch of the use of language that focuses on the speaker-meaning, the intention of an individual speaker, conversational maxims, propositional attitudes, and the illocutionary force the speaker expresses by saying something (a speech act). For our limited purpose, I will not burden us with those pragmatic aspects, but only focus on the semantic aspect of linguistic communication, which I will call informative communication:

> Cross-language (informative) communication is essentially a process of the transmission of thoughts (the literal meanings or the propositional contents) from the speaker of one language to the interpreter of the other language.

This is actually the standard model of linguistic communication adopted by most analytical philosophers since the linguistic turn.

Let us see what counts as successful communication between two disparate P-language communities in abnormal discourse based on the standard model. There are two sides to any act of linguistic communication, the speaker's side and the interpreter's side. Ideally, in a successful cross-language communication, the two sides in discourse understand a sentence from the speaker's language in the same way: The interpreter understands precisely what the speaker means. As before, I call the literal meaning of a sentence in the speaker's language 'the speaker's language-meaning' and the meaning that the interpreter ascribes to the same sentence 'the interpreter's language-meaning'. In an ideal situation, the interpreter's language-meaning and the speaker's language-meaning should be the same; in other words, the propositional content expressed by the speaker is the same as that taken in by the interpreter. If such an agreement is achieved, then we say that the act of linguistic communication between two P-language communities is successful; otherwise, we say that the communication is defective or breaks down.

Corresponding to the two sides of the act of linguistic communication, i.e., the speaker's and the interpreter's sides, we can identify conceptually two directions or two parts of the act of cross-language (informative) communication. On the one hand, the speaker of one P-language expresses a particular thought (or a propositional content) to the interpreter of another P-language. Such a speech act is performed whenever a thought is put forward to the interpreter in terms of a declarative sentence. On the other hand, the interpreter apprehends or takes in the thought put forward by the speaker. We can call the act of putting a thought to the interpreter 'the act of communicating' (in the narrow sense) and the act of apprehending a thought put forward 'the act of understanding'.

Clearly, full cross-language (informative) communication requires successful cross-language (propositional) understanding since understanding is a precondition of transferring thoughts. The failure of cross-language understanding would lead to complete cross-language communication breakdown as I have argued in chapter 12. However, the success of cross-language understanding does not guarantee the success of full cross-language communication. One (such as our Ms Jones) can understand a P-language (the Chinese medical language), but may not be able to communicate effectively to its speakers (such as our Mr Wong). This is what we shall turn to next.

4. Communication Breakdown in Abnormal Discourse

It would be an arduous task (if it is possible at all) to specify a complete list of sufficient conditions under which a successful cross-language communication is fulfilled. Fortunately, since we are only concerned with cross-language communication breakdown in the case of incommensurability, we only need to identify one significant language-dependent *necessary condition* for successful cross-language communication, namely, a *common measure* shared by two different P-languages in order for the communication between them to be successful.

According to the standard model, cross-language communication is the transmission of thoughts from the speaker of one P-language to the interpreter of another P-language in terms of declarative sentences. More precisely, we have identified the two closely related acts in the process of thought-transmission: the act of communicating and the act of understanding. For the communication to be successful, the thoughts apprehended by the interpreter have to be *the same thoughts* put forward by the speaker. Although it may be controversial as to what counts as the *same* thought (propositional contents) transferred and received, the truth-value of the thought transferred should, at least, be preserved during the transmission. That means that the interpreter should at least get the truth-value of the thought put forward by the speaker right in order to receive the same thought transferred. In particular, in formal communication such as scientific communication, we obviously have an interest in not acquiring false thoughts or propositions. As such, the primary goal of thought-transmission is, at least, to preserve the truth-value—or more stringently, to preserve the truth—of what is transferred. Therefore, truth-value-preserving thought transmission should be a necessary condition of successful cross-language communication.

Let me be more specific about the thought transmission process in cross-language communication with an eye on what is required to preserve truth-values. There is a speaker S of a P-language PL_S and an interpreter E from another P-language PL_E. Suppose that S wants to communicate to E a particular message— e.g., a thought T or a proposition p. To convey p, the speaker searches in PL_S for a sentence that is true if and only if p. Eventually, she finds a sentence A of PL_S that fits this truth condition. She utters A for E to hear or writes down A as a text for E to read, and hopes for the best. Of course, she cannot just hope that the interpreter can catch what she said. For her speech act to qualify as a successful act of communicating, the speaker is responsible to prepare the way of smooth transmission of her thoughts. More precisely, she has to express p in such a way that the interpreter is able to recognize that, first, she has *asserted* something to E when uttering A (the condition i); and second, what she asserts is p (the condition ii). To ensure the condition i (E could recognize that S has asserted something to E), the speaker needs to make sure that she has said something that is true or false to the interpreter. However, she cannot simply assume that the interpreter could somehow *conceptually* grasp its truth-value (i.e., to assume that the interpreter is a bilingual). Thus, she needs to make sure that sentence A is not only *actually* true or false (has an *actual* truth-value) from the perspective of her own language PL_S, but also true or false when A is considered within the context of the interpreter's language PL_E. In addition, to ensure the condition ii (the interpreter could understand what is asserted by A), the speaker has to make sure that the interpreter is able to establish the logical connection that A is true if and only if p, which in turn requires that the interpreter can recognize that A is *actually true* when considered within his own language PL_E. The fulfillment of both conditions i and ii requires sentence A to have an *actual* truth-value for both the speaker and the interpreter. Furthermore, based on our analysis of the concept of truth-value status in chapter 9, the interpreter could recognize that sentence A of PL_S —which is actually true or false when considered within PL_S —as *actually* true or false also when considered within PL_E only if he accepts or could accept—not just *recognize* in the case of understanding—the M-presuppositions of PL_S. This is possible only if the M-presuppositions of PL_S and PL_E are shared and compatible.

I thus conclude that shared or compatible M-presuppositions between two P-languages are necessary for carrying out a successful communication between the two distinct P-languages. In other words, shared or compatible M-presuppositions are the significant necessary *common measure* (the pre-established *semantic harmony* between the two languages) for successful cross-language (informative) communication.

This explains the awkward situation faced by our bilingual Ms Jones and Mr Wong since mutual understanding is necessary but not sufficient for successful cross-language communication. To understand a thought T put forward by the speaker S of an alien P-language PL_S, the interpreter E from another P-language PL_E has to put himself in a position to decipher and acquire T as S intended to be understood. According to the truth-value conditional theory of understanding, to comprehend T, the interpreter does not have to *accept* the M-presuppositions of PL_S. Instead, E only needs to be able to identify and comprehend them such that sentence *A* has a *conceptually* possible truth-value for him. This can be done even if PL_S and PL_E are incompatible, either by virtue of language learning or the hermeneutic interpretative understanding. But in order to communicate successfully with one another, we have concluded that the two languages involved have to be compatible. Therefore, the cross-language (informative) communication in abnormal discourses is inevitably partial.

5. Informative Communication as Mutual Understanding

One may be considering that although informative communication cannot be reduced to one-way propositional understanding, either on the interpreter's part or on the speaker's part, it could very well amount to mutual understanding. There is, in fact, a strong tendency in the related literature of confusing communication with mutual understanding. For many analytical philosophers (such as Davidson, Dummett, and others), cross-language communication is nothing but the understanding back-and-forth between the speakers from two different languages. Insofar as we are interested in communication, they believe, we are interested in mutual understanding. Such a reading of communication as mutual understanding is logically implied and widely promoted by the transmission model of communication. The goal of communication, according to the model, is simply transmission of thoughts from one side to the other. The purpose of transmitting thoughts from the speaker to the interpreter is to have the interpreter understand them. As long as the interpreter comprehends the thoughts transmitted, the goal of communication is obtained. Hence, the act of communicating could be reduced to the act of understanding; the former is only the means to the latter. Simply reversing the above process from the interpreter to the speaker, we have the act of mutual understanding. Therefore, communication equals mutual understanding.

However, as I have argued, what is involved in the act of communication is more than just mutual understanding. Mutual understanding could be achieved by two bilinguals, one on each side, such as the exchange between Ms Jones and Mr Wong. However, the fact that our bilinguals can understand one another's languages does not necessarily make them *engaged communicators*. As will become clear when we discuss the dialogue model of communication, the goal of communication is not merely mutual understanding in *the propositional sense*—simply passing on information to one side and taking in the information passed on from the other side.

To think of the act of cross-language (informative) communication as the act of transmission of information and further as mutual understanding actually reduces cross-language communication into cross-language translation. To illustrate, we can reformulate the process of thought transmission on the speaker's part by substituting 'translation' for 'transmission': To transfer a proposition p from the speaker's language to the interpreter's, the speaker first finds a sentence A from his or her own language which is true if and only if p. Second, the speaker *translates A* into a counterpart sentence B of the interpreter's language (if there is any) such that B is true if and only if p. However, it could be argued that the last step of translation cannot be carried out between two radically distinct languages since the interpreter's language does not have the necessary semantic resources (due to meaning change, reference difference, unmatchable lexical structures, or whatever other reasons) to express A. Cross-language communication breakdown turns out to be translation failure. I think the reader should have recognized by now that this reading of the transmission model brings us back to our old foe, i.e., the thesis of incommensurability as untranslatability. This is, I think, a major reason why the received interpretation of incommensurability as untranslatability sounds so natural for many, is so widespread, and is so hard to dispose of, since the interpretation is sanctioned and conditioned by the transmission model of communication.

These and other limitations of the transmission model can be attributed to its semantic foundation, namely, the monologue model of meaning. The standard transmission model of cross-language communication is propositional in essence. The model reduces the act of communication, which is supposed to be an 'alive', interactive, and dialectic process, to a 'dead', static, and monological propositional understanding: there is a fixed, self-sufficient thought in the mind of the speaker; the speaker conveys the thought to the interpreter in terms of a sentence. The sentence is self-closed and its meaning is self-contained, independent of the interpreter, simply there to be discovered. For the communication to be successful, the interpreter has to recapture the author's intention and to understand the sentence as the speaker intends and expects it to be understood.

To think beyond understanding and translation, we should ask ourselves a question: What do we want to get from communication, especially from cross-language communication? To understand each other, of course, one may answer. But why do we care whether or not we understand one another? What is the purpose of understanding anyway? Or, more precisely: What does understanding one another enable us to do? The answer cannot be that we want to understand simply for the sake of understanding. We have at least as much interest in learning from one another and coordinating our actions in a social setting through understanding. For those and other purposes, we need to hammer out disagreement and reach consensus. 'In a certain sense the question of commensurability and incommensurability with respect to competing paradigms or research traditions is the question of the possibility of objectively resolving the differences between them' (Boyd, 2001, p. 56). To do so, only passively understanding one another by passing on information is not good enough; it requires critical engagement between two sides: to respond effectively to the other side's requests, to exchange ideas effectively, and to engage in constructive dialogue and argumentation with one another. Unfortunately, this crucial aspect of genuine communication is missing from the standard model of communication.

We need to go beyond the transmission model.

Notes

1   This rather simplified statement, although it is sufficient for the time being, will be qualified later when we discuss further Gadamer's hermeneutic understanding. As we will see, for Gadamer, understanding is an open-ended process of a speaker coming-to-an-understanding and reaching-an-agreement with someone or a text about something. Such an understanding can only be achieved through conversation, which is in essence a process of communication. Therefore, no sharp distinction could be made between understanding and communication. Habermas holds a similar view. In a sense, for both Gadamer and Habermas, understanding is communication.

2   Suppose that the other side is not a bilingual, which is true in most cases of cross-language communication. But it will not affect our analysis even if both sides are bilinguals as will be illustrated in the following thought experiment.

3   By distinguishing medium from message, I do not intend to take the message, such as thoughts or propositions, to be communicated by words or a sentence (the medium) as naked and fully determined prior to any linguistic expression. The full meaning of a message is determined in the process of being expressed by some linguistic expressions. Thoughts cannot exist antecedently and independently of the vehicle of expressing them, namely, linguistic expressions. The process for a sentence to express a thought is the process of showing what the thought to be expressed is.

Chapter 15

Dialogical Communication Breakdown I: Gadamer's Conversation Model

L. Grossberg points out that 'we were living in an organization of discursive and ideological power that could be described as "the regime of communication"' (1997, p. 27). As you can guess, what Grossberg has in mind is the transmission model of communication. By calling it 'the regime of communication', Grossberg does not intend to treat the transmission model as a legitimate description of a process or a phenomenon; rather, he treats it as a certain way of thinking and talking about communication, a particular conceptual framework that M. Reddy calls the 'conduit metaphor'. Or, using M. Foucault's terminology, it is only a particular *discourse* about communication. As a dominant discourse, the transmission model does have a tremendous hold over us. We are trapped in the reality created by this way of talking. The best strategy to escape the tight grip of the regime of communication created by the discourse of the transmission model is, I think, to deploy a different set of discursive resources for the articulation of communication: the hermeneutic discourse of communication that uses conversation, rather than transmission, as the central metaphor. The two primary hermeneutic discursive resources I will draw on are Gadamer's conversation model of communication from his philosophical hermeneutics and Habermas's discourse model of communication derived from his theory of communicative action. For reasons that will become clear later, I will refer to the new model of communication based on Gadamer's and Habermas's works as the dialogue model of communication.

1. Coming to an Understanding through Conversation

Since Heidegger, in his *Being and Time*, announced his distaste for any notion of communication as mental sharing through transmitting information, no one in the Heideggerian heritage has any taste for communication as information exchange or thought transmission as described by the transmission model. Gadamer was not an exception. For Gadamer, the concept of communication no longer refers to a linear one-way transmission of some self-contained units of meanings—no matter which are ideas, thoughts, or propositional contents—from one person to another, from one language to another, or from one time or place to another, as if meanings could travel intact. Nothing 'moves' in hermeneutics. Since the term 'communication' had been so heavily associated with the transmission model, always appearing alongside terms such as 'sender', 'receiver', 'encode', 'decode', and 'transmission', Gadamer would prefer to discuss communication in the context of a different set of terms such as 'understanding', 'interpretation', and 'conversation'.

In all his work, Gadamer had been drawn to what we can learn from Plato about Socratic dialogue or conversation, which is, to Gadamer, the clue to revealing the nature of substantive hermeneutic understanding. It is Plato who made us realize the hermeneutic priority of questioning in all experience, all knowledge, and discourse. Especially, there is a close relation between questioning and understanding that is 'what gives the hermeneutic experience its true dimension' (Gadamer, 1989, p. 374). With Collingwood, Gadamer contends that, just as all knowledge starts from questions all understanding begins with questions. We can understand a text only when we have understood the question to which it is an answer.

> Thus a person who wants to understand must question what lies behind what is said. He must understand it as an answer to a question. If we go back *behind* what is said, then we inevitably ask questions *beyond* what is said. We understand the sense of the text only by acquiring the horizon of the question—a horizon that, as such, necessarily includes other possible answers. Thus the meaning of a sentence is relative to the question to which it is an reply, but that implies that its meaning necessarily exceed what is said in it. (Gadamer, 1989, p. 370; italics as original)

The real and fundamental nature of questioning is its openness. Questions always bring out the undetermined possibilities of a thing to be understood. This is the reason why understanding is always more than merely re-creating the author's intention or the text's original meaning. One's questioning of a thing to be understood opens up possibilities of its meaning. What is meaningful passes into one's own thinking on the subject in the context of one's own horizon. The fullness of meaning is constantly in the process of being redefined and can be realized only during the dialectic of question and answer (Gadamer, 1989, pp. 373, 375).

Such a dialectic interplay of question and answer, which leads to genuine understanding, is actually a reciprocal relationship of the same kind as conversation. As such, genuine understanding turns out to be a process of *reaching or coming to* an understanding through conversation. A conversation partner does not receive completed meanings from another partner. Meanings are co-created and refined as both interlocutors immerge and engage in a live conversation through questioning and answering. In contrast to the traditional binary mode of understanding—*one person* understands something *unilaterally* since the person who performs understanding has no part in meaning creation and what is to be understood cannot speak back—Gadamer in essence pushes toward a tripartite model of understanding: *one person* comes to an understanding with *another person* about *a subject matter* through conversation, the dialectic interplay of questioning and answering.

One may be wondering whether the above conversion model of understanding could be applied to understanding of written texts. Gadamer is, of course, aware of the differences between the conversation between two persons and the dialogue that we have with written texts. It is true that the text, as 'enduringly fixed expressions of life', does not speak to us in the same way as does a live interlocutor. Nevertheless, this does not mean that we cannot conduct 'conversation' metaphorically with the text. In fact, there are many common aspects shared by both situations (Gadamer, 1989, pp. 377-9, 387-8). First, interpreters can make a text speak to them by posing questions to it. Of course, the text can only 'speak' through the other partner, the interpreter. But interpreters cannot make the text speak anything they want it to speak. What the text really says 'is not an arbitrary procedure that we undertake on our own initiative but that, as question, it is related to the answer that is expected in the text. Anticipating an answer itself presupposes that the questioner is part of the tradition and regards himself as addressed by it' (Gadamer, 1989, p. 377). Second, just as we try to understand what another person speaks to us in a person-to-person conversation, so also interpreters are trying to understand what the text is *saying* to them. Both involve a heavy dose of re-interpretative efforts on the interpreter's part. 'When a translator interprets a conversation, he can make mutual understanding possible only if he participates in the subject under discussion; so also in relation to a text it is indispensable that the interpreter participates in its meaning' (Gadamer, 1989, p. 388). Third, just as interlocutors are trying to reach agreement on some subject with their partners in terms of the fusion of horizons, so also interpreters are trying to reach agreement with the text by fusing their horizon with that of the text. Gadamer concludes that it is perfectly legitimate to speak of a hermeneutic conversation with the text. A reader does not receive a pre-determined meaning from a 'dead' text. Meanings are created and recreated as interpreters engage with the text. The text comes alive only in the context of this engagement.

2. Communication as Conversation

If to understand means to come to an understanding with each other through conversation, then the further question is: What is the primary purpose of conversation? Or what does conversation enable us to achieve? For Gadamer, it is to reach agreement with one another on some subject matter.

> Understanding is, primarily agreement. ... Coming to an understanding, then, is always coming to an understanding about something. Understanding each other is always understanding each other with respect to something. From language we learn that the subject matter is not merely an arbitrary object of discussion, independent of the process of mutual understanding, but rather is the path and goal of mutual understanding itself. ... In general one attempts to reach a *substantive agreement*—not just *sympathetic understanding* of the other person—and this in such a way that again one proceeds via the subject matter. (Gadamer, 1989, p. 180; my italics)

In fact, the German term *Einverständnis*, which is closely associated with the term *Verstehen* (understanding) means 'understanding, agreement, consent'. 'Coming to an understanding with someone on something' means 'coming to an agreement with someone on something'.

> Conversation is a process of coming to an understanding. Thus it belongs to every true conversation that each person opens himself to the other, truly accepts his point of view as valid and transposes himself into the other to such an extent that he understands not the particular individual but what he says. What is to be grasped is the substantive rightness of his opinion, so that we can be at one with each other on the subject. (Gadamer, 1989, p. 385)

In order to reach substantive agreement with each other about some *subject matters* through conversation, one cannot either impose one's own point of view or tradition onto the other (the projective understanding) or place oneself into the other's horizon with the sole purpose of knowing 'objectively' the other's horizon (the adoptive, 'sympathetic' understanding). Genuine conversation is not assimilation, neither to make the other like one nor to make one like the other. In both cases, one has stopped trying to reach a genuine agreement with *one another*. In the former case, one conceals the other's otherness, thus one cannot reach agreement with the *other*. In the latter case, one absorbs one's own horizon and tradition, which makes one who one is, into the other's horizon and tradition so as to make one's own standpoint 'safely unattainable' and one's own self cannot be reached. Again, no genuine agreement can be reached between *one* and the other (Gadamer, 1989, p. 303). In contrast, to reach genuine agreement through authentic conversation, 'both partners are trying to recognize the full values of what is alien and opposed to them. If this happens mutually, and each of the partners, *while simultaneously holding on to his own arguments*, weights the counterarguments, it is possible to achieve ... a common diction and a common dictum' (Gadamer, 1989, p. 387; my italics). The process of reaching such an agreement is what Gadamer calls the process of fusing horizons: a fusion between the horizons of two parties through conversation, whereby one party's horizon is enlarged and enriched in terms of the engagement with the horizon of the other's, not *replaced* by the other's.

In his *Truth and Method*, Gadamer rarely uses the term 'communication' in discussing conversation for the reason mentioned earlier. Nevertheless, it should be clear that through his hermeneutic discourse of understanding, Gadamer not only presents a concept of understanding different from the notion of propositional understanding, but also, more precisely, he opens up a new discourse of communication that has 'conversation' or 'dialogue', instead of 'transmission', as its central metaphor.

> What characterizes a dialogue, in contrast with the rigid form of statements that demand to be set down in writing is precisely this: that in dialogue spoken language— in the process of question and answer, giving and taking, talking at cross purpose and seeing each other's point—performs *the communication of meaning* that, with respect to the written tradition, is the task of hermeneutics. (Gadamer, 1989, p. 368; my italics).

For Gadamer, the process of substantive hermeneutic understanding—the process of 'coming-to-an-understanding' and 'coming-to-an-agreement' through genuine conversation—is, in essence, communication. To distinguish it from the dominant standard discourse of communication as transmission (informative communication), I will call the new discourse about communication the hermeneutic dialogue model of communication, or briefly, dialogical communication. According to it, communication is a process of mutual creation of meanings in the flow of a live genuine conversation between two dialogists. The act of communication is co-created by both interlocutors acting and reacting to each other's utterances, with each utterance creating the conditions for the next.

In fact, it is not accurate to call Gadamer's conversation model of communication a *new* discourse. Historically, the transmission model came later than the dialogue model of communication. The standard transmission model was framed by Locke's empiricist philosophy of knowledge in the seventeenth century, further supported by the invocation of unconsciousness (E. Hartmann, F. Myers, W. James, and S. Freud) during the late nineteenth and the early twentieth centuries, enhanced by the computer metaphor of an information processing paradigm dominant within the field of modern experimental, cognitive psychology since the middle of the twentieth century, and established as a dominant, legitimate scientific model of communication by the information theory (C. Shannon, W. Weaver, and N. Wiener) around the 1950s.[1] The notion of communication theory, founded on the transmission model, is no older than the 1940s. In contrast, the dialogue model can be traced back to the ancient Greek philosophy, especially Plato's discourse on the primacy of dialectic dialogue—the art of question and answer, objection and rebuttal, argumentation and persuasion—in seeking truth and knowledge. Plato's discourse on dialogue started the exchange model of communication, according to which communication is supposed to involve interchange, mutuality, reciprocity, and engagement. Based on this Platonic tradition, a colloquial sense of communication calls for open and frank dialogue. It is not simply talk; it refers to a special kind of talk distinguished by disclosure and reconciliation: disclosing one's oneness to the other and the other's otherness to oneself (knowing oneself through knowing the other and knowing the other through knowing oneself), and reconciling oneself and the other.

Around the 1920s, we saw the revival and rehabilitation of the Platonic dialogue model in M. Heidegger's metaphysics and J. Dewey's pragmatic philosophy. Heidegger's notion of communication is neither semantic (meaning exchange), nor pragmatic (action coordination), but ontological (world disclosing and otherness openness). Communication is, for Heidegger, the interpretive articulation of our 'thrownness' into a world together with people to whom we want to open ourselves to hear their otherness. 'Communication as the revelation of being to itself through language resounds variously through those influenced by Heidegger—Sartre, Levinas, Arendt, Marcuse, Leo Strauss, Derrida, Foucault, and many more' (Peters, 1999, p. 17). With Heidegger, Dewey views language as a precondition of thought and dismisses a semantic view of language as interpersonal plumbing, carrying thought and meaning as a pipe carries water, which is the semantic foundation of the transmission model of linguistic communication. Unlike Heidegger, Dewey's notion of communication is more pragmatically orientated: communication as partaking, namely, taking part in a collective world, not simply sharing the secret of consciousness or transferring meaning. Both Heidegger and Dewey saw the dialectic dialogue as a way out of the state of alienation between people of the time. J. Peters summarizes the social context of the 1920s in which the dialogue model emerges as follows:

> Dewey took the disappearance or distortion of participatory interaction as the most alienating feature of the age. Heidegger's notion of the fall from authentic encounter was not entirely different. The notion that grace is found in dialogue was widely shared in social thinkers of the 1920s: Buber wanted to replace I-It relationships with I-Thou ones; Heidegger called for authentic confrontations; Lukács called for a joyful reconciliation of subject and object. That face-to-face dialogue or at least confrontation offered a way out from the crust of modernity is one of the key themes in thinking about communication since the 1920s, in antimodern thinkers such as Wittgenstein, Arendt, and Levinas, all of whom recognize the ultimate impossibility of dialogue, and in a host of lesser figures who do not. (1999, p. 19)

Gadamer's conversation model is a further development of this trend. It is in the hand of Gadamer that the dialogue discourse of communication reaches its maturity and universality.

3. Reaching a Common Language through a 'Fusion of Horizons'?

We have argued that the relationship between propositional understanding and informative communication is not symmetric as to the direction of message transmission. The process of informative communication (the transmission model) consists of the act of communicating (a message is put forward from the speaker to the interpreter) and the act of understanding (the message put forward by the speaker is taken in by the interpreter). Thus, propositional understanding is necessary but not sufficient for informative communication. Nevertheless, informative communication amounts to, in essence, propositional mutual understanding since both the act of communicating and the act of understanding are by nature a one-way linear transmission of message. In contrast, the relationship between Gadamer's hermeneutic understanding and dialogical communication is symmetric: The process of hermeneutic understanding (*Verstehen*)—the process of 'coming-to-an-understanding' and 'coming-to-an-agreement'—is essentially conversation or communication (*Mitteilung*). On the one hand, we can reach understanding only through conversation or communication; on the other hand, to communicate is to understand through conversation. We can say, to a certain extent, that while communication is, in the transmission model, reduced to (mutual propositional) understanding, (hermeneutic) understanding is, in Gadamer's hermeneutics, elevated to (dialogical) communication.

Although Gadamer does not want to differentiate communication from understanding and interpretation, he does try to distinguish his dialogical communication, which he calls 'authentic dialogue', from 'inauthentic dialogue', such as oral examination, certain kinds of therapeutic conversation between doctor and patient, the interrogation of a criminal, cross-language dialogue via translation, and idle talks. In all those cases, the goal is to either merely know others as *particular individuals* or kill time, not to reach genuine agreement on *some subject matters* through fusion of horizons. For example, having to rely on translation is tantamount to two people giving up their independent authority and making their own horizons and traditions unavailable. In this case, hermeneutic communication (as usual, Gadamer uses the term 'conversation' to refer to hermeneutic communication) does not really take place between two translators. This is because 'understanding how to speak is not yet of itself understanding and does not involve an interpretative process; ... Thus the hermeneutical problem concerns not the correct mastery of language but coming to a proper understanding about the subject matter, ... Mastering the language is necessary precondition for coming to an understanding in a conversation' (Gadamer, 1989, p. 385). This is exactly the case that our bilinguals Ms Jones and Mr Wong have faced repeatedly when they try to communicate through dialectic dialogue to resolve their differences. Although both of them can master the other's language, they cannot communicate effectively since they cannot engage in a productive conversation.

But why? Why cannot even two bilinguals from two distinct P-languages, who obviously understand (in the propositional sense) one another's P-language as any native speakers do, engage in a genuine/authentic dialogue—the interactive dialectic interplay of question and answer, objection and rebuttal, argumentation and persuasion—in order to reach agreement, reconciliation, or at least to effectively pin down the exact disagreement? The question can be reformulated into a less loaded one: Can two interlocutors from two incompatible P-languages communicate fully (conducting authentic conversation)? Alternatively, put into a Kantian style: Is full cross-language conversation (communication) possible in abnormal discourse? If not, does it have something to do with some semantic obstacles between two P-languages involved?

Gadamer's conversation model applies to both intra- and inter-language communication. To see how Gadamer would answer our question, we need to specify some significant necessary condition of cross-language conversation. The goal of conversation is, to Gadamer, to come to an agreement and consensus on some subject matter. But what is the precondition of reaching agreement on some subject matter between the speakers of two distinct languages? To begin with, for Gadamer, what makes conversation on any subject matter possible is language (the linguisticality of dialogue), which provides the *Mitte*, the 'medium' or 'middle ground', the 'place' where conversation takes place. Language is the *Vermittlung*, the communicative mediation that establishes common ground. This seems to be plain enough. However, not any language can serve as 'the medium' and 'middle ground'. Certainly, it cannot be one of the languages involved in a conversation. For Gadamer, it has to be a common language.

Every conversation presupposes a common language, or better, creates a common language. Hence reaching an understanding on the subject matter of a conversation necessarily means that a common language must first be worked out in the conversation. To reach an understanding in a dialogue is not merely a matter of putting oneself forward and successfully asserting one's own point of view, but being transformed into a communion in which we do not remain what we were. (Gadamer, 1989, pp. 378-9)

At first glance, Gadamer appears to bring us back to the age of logical positivism in seeking a common language underlying different scientific languages. This is not the case. The common language underlying different scientific languages that logical positivists seek is some pre-existing, fixed given language, such as a universal observation language underneath two competing theoretical languages, or an uninterpreted formal language underscoring two interpreted scientific languages, which is logically presupposed by and goes beyond the two languages involved. Such a common language is believed by logical positivists to be necessary for theory evaluation, justification, and comparison, for measurement of scientific rationality and scientific progress, and for commensurability. In contrast, although Gadamer does somehow share with logical positivism a dream of the Archimedean point, Gadamer's common language cannot be something fixed, given, and pre-existing in advance before the conversation. It cannot be established by any explicit agreement or 'social contract' that could be negotiated before conversation takes place or by any purely psychological processes of 'empathy' or 'sympathy'. Moreover, one side cannot simply accommodate the other side by adopting the other's language; nor can one side force their own language onto the other. For Gadamer, a common language can only emerge or be 'worked out' during the process of the conversation itself. As Gadamer puts it in clear terms:

Thus it is perfectly legitimate to speak of a *hermeneutical conversation*. But from this it follows that hermeneutical conversation, like real conversation, finds a common language, and that finding a common language is not, any more than in real conversation, preparing a tool for the purpose of reaching understanding but, rather, coincides with the very act of understanding and reaching agreement. ... I have described this above as a 'fusion of horizons'. We can now see that this is what takes place in conversation, in which something is expressed that is not only mine or my author's, but common. (1989, p. 388; italics as original)

In fact, the process of working out a common language during a conversation is nothing but the very process of what Gadamer speaks of as a 'fusion of horizons'. As discussed in chapter 13, Gadamer's 'horizon' is as broad as 'tradition': one's particular viewpoint formed by one's culture, language, and history, or ultimately one's whole lifeworld. Gadamer also contends that a language, more precisely, a P-language, is a worldview. One's tradition or lifeworld is essentially linguistic; it is linguistically constituted and transmitted. In this way, 'horizon' functions somewhat like 'language'; or more precisely, each language, especially our P-language, provides a unique horizon. A language is by essence perspective, 'the range of vision that includes everything that can be seen from a particular vantage point.'

Thus, Gadamer's 'common language' required for successful cross-language conversation turns out to be a 'common horizon'. Such a common language or horizon can be achieved through a fusion between two initially distinct horizons belonging to two interlocutors of a conversation, when one or both horizons undergo a shift such that a horizon is extended and enriched to make room for the object that before did not fit within it. We have thus identified a significant necessary condition for full cross-language communication according to Gadamer:

An *undistorted full* cross-language communication is possible only if a common language can be formed through a fusion of horizons.

4. The Common Language Requirement and Communication Breakdown

By identifying a common language via a fusion of horizons as a necessary condition of successful cross-language conversation, Gadamer certainly puts a strict constraint on undistorted cross-language communication. Obviously, whether cross-language communication in abnormal discourse is possible depends upon whether a common language formed through a fusion of horizons is possible between two incompatible traditions, horizons, or P-languages. However, speaking of 'a common language' through 'a fusion of horizons' in such a loose way does not help us much. We need to be more specific about the degree of fusion of horizons in order to grasp the full meaning of the common language required by Gadamer.

It has become common wisdom that any language (both natural and P-language) is an open linguistic system, open to possible modification, expansion, and evolution (both syntactically and semantically). This makes a fusion of horizons possible. In addition, for human contact between two distinct languages to be possible, some kind of *point of contact* or overlap between them has to be established. This makes a fusion of horizon desirable. Nevertheless, all these points only prove that a *partial fusion* of horizons between two languages is always possible, no matter how distant one is from the other.

To be more specific, let us consider two distinct P-languages PL_1 and PL_2. At the initial contact, the horizons of the two P-languages are distinct, H_1 and H_2, such that mutual understanding between the speakers of PL_1 and PL_2 is distorted and communication impeded. To understand the other, the speaker of PL_1 starts to extend the horizon from H_1 to $H_1(H_2)$ so as to make room for new concepts, new objects, and new beliefs that are beyond H_1's limit. Suppose that the speaker from PL_2 does the same, extending their horizon from H_2 to $H_2(H_1)$. With tremendous effort, patience, and sufficient time, the two moving horizons $H_1(H_2)$ and $H_2(H_1)$ may become fairly close such that both sides can understand one another pretty well. This process may be better seen as a fusion rather than just as an extension of horizons. At the same time the speaker of PL_1 introduces a modified language $PL_1(H_2)$ to talk about the beliefs of the speaker of PL_2 that represents an expansion in relation to PL_2. So the new language used here, $PL_1(H_2)$, opens a broader horizon, extending beyond both the original ones and in a sense combining them.

We have concluded that a *partial fusion* of two distinct horizons is not only possible and beneficial, but also feasible. Consequently, we have two *partially shared* languages, $PL_1(H_2)$ and $PL_2(H_1)$ respectively. This seems not to be controversial. The real question at stake is this: Can two radically distinct horizons determined by two incompatible P-languages be *fully* fused into *a common language* in which both sides can agree to talk undistortively of each other? Gadamer apparently believes that it is possible. He is so convinced that

> When our historical consciousness transposes itself into historical horizons, this does not entail passing into alien worlds unconnected in any way with our own; instead, they together constitute *the one great horizon* that moves from within and that, beyond the frontiers of the present, embraces the historical depths of our self-consciousness. Everything contained in historical consciousness is in fact embraced by *a single historical horizon*. (1989, p. 304; my italics)

Again:

> Transposing ourselves consists neither in the empathy of one individual for another nor in subordinating another person to our own standards; rather, it always involves rising to *a higher universality* that overcomes not only our own particularity but also that of the other. (1989. p. 305; my italics)

Unlike Davidson who attempts to establish the possibility of universal communication—especially communications between alleged incommensurables or between two radically distinct conceptual schemes, if any—through an outright rejection of the idea of a conceptual scheme, Gadamer takes a more moderate road by admitting the existence of radically distinct conceptual schemes, horizons, traditions, or language-views. Nevertheless, this is as far as Gadamer is willing to go with conceptual relativism. As I see it, Gadamer's attitude toward conceptual relativism is of two minds. On the one hand, with conceptual relativists, he believes in conceptual diversity and novelty. This I think marks the superiority of Gadamer's vision over Davidson's. On the other hand, against relativism and with objectivism, Gadamer is still dreaming of an Archimedean point, an overarching common language shared between two radically distinct P-languages or traditions (although this shared language itself evolves, and changes with the contexts of interaction between the two P-languages). For Gadamer, cross-language communication in abnormal discourse is certainly different from that in normal discourse; compared with the latter, the former involves a much more delicate dialectic interplay and back-and-forth negotiation between two P-languages or traditions (the hermeneutic circle), and will eventually work out 'the one great horizon' 'embraced' by the two distinct horizons. No matter how difficult it is, Gadamer is somewhat convinced (but not well argued) that such a common ground is attainable. Essentially, Gadamer reaches a conclusion similar to Davidson's: the universality of cross-language communication.

We have been following Gadamer step by step so far, but I have to part from him here. I agree with Gadamer that a common language is indeed necessary for a full, undistorted cross-language conversation, if not for other reasons, at least for the following logical reason. Recall that, for Gadamer, the primary goal of conversation is to 'reach a substantive agreement' through the art of conversation—argument, question and answer, objection and refutation (Gadamer, 1989, p. 180). I do not know how two interlocutors can engage in a constructive back-and-forth argumentation without first agreeing on some fundamental rules of inquiry (the rules of a language game), including logical rules and modes of justification—such as what are legitimate justifications, what are legitimate questions and acceptable answers, and so on. Furthermore, to reach a substantive agreement entails that both sides, at end, have to agree on the truth claims put forward by the other side, or at least agree on the fact that the other side has said or asserted something (to be either true or false). Therefore, a common linguistic framework consisting of those common beliefs on truth, logic, and justification has to be in place in order to carry out Gadamer's full communication as defined.

It can be argued that, unfortunately, there will never be a full fusion of horizons between two incompatible P-languages. Based on Gadamer's concept of meaning, meaning is always coming into being through the 'happening' of conversation, and conversation always happens within certain contexts. This determines that hermeneutic conversation is an open-ended process, which can never achieve finality. Accordingly, a so-called common horizon is a moving target, the ideal goal of an authentic conversation. Like conversation, a fusion of horizons is an open-ended process, which evolves with the back-and-forth interplay between the horizon of the interpreter and that of the interpreted. A fully fused horizon may never be truly actualized. But this only means, Gadamer would claim, that a common horizon is not feasible, not impossible in principle.

To me, a common language through a full fusion of the horizons of two incompatible P-languages is neither feasible, nor possible *in principle*. There is a certain limit as to how far one horizon can be extended to accommodate the other horizon without losing its own identity. For example, can the horizon of the Leibnizian language L_Z (chapter 5) be expanded to accept the existence of Newtonian absolute space and time? Can the horizon of Western medical language be enriched by the yin-yang cosmology underlying traditional Chinese medical language? As I have argued all along, those core presuppositions of each P-language are logically incompatible. They cannot be woven into one coherent theoretical framework. There is no common language possible between them. Moreover, whenever we try to understand others in a conversation, as Gadamer argues convincingly, we always carry our own tradition along. No matter how much our own horizon H_1 is fused with the other horizon H_2, the new horizon $H_1(H_2)$ is always a horizon affected by my tradition. The same happens to my interlocutor. The new fused horizon formed by her, $H_2(H_1)$, is a horizon affected by her tradition. No matter how closely the two fused horizons move toward one another, $H_1(H_2)$ is not $H_2(H_1)$. They can never merge into one common horizon $H(H_1 \& H_2)$.

To the best of my knowledge, Gadamer does not offer us many positive arguments for his conviction of the possibility of a common language through a full fusion of horizons between any two languages except the following argument from the universality of language. We can identify two aspects of Gadamer's thesis of the universality of language. First, we have the universal understandability of language: As an object of understanding, any human language, and any linguistically constituted and transmitted tradition and horizon, can be understood, no matter how alien it is to us. To be a language is to be understandable. This is, in effect, the conclusion we have reached at the end of chapter 13, that is, the openness of the incommensurables: Although the conceptual distance between two incommensurables (P-language, cultural or research traditions, paradigms, forms of life) does cause some difficulty in mutual understanding, it by no means makes mutual understanding between them impossible.

Second, we have the universal function of language. In Gadamer's view, virtually everything is linguistic, ranging from our lifeworld, experience, to our worldview, tradition, and horizon, as well as from thinking to understanding. In particular, understanding is possible only with the help of language. Understanding is in essence linguistic understanding. I think this is not controversial either. But to reach the conclusion of the possibility (or the universality) of a common language, Gadamer needs to, much more provocatively, stretch the skin of language further to make it cover all the territories supposedly controlled by universal reason:

> Its [referring to language] universality keeps pace with *the universality of reason.* Hermeneutic consciousness only participates in what constitutes the general relation between language and reason. If all understanding stands in a necessary relation of equivalence to its possible interpretation, and if there are basically no bounds set to understanding, then *the verbal form* in which this understanding is interpreted *must contain within it an infinite dimension that transcends all bounds. Language is the language of reason itself.* (Gadamer, 1989, p. 401; my italics)

After identifying language with universal reason, Gadamer is only a small step away from his dream of a universal common language. Against the relativistic-minded philosophers of language (such as B. Whorf) who insist on the uniqueness of each human language, Gadamer believes that

> In actual fact the sensitivity of our historical consciousness tells us the opposite. The work of understanding and interpretation always remains meaningful. This shows *the superior universality of reason* can rise above the limitations of any given language. The hermeneutic experience is the corrective by means of which the thinking reason escapes the prison of language, and it is itself verbally constituted. (Gadamer, 1989, p. 402)

Since universal reason can soar above any limited languages, which is only a dim reflection of the light of universal reason, the reason can build for us a common language above any two limited languages, no matter how remotely related they are.

I have to admit that I do not fully understand Gadamer's notion of universal reason, maybe because I am dazzled by the bright light of his 'superior universality of reason'. All I can see from his glorification of universal reason is the long shadow (since I am dazzled, I can only see the shadow) cast on us from modernity, especially from the dark side of the Enlightenment legacy—their desperate (and failed) attempts to discover some permanent foundations and basic constraints underneath all becoming, diversity, and temporality. Universal reason has been provoked by Descartes, Spinoza, Leibniz, Kant, and Hegel to be such an Archimedean point for justification, understanding, and knowledge. As R. Rorty (1980) has pointed out correctly, despite his own critiques of modernity and many dark sides of the Enlightenment legacy (such as Gadamer's famous critique of the Enlightenment's prejudice against prejudice), Gadamer somehow seeks to recover and preserve part of the legacy that he thinks is most vital, such as universal reason. This fits right into Cartesian rationalism and places Gadamer back in the camp of objectivism with which he has no intention of being associated. Consequently, 'despite Gadamer's incisive critiques of epistemology and the Cartesian legacy, Gadamer unwittingly is a victim of the very Cartesian legacy that he is reacting against' (Bernstein, 1983, p. 199).

When the pretension of universal reason as a universal language is unmasked, it turns out to be just another version of what has been the foundation for modern epistemology—the assumption that 'all contributions to a given discourse are commensurable'. However, hermeneutics is 'largely a struggle against this assumption' (Rorty, 1980). With Heidegger, Gadamer's hermeneutics emphasizes human finitude and facticity of hermeneutic existence, as well as historicity, contextuality, and fallibility of human knowledge and understanding. He wanted to break the modern dream of an objective, rational method that could guide understanding toward the ultimate reality and truth. Despite Gadamer's repeated protestations that the essential question of hermeneutics is not a question of such a method of understanding, Gadamer fails to be persistent with the basic spirit of his hermeneutics when he falls into the trap of universal reason. Such a move is retrogressive and does not resonate well with the rest of his hermeneutics.

We have to conclude that there is no full fusion of two distinct horizons between two incompatible P-languages. A common language above two incompatible P-languages is another modern myth, whose fate is no better than the common language dreamed by logical positivism. If a common language through a full fusion of horizons between two P-languages is required by undistorted, full communication or conversation, as it should be, then we reach the same conclusion as we did in chapter 14 from the transmission model of cross-language communication: Cross-language communication in abnormal discourse is inevitably partial.

One may be wondering how the above conclusion about communication breakdown in abnormal discourse can be reconciled with what I have argued in chapter 13, namely, the universality of hermeneutic understanding and the openness of the incommensurables. Hermeneutic cross-language understanding is in essence coming to an understanding and agreement through conversation between the interpreter from one language and the interpreted from another. Both sides could come to an understanding through conversation only in terms of a fusion of horizons. If a full fusion is unattainable in abnormal discourse, then can understanding even be possible? But we have argued, with Gadamer, that hermeneutic understanding in abnormal discourse is possible. What is going on here? This apparent inconsistency of my positions on hermeneutic understanding and on dialogical communication dissolves so long as one realizes that what we need for hermeneutic understanding is *partial fusion* of horizons. No *full fusion* of horizons is possible, nor is one needed. Accordingly, it is a partially shared language, not a common language, that makes cross-language understanding, and thus partial dialogical cross-language communication, possible.

Note

1   G. Radford traces the historical and conceptual development of the transmission model of communication in his 2005.

Chapter 16

Dialogical Communication Breakdown II: Habermas's Discourse Model

1. From Gadamer to Habermas

Contrary to what *Truth and Method* might imply, Gadamer's masterpiece turns out not to be about the relationship between truth *and* method at all. Gadamer explicitly denies that his hermeneutics, which is ontological and universal by nature, is not simply a method of understanding or interpretation. To Gadamer, the practice of understanding and interpretation is a mundane, common, ordinary feature of language that cannot be captured by methodical rules or strictures as with a scientific method. It would be a mistake to suppose that Gadamer is in the business of providing concrete rules for the carrying out hermeneutic inquiry. Nevertheless, we can see that Gadamer does offer us a very effective methodology of understanding others, especially those initially totally alien to us: to 'reach a substantive agreement' through a dialectical interplay between the horizon (tradition, prejudices) of the interpreter and that of the interpreted (the hermeneutic circle) that leads to a partial fusion of horizons. Such a 'method' of interpretation is in essence 'the art of conversation—argument, question and answer, objection and refutation'.

However, to my dismay, we do not see that truth plays any significant role at all in his hermeneutic understanding and communication. Gadamer has explicitly denied that it was his intention to play off truth against method. But this does not mean that truth should not play any significant role in his hermeneutic method. After all, 'a primary intention of *Truth and Method* is to elucidate and defend the legitimacy of speaking of the "truth" of works of arts, texts, and tradition' (Bernstein, 1983, p. 151). For Gadamer, to understand something is to grasp its 'true meaning', to let its true meaning speak to us, or more precisely, to allow the 'claim to truth', which 'makes upon us', to reveal itself. Clearly, truth is essential for his hermeneutics. Besides, according to Gadamer, one can understand and communicate effectively on some subject matter with someone only by engaging in an authentic conversation or dialogue with that person, and an authentic conversation is the process of question and answer, argument and counterargument, objection and rebuttal. Some notions of truth and truth-value are indispensable for both sides to engage in argumentative dialogue. Without a proper understanding of the role of truth in argumentative dialogue, Gadamer's conversation model of communication is substantially incomplete.

It is notorious that Gadamer's notion of truth is elusive except that we are sure that he has no taste for the traditional correspondence notion of truth and for propositional truth. For Gadamer, truth cannot be exhausted by scientific method alone. It is available to us through hermeneutic method. R. Bernstein suggests that Gadamer is in fact appealing to a concept of *discursive notion of truth*—a truth is whatever is warranted by appropriate forms of argumentation. But what distinguishes appropriate forms of argumentation from inappropriate ones? It cannot be discursive truth; otherwise we would fall into a vicious circle. What is required, in order to safeguard and promote authentic dialogue or successful communication, is a valid form of argumentation. Gadamer is properly blamed to be 'at his weakest in clarifying the role of argumentation in all the claims to truth' (Bernstein, 1983, p. 174).

We can see why the link between truth and argumentation becomes the weakest link in Gadamer's hands from a different angle: Rather than clarify the role of appropriate forms of argumentation in the notion of truth, Gadamer needs to clarify the role of truth in appropriate forms of argumentation. If the notion of truth is provoked to (as it usually does) establish valid forms of argumentation used in dialogical communication, it has to be primitive, a notion we could appeal to intuitively without justification, such as P. Horwich's minimal theory of truth—a version of the deflation notion of truth (the truth predicates, 'is true' and 'is false', exist primarily for the sake of a logical need: to ensure that we can stick to argumentative dialogue). This is in fact the notion of truth we have employed so far.

Habermas, in his theory of communicative action, brings the role of truth in cross-language communication back to the central stage. Habermas's main concern has been with such questions as: 'How is social action possible?' 'How is a social order or societal integration based on coordination of social action possible?' To address those questions, Habermas needs a comprehensive critical social theory, which in turn is based on his notion of communicative rationality. The concept of communicative rationality makes sense only against the background of Habermas's theory of communicative action. To see the central role of truth played in Habermas's view on linguistic communication, we will only focus on those parts of his theory of communicative action that focus on the crucial role of validity claims, especially truth claims, in communication.

Before we present Habermas's theory, it would be instructive to see how much Habermas, in developing his theory of communicative action and rationality, is deeply in debt to Gadamer's philosophical hermeneutics. Here are only some similarities and contrasts between them.

First, for both Gadamer and Habermas, understanding is the process of coming to understanding, which is oriented toward agreement or consensus, not simply comprehension. 'The goal of coming to an understanding [*Verständigung*] is to bring about an agreement [*Einverständnis*] that terminates in intersubjective mutuality of reciprocal understanding, shared knowledge, mutual trust, and accord with one another' (Habermas, 1979, p. 3). In fact, 'one would fail to grasp what it means to understand an utterance if one did not recognize that it is supposed to serve the purpose of bringing about an agreement (*Einverständnis*)' (Habermas,

1992, p. 78). For Gadamer, understanding is linked to coming to understanding and agreement through a fusion of horizons in conversation; for Habermas, understanding is connected with reaching understanding and consensus in terms of the evaluation of validity claims in argumentation.

Second, Habermas enthusiastically endorses the situatedness of understanding. The theme of our historicity is no less fundamental for Habermas than it is for Gadamer. With Gadamer, Habermas insists that we cannot completely escape from our own tradition and horizon (the lifeworld for Habermas) in seeking to understand what is apparently alien to us. Understanding always starts with one's own tradition or lifeworld and moves to fusion of traditions.

Third, in Gadamer's view, tradition is linguistically constituted and transmitted. Language is worldview. Understanding is no longer a sharing of consciousness and mental exchange; instead, all understanding is linguistic. Habermas's emphasis on language is, if anything, even greater. For Habermas, language is a kind of metainstitution on which all social institutions are dependent; social institutions depend on social actions and social actions are constituted in ordinary language communication. As well as serving mutual understanding and the reaching of consensus, language is also a medium of domination and social power that serves to legitimize relations of organizing power.

Fourth, both Gadamer and Habermas are concerned with the vitality of dialogue, conversation, questioning, solidarity, and community to human society. Both agree that the breakdown of communication in the form of the non-agreement of reciprocal expectations is a threat to social life.

Fifth, Gadamer's philosophical hermeneutics aims to unearth the fundamental conditions or presuppositions of human understanding (which is in essence communication). Similarly, Habermas's theory of communicative action is directed toward uncovering the universal conditions or presuppositions of all communication (understanding). We will see the connection between Habermas's notion of communicative action and Gadamer's notion of understanding when we discuss communicative action in detail in the following.

2. Habermas on Communicative Action[1]

Lifeworld and Communicative Action

Habermas identifies communicative action as the primary mode of action coordination within the lifeworld. Like Gadamer's notion of lifeworld (the world in which we live out our everyday life, which is a coherent web of meanings, beliefs, values, and norms structured and transmitted linguistically), Habermas thinks of lifeworld in linguistic terms. The lifeworld is the world structured by language and cultural tradition, more precisely, a 'culturally transmitted and linguistically organized stock of interpretive patterns' (Habermas, 1987, p. 124). Since culture is the patterns of interpretation transmitted in language, it would better to say that the lifeworld consists of a linguistically transmitted cultural stock of knowledge that is always familiar. This stock of interpretive patterns, intuitive know-how, and socially established practices serves, in the form of implicit assumptions or presuppositions, as a background to all understanding and communication. Similar to Gadamer's tradition, lifeworld is 'the horizon-forming context of communication' (Habermas, 1985), which can both restrict (as a limit) and promote (as a resource) communication. On the one hand, we can no more step outside the lifeworld than we can step outside language. On the other hand, the structure of the lifeworld lays down the forms of possible intersubjective understanding and meanings for communication. In modern societies, communication is achieved through a highly reflective mode of communicative action (in the form of critical, open-ended argumentative dialogue) against the background of plural lifeworlds.

Within the lifeworld, we coordinate our actions primarily in the form of communicative action, a form of social interaction in which the social actions of various agents are coordinated through an exchange of communicative acts that are oriented toward reaching understanding [*Verständigung*], i.e., toward reaching *both* comprehension *and* consensus/agreement [*Einverständnis*]. In its simplest terms, communicative action is the action oriented toward understanding (agreement). Clearly, not all actions are oriented toward understanding. Thus, it is important, in Habermas's view, to distinguish communicative action from nonsocial instrumental action and social strategic action (purposive-rational action). Communicative action is oriented toward understanding, coordinated through consensus, and rationalized in terms of discursive argumentation, while instrumental action and strategic action are oriented toward success, coordinated through influence, and rationalized in terms of empirical efficiency of technical means and the consistency of choices between suitable means. In Habermas's view, communicative action is the primary mechanism for social integration and purposive-rational action is merely a secondary one.

Although communicative action could be carried out by non-linguistic means, such as by extra-verbal body or facial expressions, Habermas focuses his attention on those communicative actions exhibited in the use of language, the primary mode of communicative action. Two points are worth mentioning here regarding the *use* of language in communicative action. First, according to Habermas, the use of language oriented toward understanding (the communicative mode) is the original or primary mode of language use. Other uses of language—such as the indirect mode of language use (the figurative, the symbolic) and the instrumental or the strategic mode (such as the perlocutionary effects of language use, 'Give me the money or I will shoot you', or 'May your children die before you')—are parasitic on the communicative mode of language use. Second, be aware of the fact that the language used in communicative action, according to Habermas, is not language conceived as a syntactic and semantic system, but rather language as it functions in social interaction, namely, language *in use* under certain circumstances, or simply, language as *speech act* (the utterance of a well-formed sentence in a particular situation). According to this pragmatic explanation of language, the basic unit of language is not the sentence in isolation but the sentence-as-uttered (the utterance), i.e., the speech act. Thus, communicative action is primarily a communicative speech act, i.e., the speech acts oriented toward understanding (not all speech acts are oriented toward understanding though).

Validity Claims and Communicative Action

Habermas calls the ability to perform speech acts *communicative competence*. To distinguish his own theory from other semantic theories and empirical pragmatic theories of linguistic understanding and communication, Habermas calls his theory of communicative action universal or formal pragmatics. It aims to identify the 'unavoidable' universal conditions or presuppositions underlying all communication and to reconstruct systematically the structures of communicative competence or implicit rules involved in the know-how of competent speakers in modern societies.

Habermas believes that we can reconstruct the universal conditions of communicative action through a close analysis of the formal features of our everyday communication activity. Indeed, it is not so hard to identify one precondition of successful communicative action. Imagine a professor instructs her lab assistant to fetch a frozen tube from the refrigerator by saying, 'go to get the tube for me', during a lab experiment on April 1st. The speaker, the professor, through performing the speech act by uttering the sentence under this certain situation, establishes an interpersonal relationship with her hearer, the assistant. The speech act makes an offer that the hearer can either accept or reject. Before the assistant can accept or reject the request, he has to understand it first. For this speech act to be successful (i.e., the assistant understands it and carries it out as requested) two conditions have to be met. On the speaker's side, the professor has to have some good reasons to back up the validity of her request if the assistant questions it. The professor's request could be valid in different senses. (a) It could be valid in the sense that what the professor said and what is implied are true (validity as truth): that there is a frozen tube in the refrigerator, that the refrigerator is nearby, and that the task does not require abnormal physical ability and is not hazardous to perform, and so on. (b) It may be valid in the sense that it is appropriate or normally right for the professor to make such a request (validity as rightness): The professor feels that she has a right to request her assistant to do this for her. (c) It may be valid in the sense that the professor is sincere (validity as rightfulness): she really needs the tube for the experiment and the request is sincere, and so on. On the hearer's side, the assistant, in order to understand the utterance and to want to carry out the request, has to be able to recognize what kind(s) of validity is (are) claimed (validity claims) by the request and what are the implicit reasons behind each validity claim, such that the professor would be able to defend them if questioned. This example actually reveals a defining characteristic of all speech acts: Everyday communication is connected with some validity claims that demand 'yes' or 'no' responses; and to understand a speech act is to understand the validity claims it raises.

Based on the intuition gained from the above example, we can formally introduce Habermas's *theory of validity claims*, which is at the heart of Habermas's theory of communicative action. According to Habermas, participants in communication can take up three basic attitudes—objectivating, norm-confirmative, and expressive—toward the world. Corresponding to those three different attitudes toward the world, we can relate ourselves, through communicative action, to three different (conceived) worlds: the objective world of what there are (objects, facts, and states of affairs), the social world consisting of shared norms and values used to regulate interpersonal relationships, and the subjective world of the totality of inner experiences (intentions, desires, feelings, and so on). Although a competent participant in communicative action can only take up a certain *primary attitude* to a certain world (the objectivating attitude toward the objective world, the confirmative attitude toward the social world, and the expressive attitude toward the subjective world), this does not prevent him or her from adopting different attitudes toward one and the same world. For instance, one could take up both an objectivating attitude (as the primary attitude) and an expressive attitude (as a secondary attitude) toward the objective world. For this reason, Habermas calls those different attitudes '*the performative attitude*'.

Accordingly, when performing a speech act, one can and must raise and recognize three different types of validity claims about something in each of the three worlds, as well as suppose that they can be reciprocally vindicated and redeemed if questioned, and can be accepted or rejected by the participants. (a) There are (propositional) truth claims—a claim to propositional truth—implicitly raised with and presupposed by the speech act about the objective world. For example, in our above example, in issuing the instruction to her assistant, the professor claims that there is a frozen tube in the refrigerator, and further supposes that the refrigerator with the tube stored is nearby and is easy to obtain. If the assistant replies, 'but I have only five minutes before the next class', then the professor has to either revise her implicit assumption or to make sure that he knows the truth of her claims. (b) There are (normative) rightness claims—a claim to normative rightness—about the social world. In our example, the professor assumes, as a social norm, that she has a right to request the assistant carry out her instruction and he has an obligation to obey. If the assistant complained, 'Why should I do that for you?' The astonished professor would suddenly realize that her assistant lacks basic common sense of what his duty is as a lab assistant. (c) There are (subjective) truthfulness claims—a claim to truthfulness—about the subjective world. For example, the assistant could question the truthfulness of the professor's intention by asking, 'Do you really need the tube now?' (Implication: 'You are not pulling my leg, right?').

Habermas's theory of validity claims is theoretically based on his speech-act theory of meaning. In a speech act, the speaker comes to an understanding with the hearer about something. Following K. Bühler, as we have introduced in chapter 12, Habermas claims that we can identify three dimensions of an utterance's meaning: its propositional content—which is supposed to represent states of affairs and can be defined in terms of its truth conditions; its illocutionary content—what is *used* in a speech act to enter a relationship with the hearer, and its expressive content— what is *intended* by the utterer. Every communicative speech act makes reference to each of the three dimensions of meaning simultaneously. Needless to say, those three independent and mutually irreducible dimensions of the meaning of the speech act correspond to the three types of validity claims: the propositional content to the truth claim, the illocutionary content to the rightness claim, and the expressive content to the truthfulness claim. Habermas thereby concludes that performing any speech act simultaneously raises all three types of validity claims corresponding to the three dimensions of meaning. Even so, Habermas insists that every speech act can be shown to raise only one type of claim *directly*, which is usually the primary intention of the speaker, and two other types of claims only indirectly. Thus, we can classify speech acts based on the type of the direct validity claim raised with a speech act: the constative speech acts loaded with direct truth claims, the regulative speech acts with direct rightness claims, and the expressive speech acts with direct truthfulness claims.

Closely related to the performative attitude that competent participants in communicative action take up toward each of the three worlds, the participants can (and should) adopt *a reflective relation* to each of the three worlds, namely, the participants relativize their utterances against the possibility that the validity claims behind them will be contested by others. In performing a speech act, the speaker does not blindly assume the validity of his or her validity claims and take a dogmatic stand on them (taking them as literally true or as they are); and the hearer does not blindly accept or deny those claims. Instead, those validity claims should be reciprocally contested, disputed, and defended by the giving of reasons and arguments. Therefore, with every speech act, in terms of the validity claims it makes, the speaker enters into an interpersonal relationship of mutual obligation with the hearer: The speaker is obligated to support his or her claim with reasons and the hearer is obligated to accept the claim unless he or she has good reasons not to do so. From this it follows that, to achieve the goal of understanding and agreement (mutual recognitions of the valid claims from others), participants in communicative action have to cooperate. Success of communication is not at the disposal of any individual participant.

To sum up, *mutual recognition* and *reciprocal redemption* of related validity claims made in communicative actions are a precondition (as a universal condition or presupposition of the possibility of communication) for all communicative actions. More precisely:

> The success of a communicative action depends on (i) *mutual recognition* of the validity claims raised with the action as well as on (ii) the fulfillment of *reciprocal* obligation of *redeeming* those claims. On the speaker's part, the speaker must be willing to clarify, if questioned, and be able to supply reasons to support, if challenged, the validity of the claims. On the hearer's part, the hearer must be able to identify the validity claims and to supply reasons, if questioned, for acceptance or rejection of the claims.

3. Communication as Argumentative Discourse

A communicative action is conceptually linked with the *justification* of the validity claims it raises. To make a validity claim is to assume an obligation to justify it through argumentation; and to accept or reject a validity claim is to be prepared for providing reasons for the decision. Where validity claims are disputed, attempts can be made to settle the dispute by force, by appeals to authority or tradition, or by giving good reasons. Habermas argues that the rationality internalized within our communication practice (communicative rationality) dictates that disputed validity claims be settled through justification only. More importantly, Habermas emphasizes that the justification of the validity claims cannot be done *monologically*, either from the speaker's side or from the hearer's side, without the participation from the other side. Instead, the justification should be done discursively through argumentation in dialogue between the participants.

Why does the justification of validity conditions have to be dialogical and discursive? M. Cooke (1994) identifies the two basic arguments from Habermas's writings for his intersubjective concept of justification through argumentation. According to the first one, validity claims cannot be justified independently of discussion with others. This is simply because justification, by its very nature, is discursive. 'We understand the expression of "justify" when we know the rules of the argumentation game within which validity claims can be redeemed discursively' (Habermas, 1982, p. 231). But Habermas has admitted now that the thesis of the internal connection between validity justification and discursive argumentation cannot be held universally. For example, for a significant category of truth claims, such as some straightforward empirical assertions, discursive argumentation seems not to be necessary under usual circumstances. Even so, the thesis still holds for most theoretical truth claims, such as the truth claims raised by our scientific languages. The second argument is from the fallibilist perspective of validity and justification. No validity claims, especially theoretical truth claims, can ever be conclusively verified as final 'truths'. 'Truths' (as universal validities, such as propositional truth and normative 'truth'), for Habermas, always have a moment of 'unconditionality' that transcends all spatio-temporal contexts. Thus, no validity claims are, in principle, immune to critical evaluation in argumentation. Similarly, no justification is, in principle, immune to possible revision either. Therefore, no justification is ever conclusive; it is essentially open. What counts as a good reason or a good justification can only be determined through critical evaluation with other participants in dialogue. A good justification can only emerge from and during such a process, not before. And what counts as a good reason now may not be good enough later if new evidence and methods emerge. Habermas concludes that, since fallibilism has an inescapable dialogical dimension, the justification of validity claims is dialogical in principle.

Since reciprocal justification of validity claims is essential to any communicative action and the justification is discursive in nature, communicative action is, in essence, discursive. In an 'ideal speech situation', communication is free from domination and distortion: Every validity claim is open to dispute; every participant has the same rights to bring forward reasons for or against the claims; both participants fully expect to engage in serious argument about the validity of the disputed claims; and a consensus between the participants on the disputed validity claims is eventually brought about solely through the strength of the better argument.

Calling the above undistorted communicative action an '*ideal* speech situation' implies that our actual communications do not fully comply with those 'strong idealizations'. Nevertheless, they do somehow make reference to, and attempt to bring about at least an approximation to it. In fact, the forms of argumentation involved in everyday communication could be widely different, ranging from very rudimentary—in which what counts as a good reason may be given by authorities, cultural norms, traditional rules or codes, or simply shared assumptions, many of which are not open to dispute in a certain society and culture—to open-ended and critical argumentation. Those different forms of argumentation can be distinguished based on what kinds of 'idealizing suppositions'—which are rooted in the very structures of communicative action—are materialized within certain forms of argument. Some of the important idealizing suppositions are:

(IS1)　No relevant argument is suppressed or excluded.

(IS2)　No force except that of the better argument is exerted.

(IS3)　All the participants are motivated only by concern for the better argument.

(IS4)　No validity claims are exempt from scrutiny.

(IS5)　All participants have equal rights to query any claims and to participate in debate.

(IS6)　All participants are motivated to reach a consensus (*Verständigung*: reaching-an-agreement) on the validity claims involved.

(IS7)　A rationally motivated consensus (*Einverständnis*: a well-grounded agreement) on two types of the *universal* validity claim (i.e., the truth claim and the rightness claim) is in principle possible discursively (through argumentation in dialogue).

Habermas maintains that all forms of argumentation involved in communicative action, even the very rudimentary ones, are based on some of those 'idealizing suppositions', such as IS1, IS2, and IS3, but only more advanced forms of argumentation satisfy other more stringent ones, such as IS4, IS5, and IS6. The form of argumentation that satisfies all the available idealizing suppositions, including IS1 to IS6 and maybe others, would be an ideal form of argumentation ('ideal speech situation'). Forms of argumentations coming sufficiently close to satisfying the above listed suppositions, especially IS7, could be called 'discourses': They are either *theoretical discourses* that thematize universal truth claims—the claims to propositional (empirical or theoretical) truth—and practical discourses that thematize universal claims to normative rightness. Be aware that discourses, for Habermas, are still sort of ideal forms of argumentation, which rarely exist in their pure forms.

Accordingly, we can categorize communication actions into different forms based on the forms of argumentation involved, such as basic communicative actions in which rudimentary forms of argumentation are utilized and advanced forms of communicative actions in which more demanding forms of argumentation are present, something close to discourses. What we are interested in here is the theoretical discourse since that is where scientific communication would fall. If any communicative action is a theoretical discourse at all, scientific communication is, I think, perhaps a paradigmatic theoretical discourse in action around us everyday. Scientific discourse has all the features of the theoretical discourse defined by Habermas. Since many of the following statements about the discursive feature of science are straightforward, I will state them without getting into details.

First, scientific discourse consists of scientific claims, such as 'The association of the yin and rain makes people sleepy', 'The spatial location of the body b at time t_1 is different from its location at time t_2'. Scientific claims are oriented toward scientific understanding in two senses: They are external toward nature (to understand and explain nature), and they are internal toward other participants (to communicate to or to be understood by other scientists and the general public).

Second, scientific claims aim at validity and objectivity. In other words, any scientific utterance raises truth claims (empirical or theoretical truths), both about what is *explicitly claimed* and about what is *implicitly presupposed* by the claims. Those truth claims need to be substantiated and validated in order to be understood and to be carried out. To justify the truth claims raised with a scientific utterance, one is supposed to avoid any non-argumentative means, such as tricks, force, propaganda, psychological pressure, brainwashing, appeals to authority or emotion, personal attacks, and so on (what Aristotle called *ethos* and *pathos*, as different from *logos*). Instead, only argumentation is admitted—to argue for it, to present one's best arguments, to compare one's arguments with those of one's critics, and to try to convince them by putting forward the best reasons available. In addition, no matter how strong and how convincing one's arguments and reasons for one's validity claims are, no reasons or arguments can be sustained without being subjected to counter-reasons and counter-arguments. In science, everything is fallible in principle; nothing is final, conclusive, or immune to revision, including the best evidence, reason, method, and justification at the time. Those determine that scientific justification cannot be monological, but rather dialogical, i.e., right or wrong through debate. Therefore, scientific discourse is by nature discursive.

Third, in an ideal situation, the form of scientific argumentation should be based on all essential 'idealizing suppositions' identified by Habermas, including IS1 to IS7. For example, it is an ethical duty of both parties involved in genuine scientific argumentation or debate (that is, the motivations of the argumentation are genuinely communicative) with good intention to reach agreement. To advance scientific understanding through possible consensus is, at least implicitly, inherent *telos* of scientific discourse. Of course, this does not mean that the agreement between two parties in a scientific debate could be reached in reality. But the history of science has shown us that possible convergence of different competing scientific theories is, in the long run, still possible. In fact, many scientific disciplines have developed and established some institutionalized, specialized forms of argumentation ('expert culture') which are fairly close to an 'ideal speech situation'. Be aware that to say that scientific discourse has specialized forms of argumentation does not mean that those forms have to be formal logical rules, such as *modus ponens* or *modus tollens*. Formal logic rules are only parts of the forms of argumentation used in scientific discourse. Other parts consist of informal rules, constraints, and binds, such as the idealizing suppositions given above. We can cover all those formal logical rules and informal material rules established in scientific discourse under the name, following Aristotle, 'dialectics'. In this sense, scientific discourse is not only dialogical and discursive, but also dialectic.

4. Understanding: From Comprehension to Discourse

Based on Habermas's theory of validity claims about the necessary connection between speech acts and validity claims and his discourse model of communication, we are ready to see how Habermas brings the role of truth in cross-language communication back to the central stage. To see this, the best place to start is with Habermas's pragmatic theory of understanding.

We have seen that, for both Gadamer and Habermas, hermeneutic understanding is the process of reaching understanding [*Verständigung*], which is in turn oriented toward agreement/consensus [*Einverständnis*], *not simply comprehension* (in the sense of the English word 'understanding', which roughly corresponds to the so-called 'propositional understanding'). For Gadamer, understanding is linked to reaching agreement through a fusion of horizons by engaging in genuine conversation. I have suggested that some notion of truth is indispensable for participants to engage in genuine conversation. Without a proper understanding of the role of truth in argumentative dialogue, Gadamer's conversation model of communication is substantially incomplete. My criticism of Gadamer resonates with Habermas's critique of existing pragmatic theories of meaning (such as the so-called use-theories of meaning as expounded by the later Wittgenstein). For Habermas, the main weakness of the existing pragmatic theory of meaning is that it loses sight of the connection between understanding and a context-transcendental concept of validity. By a context-transcendental validity, Habermas means a universal concept of validity that can potentially go beyond the conventional validity of a given form of life, tradition, or lifeworld. A valid claim is universal if the claim is agreeable to everyone involved, that is, everyone would agree that what is agreed upon to be valid is valid for everyone. The concept of truth would be such a universal concept of validity. Although each participant may disagree about the truth-value, even about the truth-value status of a truth claim, but truth is agreeable since all participants accept that a claim could be true or false and if it is true, it should be true for everyone. It seems to me that Gadamer makes the same mistake by removing propositional truth out of the universal conditions of understanding.

To overcome the weakness of existing pragmatic theories of understanding,

Habermas turns his attention to formal semantics (from Frege through the early Wittgenstein to Davidson and Dummett), especially truth-conditional semantics, according to which to understand an utterance is to know its truth conditions. As I see it, there are two basic characteristics of the truth-conditional theory of understanding, which are at the heart of formal semantics. First, it identifies understanding with the knowledge of *the conditions of understanding*. By the conditions of understanding, I mean the conditions that make what is to be understood (assertoric sentences or utterances) possible or that determine the semantic contents (such as meanings) of what is to be understood. To understand a sentence S is to know what makes S possible by assigning S a meaning. Second, according to the truth-conditional account, the conditions that make an assertoric sentence S possible by assigning (literal) meanings to it are its truth conditions. Therefore, to understand S is to know its truth conditions (under what conditions a sentence is true). Both assumptions work together to specify a universal condition of understanding. Briefly, to understand is to know the conditions of understanding. In other words, to understand S is to know the meaning of S; and to know the meaning of S is to know the conditions that make S's meaning possible. This basic 'innocent' intuition behind formal semantics, I believe, should serve as the foundation of any formal theory of understanding.

Habermas believes that these two basic assumptions of truth-conditional semantics should be preserved in his formal pragmatic theory of understanding. Corresponding to the first assumption of formal semantics, Habermas contends that 'we understand an utterance when we know what makes it possible' (1984, p. 297). Since what makes an utterance possible is, for Habermas, its validity claims, to understand an utterance is to understand the validity claims it raises, or more precisely, to understand *the validity conditions* (under what conditions an utterance is valid) as I will call it. More importantly, Habermas discerns from the second assumption of formal semantics what is missing from existing pragmatic theories, namely, a concept of universal validity—truth. It is necessary, Habermas argues, to re-establish the broken linkage, in both existing pragmatic theories and Gadamer's hermeneutics, between 'truth' (in a broader sense, referring to a context-transcendent concept validity, including propositional truth, normative rightness, and subjective truthfulness) and understanding. By doing so, Habermas re-establishes the central role played by truth in understanding and communication.

Of course, formal semantics is still not the formal pragmatic theory of understanding that Habermas wants because it has been guilty of 'the three abstractions': the semantics abstraction (only focusing on the meanings of sentences in isolation and ignoring the use of sentences), the cognitivist abstraction (reducing all meanings to the propositional content and ignoring both the illocutionary and expressive contents), and the objectivist abstraction (truth conditions are objective as to what makes a sentence or an utterance true independent of any subjective comprehension or knowledge). As I have argued all along, as far as our P-languages is concerned, which are basically consisting of constative utterances or assertoric sentences, we can only focus on the propositional content and put other dimensions of meanings aside.[2] For this reason, among the three abstractions, what we should be concerned about here is the so-called objectivist abstraction.

In fact, Habermas is not the first person who notices the objectivist abstraction. Not all formal semantic theories are guilty of it. The most notable truth-conditional semantics that commits itself fully to the abstraction is Davidson's truth-conditional theory of understanding. According to it, there is a necessary conceptual connection between the knowledge of *Tarskian semantic truth conditions* and understanding. Such truth conditions are objective in the sense that there are semantic procedures available for effectively deciding what the truth conditions are for every assertoric sentence and when those truth conditions are satisfied, no matter whether the interpreter is entitled to know those conditions or not. The problem with such 'objective truth conditions' is, as Dummett (1993) and many others within the formal semanticist tradition have pointed out, that the truth conditions of many sentences are either unknowable or simply not available for the interpreter to know under certain conditions, such as the truth conditions of many scientific theoretical claims. To remedy it, Dummett introduces a so-called 'epistemic turn' within formal semantics and replaces Davidson's 'objective' truth conditions with his 'intersubjective' assertability conditions (under what conditions a sentence is assertable), a verificationist version of 'truth conditions' which the interpreter is *entitled to know* when the conditions are satisfied.

Habermas applauds Dummett's 'epistemic turn'. Nevertheless, Habermas blames Dummett for not making the 'the turn' sharp enough because Dummett's assertability conditions are monological in nature, that is, the interpreter is able to know the assertability conditions (that is what Habermas desires) without engaging in dialogue with the speaker (that is what Habermas tries to avoid). In contrast, for Habermas, to understand an utterance is not just to know the conditions of understanding, namely, the conditions that make the utterance possible—such as Dummett's assertability conditions or Habermas's validity conditions—which Habermas calls the 'conditions of satisfaction', but also to be able to justify the satisfaction conditions *discursively* in dialogue with the speaker, which Habermas calls the 'conditions of validation'. In Habermas's hand, the traditional one-layer conditions of understanding, i.e., the conditions of understanding as the conditions of satisfaction, is expanded into two-layer conditions of understanding, namely, as both the conditions of satisfaction and that of validation. Therefore, to understand an utterance, we have to know both the conditions of satisfaction and the conditions of validation.

By adding the conditions of validation as a condition of understanding, Habermas wants to achieve at least three goals. One is to make the conditions of understanding (the conditions of satisfaction) *knowable* to the interpreter. That the conditions of satisfaction have to be knowable is actually presupposed by the conditions of validation since if the former are not knowable for the interpreter, then the interpreter cannot justify them. The second goal is to make the conditions of understanding *dialogical*. The conditions of understanding need to be justified in argumentation oriented toward reaching agreement. Closely related to the second goal is Habermas's attempt to substantiate his basic conviction shared with

Gadamer, that is, understanding is *a process* of reaching understanding and agreement through argumentation in dialogue. In this sense, we can say that Habermas's notion of understanding no longer simply focuses on comprehension as does propositional understanding, but rather becomes an integrated part of his formal pragmatic theory of communicative action.

To separate his version of the conditions of understanding from both Davidson's truth conditions and Dummett's asssertability conditions, Habermas names it the acceptability conditions. He claims,

> In a distant analogy to the basic assumptions of truth-conditions, I want now to explain understanding an utterance by knowledge of the conditions under which a hearer may accept it. We understand a speech act when we know what makes it acceptable. (Habermas, 1984, p. 297)

As we have mentioned, the conditions under which a hearer would accept an utterance U are its validity conditions (conditions under which U is valid, or the conditions under which the validity claims raised with U can be satisfied). But the validity conditions, as the conditions of satisfaction, need to be supplemented by the conditions of validation, that is, under what conditions the validity claims can be *justified* with good reasons. To know the validation conditions is to know the kind of reasons that a speaker would provide in support of the validity claims raised. Therefore, for Habermas,

> To understand an utterance is to know its acceptability conditions; and to know its acceptability conditions is to know its (a) satisfaction conditions (its validity conditions) and (b) its validation conditions.

For a constative utterance U, its validity conditions are its truth conditions. To know U's validity conditions is to know U's truth conditions, or to know U's propositional content. In addition, to know U's validation conditions is to know the conditions under which the speaker has convincing reasons for holding the truth claims TCs raised with and presupposed by U; and those reasons can only emerge in the process of intersubjective justification of TCs through argumentation in dialogue. However, if one knows how to justify the truth claims raised and presupposed by U, then presumably one already knows U's truth conditions. That means, in the case of constative utterances, that we can reduce the satisfaction conditions to the validation conditions. Therefore,

> To understand a constative utterance is to know how to justify, intersubjectively and reciprocally, the truth claims raised with and presupposed by it through argumentation in dialogue.

5. Truth Claims and Cross-Language Communication Breakdown

Recall that, in Gadamer's hermeneutics, the relationship between understanding and communication is symmetric: The process of coming-to-an-understanding (agreement) is in essence the process of communication (conversation). The same is true for Habermas's concepts of understanding and communication. The difference between Gadamer and Habermas seems to be this: Each starts with a different end of the equation. Starting with his notion of hermeneutic understanding, Gadamer ends with dialogical communication; while Habermas starts from his discourse model of communication, and ends with his notion of hermeneutic understanding. For Habermas, linguistic communication is an exchange of communicative speech acts oriented toward reaching understanding, i.e., toward reaching *both* comprehension *and* consensus/agreement. But understanding and agreement can only be reached through discursive argumentation in dialogue. More precisely, the process of reaching understanding and agreement is a process of reciprocal evaluation and justification of the validity claims raised through argumentation in dialogue, which is nothing but the process of communication in discourse. Therefore, for Habermas, dialogical communication turns out to be (hermeneutic) understanding.

We have seen the crucial role that truth plays in understanding of constative utterances. Since understanding is communication, based on Habermas's discourse model of communication, the connection between truth and cross-language communication is evident. This is especially true if the cross-language communication between two typical P-languages, i.e., scientific languages, is concerned. Habermas distinguishes the cognitive use from the interactive use of language. In the cognitive use of language, with the help of constative speech acts, the thematic emphasis is on the propositional content of the utterance. In contrast, the thematic emphasis in the interactive use of language, with the help of regulative speech acts, is on the relationship that the utterance establishes between the speaker and the hearer. A P-language, a scientific language in particular, is primarily used cognitively. Although a P-language could be used interactively to regulate human relationships, it is certainly not its primary function. That means that the sentences used by a P-language are primarily constative utterances (such as 'Element *a* contains more phlogiston than element *b*') with propositional contents as their primary contents (literal meanings).

When used in cross-language communication, a constative utterance of a P-language *directly* raises explicitly and presupposes implicitly one type of universal validity claims—propositional truth claims, or claims to either empirical or theoretical propositional truth (such as 'It is true that element *a* contains more phlogiston than element *b*', 'There are two elements named "*a*" and "*b*" ', and 'Phlogiston, a chemical component which is measurable in quantity, exists'). Using our terminology, part of those truth claims presupposed implicitly by an utterance of a P-language is actually what we have identified as the semantic presuppositions underlying the utterance. Among them, the most fundamental presuppositions are so-called M-presuppositions, the absolute presuppositions taken for granted by the user of the language. While Habermas seems to focus only on one type of truth claims or presuppositions, i.e., the existential presuppositions, I think that he would agree with me that other types of semantic presuppositions, at least universal principles, if not categorical frameworks, can also (and should) fit in his schema of the truth claims raised with a speech act.

In Habermas's view, communication is conceptually connected with reaching understanding and agreement through the intersubjective evaluation of validity claims in argumentation. Bear in mind that we have identified earlier two necessary conditions of successful cross-language communication:

(C1)   Mutual recognition of the truth claims raised with and presupposed by the utterances used by the other language.

(C2)   Reciprocal redemption of those truth claims through argumentation in rational dialogue.

Considering the conceptual connection between Habermas's theory of communicative action and his formal pragmatic theory of understanding (two sides of the same coin), it should be unsurprising that those two conditions of communication between two P-languages are actually the two conditions of understanding (of constative utterances). That is, (a) the knowledge of truth conditions: recognition of the conditions under which the truth claims raised with and presupposed by the utterances used by the other language (the satisfaction conditions); (b) the capacity of justifying reciprocally those truth claims through argumentation in dialogue (the validation conditions).

But there is still one more crucial necessary condition of communication (understanding) missing from the list. What is the purpose of justifying reciprocally the truth claims involved discursively? Remember that Habermas, following Gadamer, contends that communicative acts are oriented toward reaching understanding; and reaching understanding is reaching consensus, not simply comprehension. For Habermas, the very idea of engaging in serious discussion and argumentation with one another with regard to the validity of a claim makes sense only if both sides have good intention to reach consensus on the validity claim; otherwise why should we even bother to argue with one another? Habermas thinks it is absurd to assume that participants in genuine argumentation could aim at disagreement. To facilitate communication (understanding), all participants are rationally motivated to reach a consensus on the validity claims involved (idealizing supposition IS6). Furthermore, communicative rationality inherent in everyday communicative actions requires that the consensus on the validity claims can only be reached rationally and discursively through rational argumentation in dialogue, nor through other non-rational means. This seems uncontroversial. What is debatable is whether a rationally motivated consensus (*Einverständnis*: a well-grounded agreement) on two types of the *universal* validity claims (i.e., the truth claim and the rightness claim) is in principle achievable through argumentation in dialogue. Habermas is not very clear on this. It certainly cannot be achieved in most cases of communication. I would prefer to interpret Habermas as adopting IS7 as an idealizing supposition, which is achievable only in discourse, a sort of 'ideal speech situation'. Thus, we have the third condition of dialogical communication:

(C3)   In a discourse, reaching agreement *in principle* (maybe not in practice) on the validity of the truth claims in dispute discursively.

Under normal situations, especially normal intra-language communication, communication can continue smoothly so long as participants 'suppose that the validity claims they reciprocally raise are justified' (Habermas, 1979, p. 3). While all communication presupposes a background consensus that is typically taken for granted, the consensus quite often breaks down when the validity claims involved are not recognized or are questioned and disputed by the participants. Whenever this happens, argument is called for. In our example given earlier, the assistant could dispute one or two of the truth claims raised with the professor's request by questioning or challenging them. For example, 'I don't see any refrigerator *nearby*', or, 'Do you really expect me *alone* to carry that tube?' Alternatively, he can challenge the authority of his boss established by social norms: 'Hi, I am just your *academic* assistant, not your servant', or 'I really do not feel like doing anything at this moment'. The professor can then defend her request by either clarifying the facts ('See, the refrigerator is right on the corner behind the bookcase', or 'It only weights about five pounds') or restating her authority status ('Remember we are doing an experiment in the lab now. I didn't ask you to clean my house'). Of course, the assistant could be stubborn and keep disputing. Hopefully, the muddleheaded assistant could be convinced by the professor's arguments so that the communication could continue smoothly.

The intra-language communication in the above case sounds simple enough even when the validity claims are in dispute. However, for cross-language communication, especially in abnormal discourse, the process of communication would become much more complicated. Use our old friends Ms Jones (suppose this time that she is an assistant to Mr Wong for the time being) and Mr Wong as an example again. When asked why Jennifer, the patient, has a painful spleen, Mr Wong diagnoses that it is caused by an excess of the yin within her spleen (an asthenic spleen) by uttering:

(41)   Jennifer's painful spleen is caused by an excess of the yin within her spleen.

He further asks Ms Jones to administer some herbs to Jennifer to make up for the insufficiency of the yang. Remember that our Ms Jones is a bi-medical, but refuses to accept the M-presuppositions behind the Chinese medical language and for some unknown reason does not practice it. Somehow she decides to challenge the truth claims raised with Mr Wong's request in front of Jennifer, such as (7) and (5):

(7)   There is a fundamental element, force or principle in the universe, namely, the yin, and there is a pre-established connection between the human body and natural forces.

(5)   All diseases are due to the loss of the balance between the yin part and the yang part of the human body.

When challenged, Mr Wong is obligated to redeem those truth claims with reasons (at least in front of Jennifer). Of course, Ms Jones could push the debate further.

Could the two participants eventually settle their disagreement by reaching consensus? This brings us back to our primary question again: Could two P-languages with incompatible M-presuppositions communicate with one another successfully according to Habermas's discourse model of communication? To answer the question, we need to see whether the three necessary conditions of communication identified above (C1, C2, and C3) could be satisfied in abnormal discourse.

Let us consider two P-languages, PL_1 and PL_2 (such as the Western medical language L_W and the traditional Chinese medical language L_C), with two sets of incompatible M-presuppositions. For the speaker (such as Ms Jones) of PL_1 (such as L_W) to understand an utterance (such as (41)) of PL_2 (such as L_C), she has to be able to *identify* and *comprehend*[3] the truth claims raised with and presupposed by the utterance (such as (41), (7) and (5)). In other words, she needs to know under what conditions the utterance could hold to be true (to be valid). Notice that the truth claims to be identified and comprehended here not only include what is raised explicitly with (41), that is, the truth of (41), but also include what are presupposed implicitly by the utterance to be understood, i.e., (5) and (7). (7) is one of the M-presuppositions of the P-language to be understood, i.e., L_C. Therefore, one condition of communication specified by Habermas (C1)—that is, mutual recognition of the truth claims—is exactly the conclusion that we have reached in chapter 12 when we discussed effective (propositional) cross-language understanding: One can understand the sentence of an alien P-language only if one can conceptually recognize and comprehend its underlying M-presuppositions. Failure to identify and comprehend them would disturb mutual understanding.

Nevertheless, we need to be cautious with the above comparison between Habermas's conditions of understanding and the conditions of propositional understanding we identify in chapter 12. We have to realize that Habermas's notion of understanding is not propositional. Following Gadamer's innovation, Habermas moves away from the informative communication model, which is propositional by nature, by adding the second crucial condition of understanding or communication. For Habermas, only mutual recognition of the truth claims is not sufficient to understand them; in addition, those truth claims (presuppositions) have to be reciprocally justified in order for understanding or communication to be successful (By the way, in our case, mutual recognition of the truth claims is not a problem anyway since we have assumed that Ms Jones is a bi-medical who comprehends L_C).

Based on Habermas's second condition of communication (C2), for the communication between PL_1 and PL_2 to be successful, the truth claims have to be justified discursively through argumentation in rational debate between participants. Notice that in a discourse, what is in dispute is about the truth-value of the truth claims, such as the truth-value of (41). But as we have argued repeatedly, in abnormal discourse, what is at issue is not the truth-values of the claims in dispute, but rather their truth-value status. For example, (41) is neither true nor false when considered within the context of L_W. It is an established notion of validity, i.e., the belief that an utterance has to be either true or false, that is called into question in abnormal discourse. In this case, to solve the dispute, the participants have to trace the source of the truth-valuelessness to the level of more fundamental truth claims (such as (7) and (5)) presupposed by the truth claim (such as (41)). Eventually, they find themselves face to face with two radically distinct P-languages with distinct, and often mutually exclusive, cosmologies, modes of thinking and justification, and categorical frameworks.

Could we truly engage in *rational debate* about those basic M-presuppositions between two P-languages, such as L_W and L_C? To engage in rational debate on the validity of the M-presuppositions of a P-language, such as L_C, both sides have to follow certain rules of debate and argumentation. But what count as acceptable appropriate rules of justification or forms of argumentation themselves are, for many P-languages, determined by a P-language. We have argued in chapter 10 that L_C is associated with a unique mode of reasoning, which we call the mode of 'associated thinking'. It was a dominant rule of rational justification in the premodern Chinese tradition. The rule determines what facts count as evidence for justification and even what states of affairs count as permissible facts. For instance, the states of affairs about the interaction between the yin-yang parts in the human body and the yin-yang forces in Heaven count as permissible facts. For the same reason, the interaction provides good evidence for the justification of the claim that sleepiness is caused by rainfall. Similarly, I. Hacking identifies a distinct style of reasoning within the Renaissance tradition, which admits the states of affairs dealing with the mutual interaction among the human body, celestial bodies, and other chemical elements as alleged facts. This further determines that the associations between mercury, the planet Mercury, the marketplace, and syphilis count as evidence for the claim that mercury salve is good for syphilis. A Paracelsan sentence like 'Mercury salve is good for syphilis' is rationally justified based on this style of reasoning.

When those different forms of argumentation inherent in distinct P-languages are in conflict, can we locate some underlying common forms of argumentation? Habermas (1993) has admitted that there is no *metadiscourse* to which we can refer in order to justify a choice between different forms of argumentation. There is no forum for deciding when we should bring to bear arguments from other spheres of validity or for deciding which kinds of arguments are relevant. 'Habermas raises the possibility that this might be a matter for the practical judgment of individuals, only to dismiss it immediately as unacceptable. … Nor does he give any hints as to how we might solve the problem' (Cooke, 1994, p. 42).

However, the problem we face in abnormal discourse is more serious than the problem of conflicting forms of argumentation. Logically speaking, the propositions expressed by each P-language have to be in some normal bivalent logical relation (logical consequence, logical consistency, or logical compatibility) in order for both sides to engage in a rational debate. Such logical relationships between two languages can be established only if each side *adopts*, not just *recognizes*, the M-presuppositions of the other P-language. Otherwise, apparently same sentences (such as (41)), which express propositions within the context of one language (say, L_C), do not express any propositions or thoughts when considered within the other (say, L_W). In this case, there is no way to establish any logical contact between two sets of propositions expressed respectively by the two P-languages. This is the reason why Mr Wong and Ms Jones cannot engage in a rational debate. This is, in effect, also the conclusion drawn by Kuhn based on his argument from the law of non-contradiction (chapter 7). Recall that Kuhn argues that the law of non-contradiction prohibits the occurrence of *actual* truth-valueless utterances in discourse. The utterances used in communication have to be actually true or false for both the speaker and the interpreter. The utterances under consideration are actually true or false for both the speaker and the interpreter only when both hold their underlying M-presuppositions to be true (chapter 9). This is possible only when the M-presuppositions of the two languages are either shared or at least compatible. We end up with the same conclusion as we have drawn repeatedly before.

Lastly, let us turn to the third condition of communication (C3). Can the speakers PL_1 and PL_2 (such as L_W and L_C) ever reach agreement, even in principle, on the validity of the truth claims (such as (41), (5), and (7)) in dispute through rational argumentation in debate? I am not optimistic at all since both sides have lost their basic means to reach such an agreement, i.e., through rational argumentation in debate.

In conclusion, according to Habermas's discourse model, full cross-language communication in abnormal discourse is doomed to failure.

Notes

1   The following exposition of Habermas's theory of communicative action makes extensive reference to M. Cooke, 1994.
2   I do not intend to deny the existence and importance of other dimensions of meaning at all.
3   'Comprehension' used here should be understood as 'propositional understanding', namely, to know the propositional content, which is not yet the genuine 'understanding' in either Gadamer's or Habermas's sense.

Chapter 17

The Concept of Incommensurability

1. Kinds of Communication Breakdowns

We started with the observation that mutual understanding and communication between the proponents of two distinct scientific languages are difficult, problematic, and even unattainable. Following Kuhn, I suggested that incommensurability is a semantic phenomenon closely related to the problem of how two scientific language communities can effectively understand and successfully communicate with one another. Cross-language communication breakdown is the landmark of incommensurability. Intuitively, to say that two scientific languages are incommensurable is to say that a necessary common measure of some sort is lacking between them and that thereby the cross-language communication between the two language communities breaks down. By introducing the notion of P-language, I have identified such a common measure as M-presuppositions: The M-presuppositions of two P-languages have to be compatible to facilitate successful communication between them. In abnormal discourse in which the M-presuppositions of two distinct P-languages are incompatible, the communication between them is inevitably partial; in other words, communication breakdown in abnormal discourse is inevitable. In this case, the two P-languages are incommensurable. To put things in a coherent perspective, let us briefly review how we reached this conclusion in the last five chapters.

Presumably, to argue for the essential role of M-presuppositions (as a necessary condition) in cross-language understanding and communication, we need to clarify the notions of linguistic understanding and communication. In chapter 12, based on the notion of effective propositional understanding, I found that a P-language is fully intelligible only if its M-presuppositions are conceptually recognized and comprehended. Thus, the knowledge of the M-presuppositions of a P-language is necessary for effective understanding of it. In abnormal discourse, due to lack of comprehension of the M-presuppositions of a P-language to be understood, interpreters tend to project the scheme of their own language onto it. Such a projective approach to understanding in abnormal discourse often leads to the failure of understanding which then results in *complete communication breakdown* between the two language communities.

The failure of the projective understanding in abnormal discourse does not mean that mutual understanding between two P-language communities with incompatible M-presuppositions is unattainable in principle. The thesis of incommensurability does not lead to radical relativism according to which we are enclosed in a prison house of our own language. The question then becomes how to make an alien language intelligible without distorting it? In chapter 13, after rejecting the adoptive approach to understanding in abnormal discourse, I argued that we have to go beyond propositional understanding. Our attention is thereby shifted from propositional understanding to Gadamer's hermeneutic understanding, which is supposed to avoid antithetical poles and provide us with a way beyond the absolutistic–projective and the relativistic–adoptive approaches to understanding. The chapter ended with a cautious note: Although hermeneutic understanding does have a considerable advantage over propositional understanding in abnormal discourse, hermeneutic understanding, as we have argued in chapter 14, is not propositional understanding. Success of hermeneutic understanding does not guarantee the success of propositional understanding; on the contrary, it is when propositional understanding fails that hermeneutic understanding takes over. In addition, Gadamer's notion of coming to understanding oriented toward agreement through conversation, as argued in chapter 15, is in fact a process of dialogical communication. Since an *undistorted full* cross-language communication between two incompatible P-languages cannot be achieved, a *complete, full* (hermeneutic) understanding in abnormal discourse is not feasible either.

Many would argue that language learning seems to provide us an effective way of understanding the alien language in abnormal discourse so that full communication between two incompatible P-languages could be restored. I agree that *partial* understanding through language learning is possible and feasible. The process of language learning for an interpreter is, in essence, a process of hermeneutic interpretative understanding (chapter 13), which, I argued, cannot restore full propositional understanding. Another challenge may be posted based on bilingualism: A bilingual can surely understand (in the sense of propositional understanding) both P-languages as any competent native speaker does. If so, understanding should not be an insurmountable obstacle for cross-language communication, at least in the case in which both the interpreter and the speaker are bilinguals. Instead of questioning bilingualism, I decided to take it for granted for the sake of argument. To justify the conclusion that communication breakdown is inevitable in abnormal discourse, I argued in chapter 14 that it is necessary to distinguish propositional understanding as *comprehension* from informative communication based on the standard transmission model of communication; for propositional understanding is necessary but not sufficient for cross-language informative communication. I concluded that shared or compatible M-presuppositions between two P-languages are necessary for carrying out a successful communication between the two distinct P-languages. Therefore, cross-language (informative) communication in abnormal discourse is inevitably partial.

At the end of chapter 14, I found that the notion of informative communication based on the transmission model could not reveal the essence of a genuine communication in the Platonic sense. We need to go beyond the transmission model of communication. In chapter 15, I argued that, for Gadamer, the process of substantive hermeneutic understanding—the process of 'coming-to-an-understanding' oriented toward 'coming-to-an-agreement' through genuine conversation—is, in essence, communication. Through analysis and criticism of Gadamer's common language requirement in terms of a full fusion of horizons, I reached the same conclusion as I did in chapter 14: Based on Gadamer's conversation model, cross-language communication in abnormal discourse is inevitably partial.

However, Gadamer's conversation model apparently departs from the basic line of argument that I have been pursuing, namely, the essential role of M-presuppositions and truth-related claims in cross-language understanding and communication. Neither Gadamer's notion of tradition nor that of horizon could be lined up formally with my notion of M-presupposition. The notion of truth also does not play any significant role in Gadamer's conversation model, as it should. In chapter 16, I turned to Habermas's discourse model of communication to rescue these two insights lost in Gadamer's model. Habermas's theory of communicative action brings the essential roles played by both M-presuppositions and truth-related claims in cross-language communication back to the central stage. For Habermas, cross-language communication is carried out through communicative speech acts oriented toward reaching both comprehension and agreement on the status of the truth claims raised and presupposed by what is said (utterances). Such comprehension and agreement can only be reached through discursive argumentation in dialogue. Thus, for Habermas, dialogical communication is in essence (hermeneutic) understanding.

Habermas identifies three conditions of dialogical cross-language communication: (i) the knowledge of the truth claims raised and presupposed by the utterances of the other language; (ii) the redemption of those truth claims through argumentation in rational dialogue; and (iii) reaching agreement on the validity of the truth claims in dispute discursively. Since the most fundamental truth claims presupposed by the utterances of the language are actually what we call M-presuppositions, condition (i) is the same as the condition of propositional understanding I have identified in chapter 12. Failure of mutual recognition and comprehension of the truth claims raised by the other language results in complete communication breakdown between the two languages. But dialogical communication requires much more than mere propositional understanding or comprehension. One can disagree with the speaker from the other language on the status of the truth-claims raised with the language based on one's recognition and comprehension of the claims. In fact, it is precisely one's ability to comprehend the other language that explores the inability of communicating successfully with the other. To fulfill the two other conditions, I argued, two P-languages have to compatible. Again, I reached the same conclusion as I did in chapters 14 and 15: Based on Habermas's discourse model, full cross-language communication in abnormal discourse is doomed to failure. Cross-language communication in abnormal discourse is inevitably partial. It is such *partial communication breakdown* in abnormal discourse that gives the real theoretical thrust of the thesis of incommensurability as communication breakdown. Most classical cases of incommensurability identified by Kuhn, Feyerabend, and many others are cases of partial communication breakdowns.

Communication breakdown in the case of incommensurability, no matter whether it is complete or partial, occurs only in abnormal discourse. It should be distinguished from other kinds of communication breakdowns, especially communication breakdown in normal discourse. Communication between two distinct language communities in normal discourse can fall short of the ideal in many other ways. For example, a communication breakdown may occur between any two languages simply due to a divergence of the speaker's language-meaning from the interpreter's language-meaning. For example, there are cases in which the interpreter does ascribe some meaning to the sentence uttered by the speaker such that there is the interpreter's language-meaning. However, the meaning attributed to the speaker by the interpreter is not the same as the speaker's language-meaning, namely, how the speaker intends the sentences to be understood. The interpreter misunderstands the speaker. In such cases, an act of communication is not successful. However, the communication breakdown in this case is not as threatening as that due to incompatible M-presuppositions. In the case of misunderstanding, both sides agree on the truth-value status of the sentences in discourse, but specify different truth conditions for these sentences such that the interpreter attributes different meanings to the sentences in the speaker's language. Actually, these are the cases that the standard interpretation identifies as incommensurability as untranslatability due to meaning variance. This kind of communication breakdown due to misunderstanding is restorable since misunderstanding can be overcome in the process of on-going partial communication.

2. Truth-Value Gap and Communication Breakdown

We observed in chapter 5 the emergence of a (conceptually possible) truth-value gap between the Chinese medical language and the Western medical language and the occurrence of a (actual) truth-value gap between the Newtonian language and the Leibnizian language of space. By a truth-value gap between two P-languages PL_1 and PL_2, as defined formally in chapter 9, I mean the occurrence of a massive number of truth-valueless sentences of PL_1, when considered within the context of PL_2, due to the failure of the M-presuppositions of PL_1 from PL_2's point of view. Based on the distinction between conceptually possible truth-value and actual truth-value given in chapter 9, we can further distinguish two kinds of truth-value gaps. If massive numbers of truth-valueless sentences of PL_1 (say, the language of Chinese medical theory) occur because the speaker of the competing language PL_2 (say, the advocate of Western medical theory) cannot *recognize* its M-presuppositions, we say that there is a *conceptually possible* truth-value gap between the two languages. If numerous truth-valueless sentences of PL_1 (say, the Newtonian language of space) occur because its M-presuppositions *are suspended or rejected* by the advocate of the other (say, the Leibnizian), then we say that there is an *actual* truth-value gap between them.

In chapter 12, I identified a strong semantic correlate of the failure of effective understanding between two distinct P-languages—the occurrence of a conceptually possible truth-value gap between them. More precisely, if the core sentences of one P-language, when considered within the context of the other, lack conceptually possible truth-values, then it indicates the failure of effective understanding on the interpreter's side. Moreover, as argued in chapters 14, 15, and 16, compatible M-presuppositions are necessary for any successful communication between two distinct P-languages. When two P-languages are incompatible, the conflict in M-presuppositions would render massive core sentences of one language actually truth-valueless when considered within the context of the competing language; in other words, there is an actual truth-value gap between two P-languages with incompatible M-presuppositions. Then the occurrence of an actual truth-value gap between two P-languages indicates that the communication between them is in principle incomplete. In this case, even if the speaker of one language PL_1 is able to identify and comprehend the M-presuppositions of the other language PL_2, many core sentences of PL_1, which are meaningful for the speaker of PL_1, are not factually meaningful.

Therefore, the occurrence of a truth-value gap between two P-languages is a strong semantic indicator of the communication breakdown between them. Specifically, the occurrence of a *conceptually* possible truth-value gap corresponds to a *complete* communication breakdown while the emergence of an *actual* truth-value gap correlates with a *partial* communication breakdown.

Be aware that the occurrence of a truth-value gap between two competing P-languages discussed above is caused by two incompatible sets of M-presuppositions embedded in the two languages. The occurrence of a truth-value gap between two languages is measured by the occurrence of a massive number of truth-valueless sentences. However, not all occurrences of truth-valueless sentences count as a truth-value gap in the sense defined above. Many individual truth-valueless sentences could occur within a language due to purely syntactic, pragmatic, or semantic matters other than the failure of the M-presuppositions of the languages involved. For illustration, let us take a brief look at a variety of reasons responsible for the occurrence of truth-valueless (declarative) sentences.

To begin with, sentences that do not have any potential (i.e., logically possible) truth-value, such as 'Three sapdlaps sat on a bazzafrazz', are definitely truth-valueless. A sentence containing a connotationless and denotationless singular term (a proper name or description) as its subject, such as 'Florkyyzzxxm loves glork', is truth-valueless since its subject is a meaningless particle. Similarly, as to the so-called genuine or logically proper singular terms (primarily demonstratives and proper names), if B. Russell and K. Donnellan are right that these terms have a purely referential use only and have no hidden semantic structure, then we can treat them as meaningless particles if, as singular terms, they are connotationless and denotationless. Hence, a sentence containing an empty logically proper singular term, such as 'Snow White was born at midnight', is neither true nor false. In contrast, a so-called superficial singular term (primarily a description) contributes meaning to the sentence in which it occurs, irrespective of whether or not it has a referent; for the primary function of such a term is attributive, such as 'the present king of France' in the sentence, 'The present king of France is bald'. As we have discussed at length in chapter 8, truth-valuelessness of such an existential sentence is due to the failure of its semantic presuppositions, such as, 'There exists a present king of France'.

Sentences containing unspecified hidden parameters have truth-values only relative to specific contexts in which the parameters are specified. A sentence containing a vague property is a typical case for this kind of sentences. We know that many predicates, such as 'blue', 'small', 'bald', 'heap', do not have definite extensions. When such predicates are applied to objects on the borderline, the result will be sentences with no truth-values. For example, whether or not Bill's belief that Rex is big has a truth-value depends on the specification of a reference-class ('big for what' or 'big compared to what').

The truth-value status of some sentences depends on adoption of different linguistic conventions for reading the sentences. A sentence, 'Snow White was born around midnight', along the suggestion of Russell, can be read as 'There is one and only one person who is Snow White and that person was born around midnight'. That is, $\exists x$ [Born-at-Midnight(x) & $\forall y(y = $ Snow-White $\leftrightarrow x = y)$]. According to this reading, the sentence is simply false. But according to Strawson's reading, which we have adopted, the sentence is neither true nor false since its presupposition, i.e., 'Snow White is a real person' is false. A similar analysis applies to the sentences committing category mistakes. For a sentence, 'The Earth is more honest than Mars', one way of reading it is to treat it as false (W. Lycan); another way is to treat it as meaningless (Ryle); still another way is to treat it as truth-valueless (This is the reading I adopt).

Lastly, the truth-valuelessness of a sentence may be caused by the failure of one of its semantic presuppositions. The failure of any one of them will make the sentence truth-valueless. But these semantic presuppositions usually have different roles in the language containing the sentences. For illustration, imagine two chemists, Dr Bennett and Dr Braxton, who disagree on the truth-value status of sentence (30) of phlogiston theory:

(30)   Element *a* contains more phlogiston than element *b*.

Here, 'a' and 'b' stand for some definite descriptions or proper names of two chemical elements. Suppose that Braxton is working within phlogiston theory while Bennett within the framework of modern chemistry. Suppose further that for some reason, (30) is either true or false for Braxton, but is neither true nor false for Bennett. Obviously, there are many possible reasons leading to the disagreement on the truth-value status of (30) between Bennett and Braxton.

Let us first suppose that Bennett thinks that the term 'a' fails to denote anything because such a chemical element does exist. But Braxton thinks that the term really refers to a real chemical element. Assume that both persons accept Strawson's notion of semantic presupposition. Then (30) is either true or false for Braxton since the presupposition of (30), i.e., (30c), is held to be true for him.

(30c) Element *a* exists.

On the contrary, (30) is neither true nor false for Bennett since she holds (30c) to be false. However, the falsity of (30c) can be fully explained within phlogiston theory without violating any M-presuppositions of phlogiston theory. In other words, the truth-valuelessness of (30) can be derived from a factually meaningful statement within phlogiston theory itself that element *a* does not exist. Therefore, denial of (30c) does not bring any harm to the integrity of phlogiston theory.

Alternatively, let us suppose that Dr Bennett, as the speaker of the language of modern chemistry, categorically denies the truth of a M-presupposition underlying (30), namely,

. (30a) Phlogiston exists.

Be aware that Bennett's denial of (30a) is not based on her personal belief about its truth-value. Instead, her belief about the falsity of (30a) is derived from modern chemistry theory. Here we touch the heart of the debate. The failure of (30a) renders (30), when considered within modern chemistry theory, truth-valueless. However, denial of the existence of phlogiston cannot be derived from other expressible beliefs in phlogiston theory without casting doubt on the integrity of phlogiston theory as a whole. This is because (30a) is presupposed in the very linguistic setup of phlogiston theory. For this reason, rejection of (30a) cannot be carried out within phlogiston theory itself, but has to resort to another competing language, in our case, modern chemistry. There is *a (actual) truth-value gap* occurring between the two theories.

Based on the above brief survey, I think it is necessary to draw a distinction between two kinds of truth-valuelessness: the truth-valuelessness due to the failure of M-presuppositions and that due to other factors. I call the former a truth-valuelessness with *ontological significance*. Since an M-presupposition is an absolute presupposition underlying a substantial number of core sentences of a P-language, when it is suspended by the other competing language, the occurrence of truth-valuelessness is not restricted to a few individual sentences, but rather spreads to the whole language and leads to a massive number of (actual) truth-valueless sentences. Strictly speaking, only in this case can we say that there is a truth-value gap occurring between *two P-languages*. In contrast, truth-valuelessness is of no ontological significance if it is due to factors other than the rejection of M-presuppositions. Since these factors are usually associated with some isolated individual sentences only, the occurrence of this kind of truth-valueless sentence is restricted to some individual sentences.

3. The Concept of Incommensurability

The occurrence of a truth-value gap between two competing P-languages PL_1 and PL_2 due to incompatible M-presuppositions plays *a double role* in our presuppositional interpretation of the incommensurability relation between them. On the one hand, the occurrence of a truth-value gap between PL_1 and PL_2 indicates, at the epistemic level, a cross-language communication breakdown between the two language communities. On the other hand, such a gap is caused by, and can be used to indicate semantically, an ontological incompatibility between PL_1 and PL_2 at the ontological level. Because a truth-value gap between PL_1 and PL_2 plays such a double role in the emergence of incommensurability, it is more appropriate to use the occurrence of a truth-value gap caused by incompatible M-presuppositions, signified by cross-language communication breakdown, as a touchstone of incommensurability. By definition,

> Two P-languages are incommensurable (in a broad sense) when the core sentences of one language, which have truth-values when considered within its own context, lack (either actually or conceptually possible) truth-values when considered within the context of the other. The occurrence of a truth-value gap between the two P-languages indicates that the communication between the two language communities is inevitably partial.

Incommensurability so defined (in a broad sense) admits degrees since there are two kinds of truth-value gaps corresponding to two kinds of communication breakdowns. Specifically, when there is a conceptually possible truth-value gap between two P-languages, which indicates a *complete* communication breakdown between the two language communities because the speaker of one language cannot recognize and comprehend the M-presuppositions of the other, we say that two P-languages are *radically incommensurable*. I have argued that complete communication breakdown can be overcome in principle in terms of either language learning or hermeneutic understanding. That means that the radical incommensurability relation between two competing P-languages associated with a complete communication breakdown can be overcome in principle.

However, communication between two disparate P-languages with incompatible M-presuppositions is inevitably partial. When there is an actual truth-value gap between two P-languages, corresponding to a *partial* communication breakdown between the two language communities due to incompatible M-presuppositions, we say that the two P-languages are *moderately incommensurable*. Radically incommensurable languages are rare phenomena. Most classical incommensurable scientific languages identified by Kuhn, Feyerabend, and many others are moderately incommensurable. The moderate incommensurability relation between two competing P-languages, associated with a partial communication or a communication breakdown *per se*, is the incommensurability of real metaphysical significance.

Based on the notion of semantic presupposition (chapter 8), the notions of presuppositional language (chapter 9) and metaphysical presupposition (chapters 10 and 11), a truth-value conditional account of propositional understanding (chapter 12), Gadamer's notion of hermeneutic understanding (chapter 13), and the transmission model and the dialogue model of cross-language communication (chapters 14, 15, and 16), my presuppositional interpretation of incommensurability is, I believe, semantically sound, epistemologically well-established, and metaphysically profound. The interpretation enables us to explore the genuine nature and sources of incommensurability. By doing this, we establish the integrity and tenability of the notion of incommensurability, and confirm the reality of the phenomenon of incommensurability.

The presuppositional interpretation avoids many difficulties faced by the translation-failure interpretation. The translation-failure interpretation is parasitic on the concepts of meaning, reference, and translation. By sidestepping questions of meaning, reference, and translation, the presuppositional interpretation presumably avoids many intriguing and often unsolvable semantic issues concerning meaning and reference, such as the coherence and tenability of the contextual theory of meaning/reference, which are the theoretical foundation of the translation-failure interpretation. Of the four standard objections to the translation-failure interpretation (chapter 2), there is only one left that needs to be addressed here: In what sense can we say that two languages are rivals or compete with each other if they are semantically so disparate that there can be no meaning-referential continuity between them? The answer to this question is not hard to see from the perspective of presuppositional interpretation. Two P-languages are rivals or compete with one another if their underlying M-presuppositions are incompatible, which is signified by the occurrence of a truth-value gap between them.

Nevertheless, the critic might dismiss the above advantages of the presuppositional interpretation over the translation-failure interpretation by casting doubt on the independence of the former over the latter:

> Well, perhaps the presuppositional interpretation does grasp an important aspect of incommensurability, namely, the occurrence of a truth-value gap between two incommensurable languages. But truth-value gaps are exactly what incommensurable languages are supposed to lead to on the translation-failure interpretation. When there is a truth-value gap between two competing languages, truth-preserving translation between them inevitably fails; for there is no way to match up sentences with the same truth-values in two languages respectively. If so, the occurrence of truth-value gaps entails some species of untranslatability. Thus, one cannot help wondering whether the presuppositional interpretation is a real alternative to the translation-failure interpretation.[1]

If the above argument stands, it is really not necessary to introduce the presuppositional interpretation in the first place, not to mention its advantages over standard interpretation.

However, it is not clear at all how meaning-reference variance between two languages can, in the translation-failure interpretation, lead to a truth-value gap. Clearly, change in the meanings of some constituents of a sentence in itself does not make the sentence truth-valueless. For instance, suppose that, if the contextual theory of meaning was right, the term 'mass' in the sentence, 'The mass of a particle does not change with its velocity', has different meanings in Newtonian physics and relativity theory. If the sentence sounds truth-valueless when considered within the language of relativity theory, it is not because of change in the meaning of the term, but because a universal principle presupposed by the sentence, namely, that properties like shapes, masses, and periods inhere in objects and change only by direct physical interactions, is suspended by relativity theory.

I have to admit that the critic is right about one direction of the relationship between truth-value gaps and untranslatability. The existence of a truth-value gap does entail some species of untranslatability in the sense of truth-preserving translation. If two languages are incommensurable, there is necessarily a truth-value gap between them based on the presuppositional interpretation. Truth-preserving translation cannot proceed when both sides do not agree on the truth-value status of the sentences of the other language. Hence, incommensurable languages are necessarily untranslatable. This means that the presuppositional interpretation, as one will expect, can do justice to the translation-failure interpretation, which should be considered as one of its advantages. However, it is not valid to move from untranslatability to truth-value gap. Truth-preserving is necessary but not sufficient for truth-preserving translation. Untranslatability does not entail truth-value gap. Therefore, untranslatability will not necessarily, and usually will not, lead to incommensurability since the occurrence of a truth-value gap, in the presuppositional interpretation, is necessary for incommensurability. Furthermore, not all truth-value gaps lead to incommensurability. Only the truth-value gaps with ontological significance do. Therefore, the presuppositional interpretation cannot be reduced to the translation-failure interpretation.

Even if the presuppositional interpretation is distinct from the translation-failure interpretation and overcomes some typical difficulties faced by the latter, the critic may argue, it still faces its own problems. For instance, the interpretation seems to face a dilemma:

> The presuppositional interpretation seems to presuppose a *common ground* between two incommensurable languages: the speaker of one language regards the M-presuppositions of the other incommensurable language as true or false. But two languages have to have a rich enough *common language* in order for them to agree on the truth-value status of one another's M-presuppositions (when they disagree on the truth-value status of the sentences). If so, the existence of a sort of common language is inevitable between two incommensurable languages according to the presuppositional interpretation. If the above argument were sound, then the presuppositional interpretation would face a fatal dilemma: On the one hand, two incommensurable languages are supposed to lack *a common language* between them. On the other hand, the very existence of incommensurability, in the presuppositional interpretation, has to be based on the existence of a common language. Then, in what sense can we say that two languages are incommensurable *on a global scale* since they share such a common language?[2]

To respond, let me point out first that incommensurability does not *necessarily* presuppose shared truth-value status of the M-presuppositions of two P-languages involved. As I have pointed out, incommensurability admits degrees. There are two kinds of incommensurability. When two P-languages are *moderately* incommensurable, the speaker of one language can understand the other by recognizing its M-presuppositions. But the speaker does not hold them to be true. So the two sets of M-presuppositions are incompatible. In this moderate case, it is true that both sides accept the positive truth-value status of the M-presuppositions of one another's language. However, when two P-languages are *radically* incommensurable, the M-presuppositions of one language might be beyond the conceptual reach of the interpreter from the other language such that the interpreter cannot attribute any truth-value status to these M-presuppositions; for one has to understand them first in order to assign truth-value status to them. In this radical case, each side does not accept the positive truth-value status of the M-presuppositions of the other language. We might say, if you like, that the two languages are incommensurable 'on a global scale' in the sense that there is no *common language* whatsoever shared by both languages. Thus, the presuppositional interpretation does allow the possibility of incommensurability 'on a global scale'. But I have to emphasize again that radical incommensurability, or the incommensurability 'on a global scale', is a rare phenomenon and can be overcome in principle.

Return to the case of moderate incommensurability in which the speakers of two P-languages can understand the M-presuppositions of one another's language but disagree on their truth-values. In this case, there does exist some *common ground* between the two languages, namely, the shared truth-value status of each other's M-presuppositions. This conclusion is actually presupposed by the doctrine of universality of understanding that I adopt. In order for the speakers of two different languages to understand one another, there has to be some common ground between them. Otherwise mutual understanding is impossible. This sounds like a tautology. The real issue at stake is whether there exists a rich enough *common language*, not just any form of *common ground*, between two incommensurable languages so that both languages can be fully expressed. When Kuhn and Feyerabend contend that there are incommensurable languages, they do not argue against the existence of any form of common ground between them. Instead, what they strive against is the existence of a *common language* into which both incommensurable languages can be translated without loss and by which to make a point-to-point comparison between their cognitive contents. Even in our moderate case of incommensurability, there does not exist such a common language.

Actually, the existence of a common ground—in our case, shared truth-value status of the M-presuppositions—between two moderately incommensurable P-languages provides a necessary platform for theory comparison as I will address it shortly. Such a common ground should be regarded as a virtue, instead of a flaw, of presuppositional interpretation.

4. Incommensurability, Incompatibility, and Comparability

Although the concept of incommensurability has its own merits in the explication of cross-language understanding and communication, the critic might continue, the concept has stirred up many heated debates in the past four decades primarily because of its many alleged undesirable metaphysical and epistemological consequences, especially for rationalistic, realistic-minded philosophers. The thesis of incommensurability allegedly makes rational theory comparison and choice impossible, and accordingly undermines our image of science as a rational, progressive enterprise. Any theory of incommensurability would be incomplete if it did not address the issue of theory comparison, especially if it did not render rational theory comparison possible. In this respect, the presuppositional interpretation seems to fare no better than the standard one. If there is no common language between two competing P-languages, especially if there is no truth-functional connection between two languages, how can rational theory comparison and choice be possible?

Incommensurability as Incomparability

Before I clarify the position of the presuppositional interpretation of theory comparison, I have to dismiss one quite influential misunderstanding of the relation between incommensurability and theory comparison. It is commonly held that rational comparison between two theories is possible only if the statements made in each theory respectively stand in a strict logical relationship—such as logical incompatibility, inconsistency, contradiction, or logical consequence—which in turn requires that the meaning/reference-related contents of those statements overlap to some extent. According to the alleged standard argument for incommensurability based on the contextual theory of meaning/reference, no statement made in one theory is in a logical relation to any statement made in the other incommensurable one. This is because the proponents of two incommensurable theories, such as the wave and the particle theories of light, mean distinct concepts and refer to different entities by the same term, in this case 'light'. As a result, the apparently contradictory statements made in the two incommensurable theories respectively are in fact not contradictory. In other words, the incommensurable theories are not in logical conflict. In addition, since the contents of incommensurable theories are formulated in different, untranslatable languages without shared meanings and references, the cognitive contents of one theory cannot be expressed in the other. Thus their contents do not overlap. Now, in the absence of content overlap and a genuine logical relationship between two incommensurable theories, it is unclear how theory choice between them *could be made* on a rational ground. For this reason, many insist that incommensurability *amounts to* or *can be reduced to* incomparability, since the latter is a logical consequence of the translation-failure interpretation of incommensurability.[3] The thesis that theories are incommensurable turns out to be the thesis that theories may be rationally incomparable due to meaning and/or reference variance of the theoretical terms contained. This is actually another influential interpretation of incommensurability, i.e., incommensurability as incomparability due to meaning/reference variance.

Whether the incommensurables are rationally comparable has been a central issue related to incommensurability and the most controversial aspect of the thesis of incommensurability. In fact, the most persistent attacks on the thesis of incommensurability have come from those who either equate incommensurability with or reduce it to incomparability. The thesis of incommensurability so construed has been attacked both 'directly' and 'indirectly'. Many argue that the thesis *could not* be true since it directly contradicts our intuitive idea of science as a rational enterprise. By *reductio ad absurdum*, the thesis has to be rejected due to its absurd consequence, i.e., incomparability. Direct criticism, on the other hand, confronts the thesis head-on, trying to prove that the thesis is simply false since theories are rationally comparable in many effective ways.

It is well known that Kuhn and Feyerabend repeatedly denied the interpretation of incommensurability as incomparability.[4] I agree with Kuhn and Feyerabend on this. The notion of comparability itself is ill-defined and ambiguous. The term 'rational comparison' could mean totally different concepts. We can at least distinguish the following three senses of 'rational comparison' in the literature. (a) Systematic, point-by-point comparison: This is the strictest sense of rational comparison, which is possible only if there is an exact translation available between the languages of two theories to be compared (systematically identical mapping between meanings and references of the terms used by the two theories).[5] (b) Meaning/reference-related content comparison: Theory comparison means to compare meaning/reference-related contents—such as cognitive or empirical contents—of two theories, for instance, consequence classes, empirical predictions, observational evidences, truthfulness or truthlikeness, or any other such considerations relating to meaning/reference-contents. This kind of comparison is possible only if there is enough meaning/reference overlap or continuity between the terms of two theories. (c) Pragmatic comparison: Comparison between two theories can be carried out by appealing to other non-semantic aspects of the theories, such as cognitive norms or values, methodological considerations, pragmatic efficiency, etc.

As Kuhn and Feyerabend point out repeatedly, the thesis of incommensurability as untranslatability does not *amount to* or cannot be reduced to *incomparability in general*. There are many effective ways to make rational comparison between two incommensurable theories. Not only is pragmatic comparison feasible between two incommensurable theories, but also are some types of content comparison.[6] What Kuhn and Feyerabend deny is point-by-point comparison between the incommensurables and some content comparison that depends upon the invariance of meaning and/or reference. Therefore, incommensurability, in the translation-failure interpretation, does not *amount to* or cannot be reduced to rational incomparability in general. It at most leads to *some kinds* of incomparability, such as the impossibility of point-by-point comparison.

Therefore, incommensurability and incomparability are two separate issues; the latter is not a necessary ingredient of the former. Incomparability has nothing whatsoever to do with *the nature* of incommensurability. It is at most an epistemological consequence (not a logical consequence) of incommensurability, not the very definition of it. For me, it is an open question whether two incommensurable theories are rationally comparable. It is open because the answer depends upon many related notions, such as the concept of incommensurability, the concept and standards of rational comparison, the concept of semantic relation between theories, and even the notion of theory.[7] It is my intention to leave it open whether two incommensurable theories, in the presuppositional interpretation, are rationally comparable.

Possibility of Semantic Comparison

It is obvious that rational comparison between the incommensurables differs from and is more problematic than that between the commensurables. Classical point-by-point comparison, which requires the existence of a common language into which both languages to be compared can be translated without loss, has to be given up in the case of incommensurability. Similarly, classical content-comparison based on the sameness of meaning/reference cannot be carried out between two incommensurable theories. Unfortunately, the critic quickly jumps from this observation to the conclusion that the thesis of incommensurability not only makes classical rational comparison problematic, but also makes rational comparison in general impossible.

Notice that the question of concern here is the *possibility*, not the *feasibility*, of rational comparison. In fact, to say that two theories are rationally comp*arable* is to say that it is *possible* to make a rational comparison between them. So what we want to know is whether rational comparison between the incommensurables is possible. If it is not, then the thesis of incommensurability, no matter how it is interpreted, indeed threatens to undermine our image of science as a rational enterprise.

Furthermore, as I have shown above, the debate over the issue of comparability does not concern whether rational comparison *in general* between the incommensurables is possible, but rather concerns *what kind of* rational comparison is possible. Especially, the controversy arises regarding whether *semantic comparison* between two incommensurable theories is *possible*. By semantic comparison, I mean the rational comparison between *semantic contents* of the theories involved. Both point-by-point comparison and content comparison are two classical types of semantic comparison. But the concept of semantic comparison I adopt here is broader than these two. The semantic contents to be compared in semantic comparison not only include meaning/reference-related contents at the theoretical level, such as consequence classes, empirical predictions, observational evidences, truthfulness or truthlikeness, but also include semantic-presupposition-related content at the *meta-theoretical level* as I will clarify shortly. Value-related contents, such as cognitive norms or values, methodological considerations (simplicity, coherence, etc.), or pragmatic efficiency (problem-solving ability, predictability, etc.), are dealt with in pragmatic comparison, not in semantic comparison. So we might divide rational comparison into two subcategories, semantic comparison versus pragmatic comparison.

Now it should become clear that the question of comparability related to the thesis of incommensurability is a Kantian one. It is a question of 'possibility': Is *semantic comparison* between the incommensurables *possible*? It is not the question of 'how to': How to make a rational comparison between two incommensurable theories?

Semantic comparison is in principle possible if there is a certain semantic relationship, whatever it is, holding between the languages of two theories to be compared. There is nothing in principle that prevents pooling together all the potential semantic resources of two competing theories, no matter how disparate they are, so as to bring them into some semantic relation. Therefore it is my conviction that semantic comparison is always possible between any two competing theories. Otherwise, it does not even make sense to say that they are competing with one another if they do not stand in any semantic relation with one another whatsoever. That means that some semantic comparison between the incommensurables should be in principle possible. The real trouble for semantic comparison between the incommensurables is that the languages of two incommensurable theories may be related to one another semantically in such a hidden way at *a meta-language level* which is normally ignored. In this case, whether two incommensurable languages can be semantically connected depends on how to locate such a crucial semantic connection between the incommensurable P-languages.

To confirm the above intuition, we need to specify a certain kind of semantic relationship needed for semantic comparison between competing theories. Presumably, semantic incompatibility is such a semantic relation we need. If two theories are semantically incompatible, then we can identify a situation or a possible world in which they cannot both be true. In such a case, it is possible to make a rational comparison about their relative merits based on some commonly accepted criteria of comparison. Thus, two theories are semantically comparable if the languages of the theories are semantically incompatible. Now our task is to identify a semantic incompatibility relationship between two incommensurables.

There are at least two types of semantic incompatibility of concern here. One is what I will call *truth-theoretical incompatibility*; the other is what I will call *presuppositional incompatibility*. Truth-theoretical incompatibility includes logical and quasi-logical incompatibility. Logical incompatibility is a strict logical relation defined as follows:

Suppose: (a) Two theories T_1 and T_2 have predicates F_1 and F_2. (b) Terms a_1 and a_2 are in the core paradigmatic part of T_1 and T_2, respectively. (c) There exist some theory of meaning M and theory of reference R shared by both T_1 and T_2 so that Meaning (F_1) = Meaning (F_2), Reference (F_1) = Reference (F_2), Meaning (a_1) = Meaning (a_2), Reference (a_1) = Reference (a_2). If $\{T_1, T_2, M, R\} \vdash \neg[\text{True}(F_1(a_1)) \land \text{True}(F_2(a_2))]$, then T_1 and T_2 are logically incompatible.

I will call the semantic comparison based on logical incompatibility defined above logical comparison, which is actually the counterpart of point-by-point comparison.

Clearly, logical incompatibility requires sameness of meaning and reference. According to the translation-failure interpretation, logical comparison between two incommensurable theories is impossible due to variance of meaning and reference.

In contrast, quasi-logical compatibility allows difference of meanings as long as there is a partial overlap of reference. By definition:

> Suppose: (a) Two theories T_1 and T_2 have predicates F_1 and F_2. (b) Terms a_1 and a_2 are in the core paradigmatic part of T_1 and T_2, respectively. (c) [Overlapping-reference hypothesis H]: The referents of F_1, F_2, a_1, and a_2 overlap somehow according to an available theory of reference R: Reference $(F_1) \cap$ Reference $(F_2) \neq \emptyset$, Reference $(a_1) \cap$ Reference $(a_2) \neq \emptyset$. If $\{T_1, T_2, R, H\} \vdash \neg[\text{True}(F_1(a_1)) \wedge \text{True}(F_2(a_2))]$, then T_1 and T_2 are quasi-logically incompatible.

The critics of Kuhn and Feyerabend who take the referential alternative to the semantic values of scientific terms contend that sameness of meaning and reference is not required for semantic comparison. Semantic comparison between two incommensurable theories is possible as long as they are quasi-logically incompatible. Since such a relation only requires overlap of reference, not sameness of meaning and reference, the semantic comparison between incommensurables is possible.

However, according to my presuppositional interpretation, semantic comparison based on quasi-logical incompatibility, which I will call quasi-logical comparison, between the incommensurables is also impossible. One reason is that, as I have argued in chapter 2, the critic is unsuccessful in providing us with the referential continuity required by quasi-logical incompatibility. Another reason is more significant. I find that both logical incompatibility and quasi-logical incompatibility exclusively focus on the meaning-referential relationship between theories. They are intended to be used to represent the confrontation of truth-values of the statements of two incommensurable theories. This in turn presupposes that the same truth-value status can cross the boundary of two incommensurable P-languages employed by the theories. Then, we can reduce both logical and quasi-logical incompatibility to a broader framework, namely, truth-theoretical incompatibility:

> Suppose (a) two theories T_1 and T_2 and their languages $L(T_1)$ and $L(T_2)$, (b) a metalanguage $M \cap \{L(T_1) \wedge L(T_2)\}$, (c) a truth theory TR formulated in M, and (d) core sentences S_1 and S_2, $S_1 \in L(T_1)$ and $S2 \in L(T_2)$. T_1 and T_2 are truth-theoretically incompatible if $\{T_1, T_2, L(T_1), L(T_2), M, TR\} \vdash \neg[\text{True}(S_1) \wedge \text{True}(S_2)]$.

The concept of truth-theoretical incompatibility assumes that a sufficient neutral metalanguage can be found to establish a unitary truth theory accepted by both theories. We can identify a confrontation on truth-values since truth-value status is preserved across the theories. But this assumption cannot be held in the case of incommensurability based on the presuppositional interpretation. I have argued that there is a truth-value gap between the languages of two incommensurable theories. Therefore, we simply cannot compare the truth-values of the core sentences in two incommensurable languages since they differ on truth-value status. We cannot identify the confrontation of truth-values between core sentences of two incommensurable theories. Consequently, the truth-theoretical incompatibility relation is not available between them. That means that the semantic comparison based on truth-theoretical incompatibility, which I will call truth-theoretical comparison, is impossible.

One obvious objection to the presuppositional interpretation emerges immediately: 'Your presuppositional interpretation actually brings more damage to the image of science as a rational enterprise than the translation-failure interpretation does. Based on the translation-failure interpretation, the semantic comparison between the incommensurables is still possible as long as a quasi-logical incompatibility relationship between them remains. But in the presuppositional interpretation, even quasi-logical incompatibility between the incommensurables is excluded *a priori*. Then the presuppositional interpretation implies that semantic comparison is impossible between the incommensurables. It is more radical than the traditional interpretation.'

The critic here is too hurried to jump from the impossibility of truth-theoretical incompatibility to the impossibility of semantic incompatibility between the incommensurables. Truth-theoretical incompatibility is neither the only nor the most interesting aspect in which two competing theories can be incompatible. I have argued that it is the truth-value functional relation between competing theories, instead of the meaning-referential relation, that constitutes the dominant semantic relation between two incommensurable theories. The disagreement between two incommensurable theories is not in what counts as truth, but rather in what counts as truth-or-falsity. This is in turn determined by the M-presuppositions underlying each theory. When the M-presuppositions of two competing theories are incompatible, the two theories are incommensurable. The incompatibility between the M-presuppositions of two incommensurable theories is the fundamental semantic incompatibility needed for semantic comparison. By definition:

Suppose (a) two theories T_1 and T_2 and their languages $L(T_1)$ and $L(T_2)$, (b) two built-in truth theories TR1 and TR2 for $L(T_1)$ and $L(T_2)$, respectively, and (c) the M-presuppositions MP_1 and MP_2 underlying $L(T_1)$ and $L(T_2)$, respectively. T_1 and T_2 are presuppositionally incompatible if $\{T_1, T_2, L(T_1), L(T_2), TR_1, TR_2\} \vdash_L \neg [\text{True}(MP_1) \wedge \text{True}(MP_2)]$.

The subscript '$_L$' indicates that both MP_1 and MP_2 are considered within the context of one language, either $L(T_1)$ or $L(T_2)$. We have identified many presuppositionally incompatible scientific languages before. I will not belabor them here.

Since the presuppositional incompatibility relation exists between two incommensurable theories, semantic comparison based on it is possible. I will call such comparison *presuppositional comparison*. The virtues of presuppositional comparison are obvious. Compared with much stronger requirements of logical or quasi-logical comparison, presuppositional comparison asks much less. It does not require that there exist a neutral language into which both theories can be translated without loss; it does not require that there exist a unitary truth theory accepted by both languages; it does not require sameness or overlap of meaning or reference. Actually, it sidesteps many problems caused by meaning, reference, and translation. As long as we can show that the M-presuppositions of two incommensurable theories, when considered within the context of one theory, cannot both be held to be true, we know that they are semantically incompatible, which makes semantic comparison between the two theories possible. Unlike truth-theoretical comparison which focuses on the contents of corresponding statements of two incommensurable theories, presuppositional comparison starts from the bottom up, namely, to compare and evaluate the incompatible M-presuppositions of two incommensurable theories at the meta-theoretical level. The most effective way of pooling together two incommensurable theories is to bridge the ontological gap caused by incompatible M-presuppositions. Presuppositional comparison provides a promising way to do this.

Notes

1   Both Ruth Millikan and Austen Clark pointed out this potential connection between the two interpretations.
2   This potential dilemma was pointed out by Samuel Wheeler III.
3   I. Scheffler, 1967, pp. 81-6; C. Kordig, 1971, pp. 22, 52; L. Laudan, 1977, pp. 139-42; 1990, pp. 121-3; M. Devitt, 1979, p. 29; W. Newton-Smith, 1981, p. 148; D. Shapere, 1984, pp. 66, 83; D. Pearce, 1987, p. 3.
4   Kuhn, 1976, pp. 191, 198; 1970b, pp. 234, 266; 1983b, p. 671; 1977a, pp. 320-39. Feyerabend, 1970, 1977.
5   For a clear illustration of one-by-one comparison, please refer to W. Balzer, 1989.
6   Feyerabend, in his 1965a, lists a few ways to make content comparison.
7   According to W. Stegmuller's (1979) structuralist interpretation of incommensurability, a pair of incommensurable theories may be semantically incomparable if they are viewed as classes of sentences or statements, since they are not logically related so construed in the case of incommensurability. However, when those theories are construed as structure of a certain sort, not as statements, they are semantically comparable since the structures of two incommensurable theories can be canonically related where statements cannot.

Bibliography

Atlas, J. (1989), 'A Note on A Confusion of Pragmatics and Semantic Aspect of Negation', *Linguistic and Philosophy*, Vol. 3, pp. 411-14.

Baghramian, M. (1998), 'Why Conceptual Schemes?' *The Proceedings of the Aristotelian Society*, Vol. 98, pp. 287-306.

Balzer, W. (1989), 'On Incommensurability', in K. Gavroglu, et al. (eds), *Imre Lakatos and Theories of Scientific Change*, Kluwer Academic Publishers, Boston, pp. 287-304.

Barsalou, L. and Sewell, D. (1984), *Constructing Representations of Categories from Different Points of View*, Emory Cognition Project Report #2, Emory University, Atlanta.

Barsalou, L. (1987), 'The Instability of Graded Structure: Implication for the Nature of Concepts', in U. Neisser (ed.), *Ecological and Intellectual Factors in Categorization*, Cambridge University Press, Cambridge, pp. 101-40.

Beijing College of Traditional Chinese Medicine (1978), *Traditional Chinese Medicine*, Shanghai.

Bergmann, M. (1981), 'Presupposition and Two-Dimensional Logic', *The Journal of Philosophical Logic*, Vol. 10, pp. 27-53.

Bernstein, R. (1983), *Beyond Objectivism and Relativism: Science, Hermeneutics, and Praxis*, The University of Pennsylvania Press, Philadelphia.

Bertalanffy, L. von (1955), 'An Assay On the Relativity of Categories', *Philosophy of Science*, Vol. 22, pp. 243-63.

Biagioli, M. (1990), 'The Anthropology of Incommensurability', *Studies in History and Philosophy of Science*, Vol. 21, pp. 183-209.

Böer, S. and Lycan, W. (1976), *The Myth of Semantic Presupposition*, Indiana University Linguistic Club, Bloomington.

Borges, L. (1966), *Other Inquisitions 1937-1952*, Washington Square Press, New York.

Boyd, R. (2001), 'Reference, (In)commensurability and Meanings: Some (Perhaps) Unanticipated Complexities, in P. Hoyningen-Huene and H. Sankey (2001), pp. 1-64.

Brown, H. (1983), 'Incommensurability', *Inquiry*, Vol. 26, pp. 3-29.

Bühler, K. (1934), *Sprachtheorie*, Fischer.

Carey, J. (1992), *Communication as Culture: Essays on Media and Society*, Routledge, New York.

Cavell, S. (1979), *The Claim of Reason: Wittgenstein, Skepticism, Morality, and Tragedy*, Oxford University Press, New York.

Chan, W. (1963), *A Source Book in Chinese Philosophy*, Princeton University Press, New Jersey.

Collingwood, R. G. (1940), *An Essay On Metaphysics*, the Clarendon Press, Oxford.

Cooke, M. (1994), *Language and Reason: a study of Habermas's pragmatics*, The MIT Press, Boston.

Davidson, D. (1984), *Inquires into Truth and Interpretation*, Oxford University Press, New York (including 'On the Very Idea of a Conceptual Scheme', pp. 183-98).

Davidson, D. (2001a), 'The Myth of the Subjectivity', in his *Subjective, Intersubjective, Objective*, Clarendon Press, Oxford, pp. 39-52.

Davidson, D. (2001b), 'A Coherent Theory of Truth and Knowledge', in his *Subjective, Intersubjective, Objective*, Clarendon Press, Oxford, pp. 137-57.

Devitt, M. (1979), 'Against Incommensurability', *Australasian Journal of Philosophy*, Vol. 57, pp. 29-50.

Devitt, M. (1984), *Realism and Truth*, Princeton University Press, New Jersey.

Doppelt, G. (1978), 'Kuhn's Epistemological Relativism: An Interpretation and Defense', *Inquiry*, Vol. 21, pp. 33-86

Dummett, M. (1993), *The Sea of Language*, Clarendon Press, Oxford.

Englebretson, G. (1973), 'Presupposition, Truth and Existence', *Philosophical Papers*, Vol. 2, pp. 39-40.

English, J. (1978), 'Partial Interpretation and Meaning Change', *The Journal of Philosophy*, Vol. 75, pp. 57-76.

Evans, G. (1973), 'The Causal Theory of Names', in A. Martinich (1996), pp. 271-83.

Evans, G. and McDowell, J. (eds), (1976), *Truth and Meaning: Essays in Semantics*, Clarendon Press, Oxford.

Evans-Pritchard, E. (1964), *Social Anthropology*, Cohen & West, London.

Favretti, R., Sandri, G. and Scazzieri, R. (eds), (1999), *Incommensurability and Translation: Kuhnian Perspectives on Scientific Communication and Theory Change*, Edward Elgar, Nothampton.

Feyerabend, P. (1962), 'Explanation, Reduction, and Empiricism', in F. Feigh and G. Maxwell (eds), *Scientific Explanation, Space, and Time*, The University of Minnesota Press, Minneapolis, pp. 28-97.

Feyerabend, P. (1965a), 'Reply to Criticism', in his (1981), pp. 104-31.

Feyerabend, P. (1965b), 'On the "Meaning" of Scientific Terms', *The Journal of Philosophy*, Vol. 62, pp. 266-74.

Feyerabend, P. (1965c), 'Problems of Empiricism', in R. Colodny (ed.), *Beyond the Edge of Certainty*, Prentice-Hall.

Feyerabend, P. (1970), 'Consolations for the Specialist', in I. Lakatos and A. Musgrave (eds), *Criticism and Growth of Knowledge*, Cambridge University Press, New York, pp. 197-229.

Feyerabend, P. (1977), 'Changing Patterns of Reconstruction', *The British Journal for the Philosophy of Science*, Vol. 28, pp. 351-69.

Feyerabend, P. (1978), *Against Method*, Verso, London.

Feyerabend, P. (1981), *Realism, Rationalism and Scientific Method, Philosophical Paper I*, Cambridge University Press, Cambridge.

Feyerabend, P. (1987), 'Putnam on Incommensurability', *The British Journal For the Philosophy of Science*, Vol. 38, pp. 75-82.

Field, H. (1973), 'Theory Change and the Indeterminacy of Reference', *The Journal of Philosophy*, Vol. 70, pp. 487-509.

Forster, M. (1998), 'On the Very Idea of Denying the Existence of Radically Different Conceptual Schemes', *Inquiry,* Vol. 41, pp. 133-85.

Fraassen, B.C. van (1970), 'On the Extension of Beth's semantics of Physical Theories', *Philosophy of Science*, Vol. 37, pp. 325-39.

Fraassen, B.C. van (1989), *Laws and Symmetry*, Oxford University Press, Oxford.

Fu, D. (1995), 'Higher Taxonomy and Higher Incommensurability', *Studies in History and Philosophy of Science*, Vol. 26, pp. 273-94.

Fung, Y. (1947), *The Spirit of Chinese Philosophy*, trans. by R. Hughes, Kegan Paul, London.

Fung, Y. (1952 and 1953), *A History of Chinese Philosophy*, 2 Vols., trans. by D. Bodde, Princeton University Press, New Jersey.

Gadamer, H.-G. (1975), *Truth and Method*, trans. and ed. G. Barden and J. Cumming, Seabury Press, New York.

Gadamer, H.-G. (1976), *Philosophical Hermeneutics*, trans. by D. Linge, University of California Press, Berkeley.

Gadamer, H.-G. (1979), 'The Problem of Historical Consciousness', in P. Rabinow and W. Sullivan (eds), *Interpretive Social Science: A Reader*, University of California Press, Berkeley.

Gadamer, H.-G. (1985), 'Grenzen der Sprache', *Evolution und Sprache: Uber Entstebung und Wesen der Sprache,* Herrrenalber Texye 66t, pp. 89-99.

Gadamer, H.-G. (1986), *Gesammelte Werke,* Vol. 2: *Hermeneutik II: Wahrheit Mrthode*, Erganzungen, Registert.

Gadamer, H.-G. (1989), *Truth and Method*, second revised edition, trans. by J. Weinsheimeer and D. Marshall, Continuum, New York.

Gaifman, H. (1975), 'Ontology and Conceptual Frameworks', Part I, *Erkenntnis*, Vol. 9, pp. 329-53.

Gaifman, H. (1976), 'Ontology and Conceptual Frameworks', Part II, *Erkenntnis*, Vol. 10, pp. 21-85.

Gaifman, H. (1984), 'Why Language?', in W. Balzer et al. (eds), *Reduction in Science,* D. Reidel Publishing Company, Boston, pp. 319-30.

Garfield, J. and Kiteley, M. (eds), (1991), *Meaning and Truth: The Essential Readings in Modern Semantic*, Paragon House, New York.

Gazdar, G. (1979), *Pragmatics: Implication, Presupposition, and Logical Form*, Academic Press, New York.

Geertz, C. (1979), 'From the Native's Point of View: On the Nature of Anthropological Understanding', in P. Rabinow and W. Sullivan (eds), *Interpretive Social Science: A Reader*, University of California Press, Berkeley, pp. 225-41.

Grossberg, L. (1997), *Bringing it All Back Home: Essays on Cultural Studies*, Duke University Press, Durham.

Habermas, J. (1979), 'What Is Universal Pragmatics?' in his *Communication and the Evolution of Society*, trans. by T. McCarthy, Beacon Press, Boston.

Habermas, J. (1982), 'A Reply to My Critics', in J. Thompson and D. Held (eds), *Habermas: Critical Debates*, Macmillan.

Habermas, J. (1984), *The Theory of Communication Action*, Vol. 1, trans. by T. McCarthy, Beacon Press, Boston.

Habermas, J. (1985), 'Remarks on the Concept of Communicative Action', in G. Seebaβ and R. Tuomela (eds), *Social Action*, Reidel, Boston.

Habermas, J. (1987), *The Theory of Communication Action*, Vol. 2, trans. by T. McCarthy, Beacon Press, Boston.

Habermas, J. (1992), *Postmetaphysical Thinking*, trans. by W. Hohengarten, The MIT Press, Cambridge.

Habermas, J. (1993), *Justification and Application: Remarks on Discourse Ethics*, The MIT Press, Cambridge.

Hacker, S (1996), 'On Davidson's Idea of a Conceptual Scheme,' *The Philosophical Quarterly*, Vol. 46, pp. 289-307.

Hacking, I. (1982), 'Language, Truth and Reason', in M. Hollis and S. Lukes (eds), *Rationality and Relativism*, The MIT Press, Cambridge, pp. 48-66.

Hacking, I. (1983), *Representing and Intervening*, Cambridge University Press, New York.

Hacking, I. (1992), '"Style" for Historians and Philosophers', *Studies in the History and Philosophy of Science,* Vol. 23, No. 1, pp. 1-20.

Hacking, I. (1993), 'Working in a New World: The Taxonomic Solution', in P. Horwich (1993), pp. 275-309.

Harman, G. (1984), 'Conceptual Role Semantics', *Notre Dame Journal of Formal Logic*, Vol. 23, pp. 242-56.

Heidegger, M. (1962), *Being and Time*, trans. by J. Macquarrie and E. Robinson, Happer & Row Publisher, New York.

Heidegger, M. (1975), *Gesamtausgade*, Frankfurt.

Henderson, D. (1994), 'Conceptual Schemes After Davidson', in G. Preyer *et al.* (eds), *Language, Mind and Epistemology*, Kluwer Academic Publishers, pp. 171-97.

Hintikka, J. (1988), 'On the Incommensurability of Theories', *Philosophy of Science*, Vol. 55, pp. 25-38.

Horwich, P. (1990), *Truth*, Basil Blackwell, Cambridge.

Horwich, P. (ed.) (1993), *World Change: Thomas Kuhn and the Nature of Science*, The MIT Press, Cambridge.

Hoyningen-Huene, P. (1990), 'Kuhn's Conception of Incommensurability', *Studies in History and Philosophy of Science*, Vol. 21, pp. 481-92.

Hoyningen-Huene, P. (1993), *Reconstructing Scientific Revolutions*, The University of Chicago Press, Chicago.

Hoyningen-Huene, P. and Sankey, H. (eds), (2001), *Incommensurability and Related Matters*, Kluwer Academic Publishers, Boston.

Hung. H. (1981a), 'Theories, Catalogues, and Languages', *Synthese*, Vol. 49, pp. 375-94.

Hung. H. (1981b), 'Nomic Necessity is Cross-theoretic', *British Journal for the Philosophy of Science*, Vol. 32, pp. 219-36.

Hung. H. (1987), 'Incommensurability and Inconsistency of Languages', *Erkenntnis*, Vol. 27, pp. 323-52.

James, W. (1909), *A Pluralistic Universe*, University of Nebraska Press, Lincoln.

Kaplan, D. (1975), 'What is Russell's Theory of Descriptions?', in D. Davidson (ed.), *The Logic of Grammar,* Dickenson, California, pp. 210-17.

Kempson, R. (1975), *Presupposition and the Deliminatioion of Semantics*, Cambridge University Press, New York.

Khalidi, M. (1991), *Meaning-Change and Theory-Change*, Dissertation, Columbia University.

Kitcher, P. (1978), 'Theories, Theorists, and Theoretical Change', *Philosophical Review*, Vol. 87, pp. 519-47.

Kordig, C. (1971), *The Justification of Scientific Change*, D. Reidel Publishing Company, Holland.

Kripke, S. (1972), 'Naming and Necessity', in A. Martinich (1996), pp. 255-70.

Kuhn, T. (1970a), *The Structure of Scientific Revolutions*, second edition, the University of Chicago Press, Chicago.

Kuhn, T. (1970b), 'Reflection on My Critics', in I. Lakatos and A. Musgrave (eds), *Criticism and Growth of Knowledge*, Cambridge University Press, New York, pp. 231-78.

Kuhn, T. (1976), 'Theory-Change as Structure-Change: Comments on the Sneed Formalism', *Erkenntnis*, Vol. 10, pp. 179-99.

Kuhn, T. (1977a), *The Essential Tension* (including 'Objectivity, Value Judgment, and Theory Change', originally published in 1973, and 'Logic of Discovery of Psychology of Research', originally published in 1970), the University of Chicago Press, Chicago.

Kuhn, T. (1977b), 'Second Thoughts on Paradigms', in F. Suppe (ed.), *The Structure of Scientific Theories,* 2nd edition, The University of Illinois Press, pp. 459-517.

Kuhn, T. (1979), 'Metaphor in Science', in A. Rotony (ed.), *Metaphor and Thoughts*, Cambridge University Press, Cambridge, pp. 409-19.

Kuhn, T. (1983a), 'Rationality and Theory Choice', *The Journal of Philosophy*, Vol. 80, pp. 563-70.

Kuhn, T. (1983b), 'Commensurability, Comparability, Communicability', *PSA 1982*, Vol. 2, pp. 669-88.

Kuhn, T. (1987), 'What are Scientific Revolutions?' in L. Krüger, G. Gigerenzer and M. S. Morgan (eds), *The Probabilistic Revolution*, Vol.2: *Ideas in the sciences*, The MIT Press, Cambridge, pp. 7-22.

Kuhn, T. (1988), 'Possible Worlds in History of Science', in S. Allen (ed.), *Possible Worlds in Humanities, Arts and Sciences*, Walter de Gruyter, Berlin, pp. 9-32.

Kuhn, T. (1991), 'The Road Since Structure', *PSA 1990*, Vol. 2, pp. 3-13.

Kuhn, T. (1992), "The Natural and the Human Sciences', in D. Hiley, J. Bohman, and R. Shusterman (eds), *The Interpretative Turn: Philosophy, Science, and Culture*, Cornell University Press, pp. 17-24.

Kuhn, T. (1993a), 'Afterwords', in P. Horwich (1993), pp. 311-41.

Kuhn, T. (1993b), 'Foreword', in P. Hoyningen-Huene (1993), pp. xi-xiii.

Kuhn, T. (1999), 'Remarks on Incommensurability and Translation', in R. Favretti et al. (1999), pp 33-7.

Kuhn, T. (2000), *The Road Since Structure*, The University of Chicago Press, Chicago.

La Barre, W. (1954), *The Human Animal*, The University of Chicago Press, Chicago.

Lan, Z. (1988), 'Incommensurability And Scientific Rationality', *International Studies in the Philosophy of Science*, Vol. 2, pp. 227-36.

Laudan, L. (1977), *Progress and Its Problems: Towards a Theory of Scientific Growth*, University of California Press, Berkeley.

Laudan, L. (1990), *Science and Relativism*, The University of Chicago Press, Chicago.

Lehmann, S. (1994), 'Strict Fregean Logic', *Journal of Philosophical Logic*, Vol. 23, pp. 307-36.

Leplin, J. (1979), 'Reference and Scientific Realism', *Studies in History and Philosophy of Science*, Vol. 10, pp. 265-84.

LePore, E. (ed.) (1986), *Truth and Interpretation: Perspectives on the Philosophy of Donald Davidson*, Basil Blackwell Ltd, Cambridge.

Lewis, C.I. (1929), *Mind and the World Order*, Dover Publications, New York.

Locke, J. (1996), *An Essay Concerning Human Understanding*, abridged and edited by K. Winkler, Hackett Publishing Company, Cambridge.

Lycan, W. (1984), *Logical Form in Natural Language*, the MIT Press, Cambridge.

Lycan, W. (1987), 'Projection Problem for Presupposition', *Philosophical Topics*, Vol. 15, pp. 169-75.

Lycan, W. (1994), *Modality and Meaning*, Kluwer Academic, Boston.

Lynch, M. (1998), 'Three Modes of Conceptual Schemes', *Inquiry*, Vol. 40, pp. 407-26

MacIntyre, A. (1985), 'Relativism, Power and Philosophy', *Proceedings and Addresses of the American Philosophical Association*, Vol. 59, pp. 5-22.

Malinowski, B. (1965), *The Language of Magic and Gardening*, Indiana University Press, Bloomington.

Malone, M. (1993), 'Incommensurability and Relativism', *Studies in the History and Philosophy of Science*, Vol. 24, pp. 69-93.

Martin, J. (1975), 'A Many-Valued Semantics for Category Mistakes', *Synthese*, Vol. 31, pp. 63-83.

Martin, J. (1979), 'Some Misconceptions in the Critique of Semantic Presupposition', *Theoretical Linguistics*, Vol. 6, pp. 235-82.

Martin, M. (1971), 'Referential Variance and Scientific Objectivity', *The British Journal For the Philosophy of Science*, Vol. 22, pp. 17-26.

Martinich, A. (ed.), (1996), *The Philosophy of Language*, 3rd edition, Oxford University Press, New York.

McDowell, J. (1994), *Mind and World*, Harvard University Press, Cambridge. ,

Needham, J. (1956), *Science and Civilization in China*, Vol. 2, Cambridge University Press, New York.

Nersessian, N. (1984), *Faraday to Einstein: Constructing Meaning in Scientific Theories*, Martnus Nijhoff Publishers, Boston.

Nersessian, N. (1989), 'Scientific Discovery and Commensurability of Meaning', in K. Gavroglu, et al. (eds), *Imre Lakatos and Theories of Scientific Change*, Kluwer Academic Publishers, Boston, pp. 323-34.

Newton-Smith, W. (1981), *The Rationality of Science*, Routledge.

Orenduff, J. (1979), 'Why the Concept of Presuppositions is Otiose', *Southwest Philosophical Studies*, Vol. 4, pp. 73-7.

Pearce, D. (1987), *Roads to Commensurability*, D. Reidel Publishing Company, Boston.

Peters, J. (1999), *Speaking Into the Air: a history of the idea of communication*, The University of Chicago Press, Chicago.

Przelecki, M. (1979), 'Commensurable Referents of Incommensurable Terms', in I. Niiniluoto and R. Tuomela (eds), *The Logic and Epistemology of Scientific Change,* North-Holland Publishing Company, Amsterdam, pp. 347-65.

Putnam, H. (1973), 'Meaning and Reference', in A. Martinich (1996), pp. 284-91.

Putnam, H. (1981), *Reason, Truth and History*, Cambridge University Press, New York.

Quine, W. V. (1960), *Word and Object*, The MIT Press, Cambridge.

Quine, W. V. (1969), *Ontological Relativity and Other Essays,* Columbia University Press, New York.

Quine, W. V. (1970), 'Philosophical Progress in Language Theory', *Metaphilosophy*, Vol.1, pp. 2-19.

Quine, W. V. (1980), 'Two Dogmas of Empiricism', in his *From a Logical Point of View*, 2nd edition, Harvard University Press, Cambridge, pp. 20-46.

Quine, W. V. (1981), 'On the Very Idea of a Third Dogma', in his *Theories and Things*, Harvard University Press, Cambridge, pp. 39-42.

Radford, G. (2005), *On the Philosophy of Communication*, Wadsworth, Belmont.

Ramberg, B. (1989), *Donald Davidson's Philosophy of Language*, Basil Blackwell Press, Cambridge.

Rescher, N. (1980), 'Conceptual Schemes', in P. French et al. (eds), *Studies in Epistemology*, University of Minnesota Press, Minneapolis, pp. 323-45.

Rorty, R. (1979), *Philosophy and the Mirror of Nature*, Princeton University Press, Princeton.

Rorty, R. (1980), 'Pragmatism, Relativism, and Irrationalism', *Proceedings and Addresses of the American Philosophical Association*, Vol. 53, pp. 719-38.

Rorty, R. (1982), 'The World Well Lost', in his *Consequences of Pragmatism*, University of Minnesota Press, Minneapolis, pp. 3-18.

Rosch, E. (1973), 'On the Internal structure of perceptual and semantic categories', in T. Moore (ed.), *Cognitive Development and the Acquisition of Language*, Academic Press, New York.

Rosch, E. (1975), 'Universals and Cultural Species in Human Categorization', in R. Brislin, S. Bochner, and W. Lonner (eds), *Cross-Cultural Perspectives on Learning*, John Wiley & Sons Inc, Mississauga, Ontario, pp. 177-206.

Rosch, E. (1978), 'Principles of Categorization', in E. Rosch and B. Lloyd (eds), *Cognition and Categorization*, Lawrence Erlbaum Associates Publishers, pp. 27-48.

Rosch, E. and Mervis, C. (1975), 'Family Resemblance: Studies in the Internal Structure of Categories', *Cognitive Psychology*, Vol. 7, pp. 573-605.

Russell, B. (1905), 'On Denoting', in A. Martinich (1996), pp. 199-207.

Russell, B. (1957), 'Mr. Strawson on Referring', in J. Garfield and M. Kiteley, (1991), pp. 130-35.

Salmon. N. (1986), 'Reflexivity', *Notre Dame Journal of Formal Logic*, Vol. 27, 3, pp. 401-29.

Sankey, H. (1991), 'Incommensurability and the Indeterminacy of Translation', *Australasian Journal of Philosophy*, Vol. 69, pp. 219-23.

Sankey, H. (1994), *The Incommensurability Thesis*, Aveburg, Sydney.

Sankey, H. (1997), *Rationality, Relativism and Incommensurability*, Averburg, Sydney.

Scheffler, I. (1967), *Science and Subjectivity*, The Bobbs-Merrill Company, Indianapolis.

Schlick, M. (1991), 'Positivism and Realism', in R. Boyd et al. (eds), *The Philosophy of Science*, The MIT Press, Cambridge, pp. 37-55.

Searle, J. (1958), 'Proper Names', in A. Martinich (1996), pp. 249-54.

Searle. J. (1995), *The Construction of Reality*, Penguin, London.

Sellars, W. (1954), 'Presupposing', *Philosophical Review*, Vol. 63, pp. 197-215.

Shapere, D. (1984), *Reason and the Search for Knowledge*, D. Reidel Publishing Company, Boston.

Stegmüller, W. (1979), *The Structuralist View of Theories*, Berlin-Heidelberg, New York.

Strawson. P. (1950), 'On Referring', in A. Martinich (1996), pp. 215-30.

Strawson. P. (1952), *Introduction to Logical Theory*, Methuen, London.

Strawson. P. (1969), 'Meaning and Truth', in A. Martinich (1996), pp. 104-14.

Strawson, P. (1992) *Analysis and Metaphysics*, Oxford University Press, Oxford.

Tarski, A. (1949), 'The Semantic Conception of Truth', in H. Fiegle and W. Sellars (eds), *Readings in Philosophical Analysis,* Appleton-Century-Crofts, New York, pp. 52-84.

Taylor, C. (1982), 'Rationality', in H. Martin and S. Lukes (eds), *Rationality and Relativism*, The MIT Press, Cambridge, pp. 95-6.

Taylor, C. (2002), 'Understanding the Other: A Gadamerian View on Conceptual Schemes', in J. Malpas, U. Armswald, and J. Kertscher (eds), *Gadamer's Century*, The MIT Press, Cambridge, pp. 279-97.

Wittgenstein, L. (1969), *On Certainty*, Harper Torchbooks, New York.

Whorf, B. (1956), *Language, Thought and Reality*, The MIT Press, Cambridge.

Wilson, D. (1975), *Presupposition and Non-truth-conditional Semantics*, Academic, London.

Winch, P. (1958), *The Idea of Social Science*, Humanities Press, New York.

Winch, P. (1979), 'Understanding a Primitive Society', in B. Wilson (ed.), *Rationality*, Harper & Row, Evanston, pp. 100, 104-5.

Wong, D. (1989), 'Three Kinds of Incommensurability', in M. Krausz (ed.), *Relativism: Interpretation and Confrontation*, The University of Notre Dame Press, Notre Dame, pp. 140-58.

Index